Nov 2017

# EXACT THINKING
IN DEMENTED
TIMES

## Also by Karl Sigmund

*The Calculus of Selfishness*

*Evolutionary Games and Population Dynamics*
(co-authored with Josef Hofbauer)

*Games of Life: Explorations in Ecology, Evolution, and Behaviour*

# EXACT THINKING

## IN DEMENTED

## TIMES

*The Vienna Circle and
the Epic Quest for the
Foundations of Science*

## KARL SIGMUND

BASIC BOOKS

NEW YORK

Basic Books
Hachette Book Group
1290 Avenue of the Americas, New York, NY 10104
www.basicbooks.com

Printed in the United States of America
Originally published in hardcover and ebook by Springer Spektrum in May 2015 in
Germany
First United States Edition: December 2017
Published by Basic Books, an imprint of Perseus Books, LLC, a subsidiary of Hachette
Book Group, Inc.

The Hachette Speakers Bureau provides a wide range of authors for speaking events.
To find out more, go to www.hachettespeakersbureau.com or call (866) 376-6591.
The publisher is not responsible for websites (or their content) that are not owned by
the publisher.

Additional copyright/credits information is on page 389.

*Print book interior design by Linda Mark.*

Library of Congress Cataloging-in-Publication Data

Names: Sigmund, Karl, 1945– author.
Title: Exact thinking in demented times : the Vienna Circle and the epic quest for the
    foundations of science / Karl Sigmund.
Description: New York : Basic Books, [2017] | Includes bibliographical references and index.
Identifiers: LCCN 2017037770 | ISBN 9780465096954 (hardcover) |
    ISBN 9780465096961 (ebook)
Subjects: LCSH: Science—Philosophy—History—20th century. | Vienna circle. |
    Logical positivism.
Classification: LCC Q175 .S5554 2017 | DDC 501—dc23
LC record available at https://lccn.loc.gov/2017037770

LSC-C

10  9  8  7  6  5  4  3  2  1

*I am sitting with a philosopher in the garden; while pointing at a nearby tree, he says several times, "I know that that's a tree." Another person arrives and hears this, and I say: "This fellow isn't nuts. We're just philosophizing."*

—LUDWIG WITTGENSTEIN

*If we were to open the window so that passersby could hear us, we would wind up either in jail or in the loony bin.*

—HANS HAHN

# Contents

# *Preface*

by Douglas Hofstadter

ONE EVENING IN THE EARLY FALL OF 1959, WHILE I WAS BROWS-
ing at random in Kepler's bookstore in Menlo Park, I stumbled across
a slim paperback book called *Gödel's Proof,* co-written by Ernest Nagel
and James R. Newman. Aged fourteen at the time, I had never heard of
Gödel, but I liked the exotic dots floating above his name, and in my
high school math class, I had recently become spellbound by the notion of mathematical proof, so my curiosity was piqued. Upon flipping
through the book's pages, I was quickly hooked. Here was a book about
many things, including logic, the nature of mathematics, language and
symbols, truth and falsity, proofs about provability, and perhaps best
of all, paradoxes and self-referential statements. All these topics were
amazingly alluring to me. I had to buy this book!

My father, a physics professor at Stanford, was with me that evening,
and as we were paying for our purchases, he saw the cover of my book
and told me with delight that he knew Ernest Nagel quite well. I was
bowled over. In fact, he had taken a philosophy course from Nagel in
New York City in the early 1930s, and they had become friends as a

result, although it had been many years since they had seen each other. This friendship, so unexpected by me, was certainly a welcome confirmation of my book choice.

Unbeknownst to either of us, Ernest Nagel, who had long been a professor of philosophy at Columbia University, had just arrived at Stanford a couple of weeks earlier to spend a sabbatical year "out west" with his family. And not long thereafter, my father ran into his old friend on the Stanford campus by chance, and they had a happy reunion. One thing led to another, and pretty soon my father took me over to the Nagels' rented house on the Stanford campus. There I met Ernest Nagel and his wife, Edith, who taught physics at the City College of New York, and their two sons, Sandy and Bobby, who were fascinated by math and science, as I was. Not only were all four Nagels sparkling intellects, but they were also among the kindest, warmest people I had ever met in my life. There was an immediate resonance among us all, and thus began a lifelong friendship.

During that wonderful year, Ernest told me many stories about interesting characters whom he had met in Europe and the United States, such as Rudolf Carnap, Moritz Schlick, Carl Hempel, and others. And in my frequent browsings at Kepler's bookstore, I kept on running into books by various people Ernest had mentioned to me. One of my favorites was *Introduction to Mathematical Thinking* by Friedrich Waismann, from which I learned a great deal.

First through Ernest's stories and then through books, I learned about the Vienna Circle and the ambitious philosophical movement that it launched, called *logical positivism.* This group of roughly a dozen people, who were fascinated by issues of philosophy, linguistics, physics, mathematics, logic, social reform, education, architecture, and communication, had the idealistic goal of forging a great unification of human knowledge. They were working on this grandiose plan during a period of tremendous economic and political upheaval in Austria and Germany—right between the two world wars. That was a tough time to be thinking idealistic thoughts!

I will always remember my extreme curiosity, even exitement, when I espied the provocative series of paperbacks called the International Encyclopedia of Unified Science on the shelves at Kepler's. As I browsed through these volumes, I got the clear impression that the greatest questions of all time were being answered at this moment in history by deep thinkers who had once belonged to the now-defunct Vienna Circle and by their close colleagues.

When I was fifteen, I spotted Rudolf Carnap's *The Logical Syntax of Language* in one of my favorite bookstores, the Princeton U-Store, and snapped it up (for $1.15). This book, filled to the brim with long, mysterious-looking formulas using exotic fonts, oozing with references to Gödel, Hilbert, Tarski, Frege, Russell, and others, and featuring lengthy discussions of languages, metalanguages, proof schemas, syntactical antinomies, and so on, practically set my young brain on fire. Why? Because at that tender age, I had gotten fully caught up in the exciting notion that human thinking and pure deductive logic were one and the same thing. Even though Carnap's book was mostly opaque to me, it inspired in me indescribable feelings of depth. I was, after all, only fifteen. . . .

Around that same time, I also ran into the mythical Ludwig Wittgenstein and his imposing-sounding *Tractatus Logico-Philosophicus*, his *Brown Book* and his *Blue Book*, and other volumes. These were praised to the skies by renowned authorities, such as Bertrand Russell. I had to check them out! At first I was intrigued by Wittgenstein's pithy numbered sayings, but after trying hard to figure them out, I couldn't make a great deal of sense of them. Even so, I remained impressed; after all, so many people to whom I looked up seemed to think that they were works of great genius. However, after a while I started to be a bit more self-confident about my own opinions, and at some point, I grew skeptical of Wittgenstein's oracular tone and cryptic phrasings. His sentences struck me more as pretentious obscurantism than as insightful clarity. Eventually I lost patience and decided that whether or not he had something important to say, his way of communicating was pretty orthogonal to my own, and so I dropped him like a hot potato.

Well, all that was a long, long time ago. We now fast-forward nearly sixty years. It is June 2016, and I am in Stockholm, Sweden, participating in a small two-day symposium on philosophy and science, organized by the writer and publisher Christer Sturmark, and there I meet quite a few interesting people, including Björn Ulvaeus (once one of the stars in the Swedish pop group ABBA), Anton Zeilinger (a pioneering quantum physicist from Vienna), and Karl Sigmund (a Viennese mathematician who once wrote a biography of Gödel). We are strolling across the charming park called Skansen shortly after lunch, and the easygoing Professor Sigmund tells me that he has just completed a book about the Vienna Circle. My ears perk up with interest, since that is a set of thinkers whom I have known my whole life, at least indirectly, and a couple of them even exerted monumental influences on me. I ask him what led him to write the book, and he explains that he grew up in the shadow of the Vienna Circle, so to speak, always haunted by their presence wherever he went in his hometown.

In many ways, Karl Sigmund's reasons for being interested in the Circle were like mine, except raised to the $n$th power. Of course he had had to write such a book—it was practically his destiny! As we talked, he sensed my genuine enthusiasm and told me he would be happy to send me a copy when he got home to Vienna. I was gung ho! And indeed, a few weeks later, I received in the mail a copy of *Sie nannten sich Der Wiener Kreis: Exaktes Denken am Rand des Untergangs.* The moment I opened it up, I was stunned by the profusion of photographs of people and places, the reproductions of handwritten letters, book covers, sales receipts, ticket stubs, and God knows what all else. The book was an amazing historical museum! I could hardly wait to read it. It didn't hurt that I was planning on spending the first part of my upcoming sabbatical year in Vienna, so the prospect of plunging myself into Vienna's intellectual history was also delicious.

It took me about a month to read the book from start to finish. In doing so, I learned a great deal about the Vienna Circle and its intellectual roots and contributions. Of course I already knew about Kurt Gödel's incompleteness theorems, but I also discovered that among the many diverse creations rooted in the Vienna Circle were Otto and Marie Neurath's pathbreaking use of icons in communication, Karl Menger's brilliant invention of dimension theory, Hans Hahn's pioneering ideas about functional analysis, Ludwig Wittgenstein's cryptic declarations, Karl Popper's influential ideas about falsifiability in science, and Rudolf Carnap's heroic attempt to unify all the sciences with logic.

I also learned more than I had ever wanted to know about the horrible turmoil that was reigning all over Eastern Europe at the time these intellectual developments were taking place. This turmoil of course affected everyone in the Vienna Circle, leading to the cold-blooded murder of its leader, and eventually causing most of its members to flee Austria. That, of course, is why Karl Sigmund had given his book the subtitle "Exact Thinking at the Brink of Doom" (or words roughly to that effect).

While I was reading the book, I busily penciled all sorts of notes into the margins. Mostly these were just literal translations of German words and idioms, but some were thoughts about how to render the ideas colorfully and idiomatically in English. Why was I writing such marginal notes to myself? Well, when I had gotten only a chapter or two into the book, it suddenly occurred to me that during my upcoming stint in Vienna, I could translate this book into English. What could possibly give me a more deeply Viennese experience than that?

I had already translated several books into English, but none from German. Luckily, though, my knowledge of German was pretty decent, since I had studied it in college, and later, in the mid-1970s, as a graduate student in physics, I had spent time at the University of Regensburg, where I read novels in German, talked for hundreds of hours with German students and professors, and even taught a physics lab course in German. Forty years later, my German was a bit rusty but still passable. So what better way could there be to renew my old engagement with the German language than through translating this book into English?

As soon as I had finished reading *Sie nannten sich Der Wiener Kreis,* I wrote an email to Karl Sigmund telling him how much I had enjoyed it, and saying that it would be an honor to translate it into English, if he would like that. To my surprise, he replied not from Vienna but from the island of Mauritius, where he was on vacation, and to my greater surprise, he wrote the following: "The idea of having you translate my book is something so extraordinary that I am still reeling! I feel that I have missed the opportunity of my lifetime." That really threw me!

He then explained to me that in fact he himself had already translated it into English, and that his manuscript was currently in the process of being proofread and finalized by two native speakers of English. And then came a most curious coincidence: his publisher was Basic Books in New York City—also my publisher, ever since 1978—and the editor he was dealing with there was TJ Kelleher, who is also my editor at Basic. All this made me smile.

I was of course most flattered that Karl (by now we were on a first-name basis) had said something so generous about the missed chance of my translating his book, but in my reply I told him that I thought it was probably better that he had done the job himself, since he knew exactly what he meant by each sentence and each word choice, and no one else could get all the nuances across in the same way as the author could. And in case he was worried about the idiomaticity of the English, well, he had two native speakers helping to fix up any language glitches.

A few days later, though, as I continued to mull over Karl's keen sense of loss, I had an idea. I wrote to him again, saying that if he was still interested in having me participate in the creation of an English-language version of his book, I would be delighted to read through the proofs and to make suggestions here and there, in order to make the prose flow as smoothly and as vividly as possible. I pointed out that I had the advantage of having just read the original work in German with a fine-tooth comb, that I knew math and logic well, that I had been familiar with the Vienna Circle for most of my life, and that I had written scads of marginal notes to myself while daydreaming of translating it during my sabbatical. In sum, I said that it would be both a pleasure and a privilege to help put finishing touches on the English-language version of his book, if that was of interest to him.

Well, Karl was very taken with my offer, and TJ approved the idea as well, although he told us there was considerable time pressure, so I had to promise to make it snappy. As soon as we'd all agreed that I would take this job on (and would make it snappy), Karl emailed me all his files, and thus began an intense several-week adventure. I had the fascinating experience of living intimately with the Vienna Circle once again, with this second go-round taking place in English (though of course I was constantly revisiting the German book as well), and also encountering all sorts of new episodes that Karl had thrown into this version.

As I carried out my editing task, I had the pleasure of inserting appropriate idiomatic phrases into the text here and there (such as "they weren't a dime a dozen" and "he put physics on the back burner for a while"), and of throwing in quite a few other vivid turns of phrase. However, I also quickly realized that Karl had a wonderful command of English, with an extremely rich vocabulary and a superb mastery of idioms. Although in my painstaking, microscopic labors over the next few weeks, I sometimes added words here and subtract words there, my changes were always made with a profound respect for the untold thousands of highly intelligent, well-considered choices that had taken place behind the scenes.

Of course Karl had total veto power over any of my suggestions, and he often exercised it, since I sometimes went a little overboard with my ways of expressing things. Also, I should point out that most of the idioms in this book originated with Karl, not with me. He really uses words skillfully and colorfully! And lastly, if the reader should detect an overabundance of *indeeds* and *after alls* in these thirteen chapters, well, I take the blame for that—it is entirely my fault!

It was both informative and touching to me to deepen my familiarity with the many colorful characters in Karl's book, some of whom were official members of the Circle, others of whom were "associates" or else marginal characters of one sort or another. For instance, I grew fond of, and then quite exasperated with, and then once again fond of, elephant-loving, statistics-loving, and woman-loving Otto Neurath. I felt deep pity for poor Friedrich Waismann, so long exploited by the capricious and insensitive Ludwig Wittgenstein. I felt admiration for faithful Adele Nimbursky, who so staunchly backed her brilliant but tormented husband Kurt Gödel. I felt shocked by Albert Einstein's friend, the maniacal Friedrich Adler, who turned out to be just as evil as Johann Nelböck, the killer of the Circle's founder, Moritz Schlick. I felt compassion for the long-suffering Rose Rand—and so forth and so on.

Two figures stood out as particularly troubling to me, one of them being the philosopher Paul Feyerabend, who rose to the rank of lieutenant in Hitler's army and who then, after the war, left behind his Nazi service, got himself a PhD in philosophy, and soon became world-famous for spouting random nonsense about how science is supposedly done. I couldn't stand any of this, and I had the cheek to insert a few cynical words into Karl's text that reflected my own personal take on Feyerabend, but Karl vetoed my harsh words, writing a kind and intelligent note to me that ran as follows: "Slight change, to make it less accusatory. Please understand me: so many Austrians and Germans are nowadays pointing the finger. It is easy. But what would *they* have actually done? Most would not have been in the *Widerstand* [the Resistance]. Statistics preclude it. Heroes are rare. And what would *I* have done?" I greatly respected Karl's reflections, and stood corrected.

The other person whom I could not abide was the two-faced philosopher Martin Heidegger, who, when Hitler came to power, became the rector of the University of Freiburg and as such, gave rabble-rousing speeches wearing stormtrooper shirts and shouting "Heil Hitler!" What completely flummoxed me was that my adored uncle Albert Hofstadter, for many years an esteemed colleague of Ernest Nagel's in Columbia's philosophy department, was a huge admirer of Heidegger's ideas and even translated two of Heidegger's books into English. For me, though, not only was Heidegger rotten to the core, but his writings seemed incomprehensible from start to finish. What earthly good had dear old Uncle Albert ever seen in him? I guess I'll never know. Heidegger, of course, was never a member of the Vienna Circle but someone whose philosophy was so diametrically opposed to its ideas that he represents, in some sense, the loyal opposition, and several Circle members explicitly heaped scorn on his opaque writings.

Well, I have come a long way since my teenage infatuation with the vision of mathematical logic as the crux of human thinking. Today such an idea strikes me as deeply implausible. And yet I still vividly recall how that idea consumed me for years, inspiring me to think as hard as I possibly could about what thought was. In that sense, my teenage addiction to the writings of a few Vienna Circle members was not a bad thing at all for me—in fact, it kick-started my fascination with the amazingly subtle nature of human thinking, which has lasted my entire life.

And now, having just read Karl Sigmund's book so carefully in two languages, I have realized that the Vienna Circle's philosophical vision, though idealistic, was also quite naïve. The idea that pure logic is the core of human thought is certainly tempting, but it misses virtually all of the subtlety and depth of human thinking. For instance, the Circle's claim that the act of induction—moving from specific observations to broad generalizations—plays no role at all in science is one of the silliest ideas I have ever heard. The way I see it, induction is the seeing of patterns, and science is the seeing of patterns *par excellence*. Science is nothing if not a grand inductive guessing game, where the guesses are constantly rigorously tested by careful experiments. Contrary to the Vienna Circle's viewpoint, science has everything to do with induction, and precious little to do with syllogistic reasoning or any other type of strict, mathematical reasoning.

The Vienna Circle had a deeply idealistic view of the world of thinking and of politics, but ultimately it became a victim of the tragedy of its times. Fascism and Nazism blasted to smithereens the great cultures of Austria, Germany, and Italy for a few decades, and much of this book is about that horrible destruction. The Circle was a salient counterforce to those forces of evil. It was a noble dream, some of whose colorful shards remain with us today, greatly enriching the complex mosaic of thoughts and personalities that is our collective intellectual heritage from previous generations.

Though it is long gone and not so often talked about today, there is no doubt that *Der Wiener Kreis* was an assemblage of some of the most impressive human beings who have ever walked the planet, and Karl Sigmund's book tells its story, and their stories, in a gripping and eloquent fashion. It is a wonderful historical document, and perhaps it will inspire some readers to dream great dreams in the way that they were dreamt in the Vienna of those far-off days.

# Bringing the Vienna Circle into Focus

## MIDNIGHT IN VIENNA

To do full justice to the story of the Vienna Circle, I'd need to be an artist. But alas, an artist I am not.

I just wish I had the magic of Woody Allen to lure you into a taxicab and give you my vision of *Midnight in Vienna,* dropping in spontaneously on various moments in the rich past of my hometown. Most of the time, on climbing out of the cab, you would find yourself somewhere in the interwar years, but now and then you would be somewhere in the wake of World War II, with the theme from *The Third Man* vaguely humming in the background. And to begin properly, I really would have to carry you all the way back to the years before World War I, with a waltz from *The Merry Widow* on the soundtrack.

Unfortunately, I cannot introduce you to Gustav Klimt, Egon Schiele, and Oskar Kokoschka, or to Otto Wagner and Adolf Loos, or to Dr. Freud and Dr. Schnitzler. You'll just get fleeting glances of

FIGURE 1.1 (LEFT)  Moritz Schlick (1882–1936).
FIGURE 1.2 (MIDDLE)  Hans Hahn (1879–1934).
FIGURE 1.3 (RIGHT)  Otto Neurath (1882–1945).

them—cameo appearances—through the windows of a brilliantly lit coffeehouse. Most of the cast of my film—and please don't let this turn you off!—consists just of philosophers. They come in many stripes, but what links them all is one absorbing interest—namely, science.

If you are still with me after this disclosure, then let me briefly sketch the plot:

In 1924, philosopher Moritz Schlick, mathematician Hans Hahn, and social reformer Otto Neurath joined forces to launch a philosophical circle in Vienna. At that time, Schlick and Hahn were professors at the University of Vienna, and Neurath was the director of the Vienna Museum for Social and Economic Affairs.

From that year on, the circle met regularly on Thursday evenings in a small university lecture hall on a street named after the Austrian physicist Ludwig Boltzmann, where they discussed philosophical questions such as: What characterizes scientific knowledge? Do metaphysical statements have any meaning? What makes logical propositions so certain? Why is mathematics applicable to the real world?

The manifesto of the Vienna Circle proclaimed: "The scientific worldview is characterized not so much by theses of its own, but rather by its basic attitude, its points of view, its direction of research."

FIGURE 1.4 (LEFT) Ernst Mach (1838–1916).
FIGURE 1.5 (RIGHT) Ludwig Boltzmann (1844–1906).

The circle sought to create a purely science-based philosophy without any highbrow talk of unfathomable depths and without any otherworldly obscurantism: "In science there are no 'depths'; instead, there is surface everywhere. All experience forms a complex network, which cannot always be surveyed in its totality and which often can only be grasped in parts. Everything is accessible to Man; and Man is the measure of all things."

The Vienna Circle forged ahead in the tradition of Ernst Mach and Ludwig Boltzmann, two towering physicists who had made great discoveries and had taught philosophy in turn-of-the-century Vienna. The other main guiding lights of the small band of thinkers were the physicist Albert Einstein, the mathematician David Hilbert, and the philosopher Bertrand Russell.

Before long, a thin volume that had just been published came to dominate the discussions of the Vienna Circle. This was the *Tractatus Logico-Philosophicus*, written by Ludwig Wittgenstein during his

FIGURE 1.6 Ludwig Wittgenstein (1889–1951).

military service in the trenches of World War I. Wittgenstein, after renouncing his huge inheritance, had become a teacher in an elementary school in rural Lower Austria. After a while, however, he began talking with a few members of the Vienna Circle, and this link gradually brought him back to philosophy.

The Vienna Circle wanted to have nothing to do with hallowed (and often stuffy) philosophical traditions: "The scientific worldview knows no unsolvable riddles. Clarification of the traditional philosophical problems sometimes leads to their unmasking as pseudo-problems, and other times converts them into empirical problems, which can thereby be subjected to the methods of experimental science. The task of philosophical work lies in this type of clarification of problems and statements, rather than in the crafting of special 'philosophical' statements."

Brilliant newcomers joined the group, such as philosopher Rudolf Carnap, mathematician Karl Menger, and logician Kurt Gödel. These three in particular were eventually to radically redefine the border regions between philosophy and mathematics. The philosopher

FIGURE 1.7 (LEFT) Rudolf Carnap (1891–1970).
FIGURE 1.8 (MIDDLE) Karl Menger (1902–1985).
FIGURE 1.9 (RIGHT) Kurt Gödel (1906–1978).

Karl Popper, too, became closely connected with the Vienna Circle, although he never was invited to its meetings.

The circle quickly became the world center of the movement called *logical empiricism*. Leading thinkers in Prague, Berlin, Warsaw, Cambridge, and Harvard picked up the threads of its discussions.

In 1929, the Vienna Circle embarked on a new public presence, through its own journals, conferences, books, and lecture series. This significant shift was heralded by a manifesto titled *The Scientific Worldview*.

The manifesto was not so much a birth certificate—after all, Schlick's circle had already existed for five years—as it was a christening. The name *Vienna Circle*, proposed by Otto Neurath, was new, and was meant to evoke positive associations, such as *Vienna Woods* or *Viennese Waltz*, and also was intended to serve as a trademark. The content of the manifesto was a signpost announcing not only a new philosophical school, but also a new social and political agenda. "The scientific worldview serves life, and life embraces it."

The authors of the manifesto belonged to the left wing of the small group, and they made no secret of their ardent wish to thoroughly reform society. The Ernst Mach Society, founded by members of the Vienna Circle in 1928, was dedicated to "the spreading of the scientific

FIGURE 1.10  Karl Popper (1902–1994).

worldview." It sided with Social Democratic *Red Vienna* in the political struggle for reforms, especially in housing and education. (Despite its name, Red Vienna, or *Rotes Wien* in German, was not a communist movement but simply the nickname for Vienna in the period under the Social Democrats, which lasted from 1918 through 1934.)


## COFFEE AND CIGARS

It wasn't long before the Vienna Circle and the Ernst Mach Society had become favorite targets for the anti-Semitic and right-wing currents that existed in Vienna. The political ambience was growing increasingly menacing. During this second, public phase, the Vienna Circle slowly disintegrated.

Carnap moved to Prague, Wittgenstein to Cambridge. After the Austrian civil war of 1934, Neurath was barred from returning to Austria. In that same year, Hahn died unexpectedly. Young Gödel repeatedly had to spend time in psychiatric hospitals. In 1936, Schlick was assassinated on the steps of the main building of the university by a former student. Soon after that, Menger and Popper, disgusted by the prevailing public mood, opted for emigration. Most members of the Vienna

Circle left Vienna well ahead of the so-called cleansing following the *Anschluss* (the annexation of Austria by the Third Reich), but not all of them. In the war year of 1940, Kurt Gödel, against all odds, finally arrived in the United States, as a straggler. He had had to get there the long way round, via Siberia, Japan, and the vast Pacific.

The Vienna Circle, by that time very famous, had lost its Viennese roots, and it did not regain them after World War II. However, it was able to find shelter in Anglo-Saxon countries, and from there it exerted a seminal influence on the intellectual and scientific history of the twentieth century, decisively shaping analytical philosophy, formal logic, and economic theory. For example, the algorithms and computer programs that pervade our daily lives can be traced all the way back to the abstract investigations of Russell, Gödel, and Carnap into symbolic logic and computability.

Tales of murder and suicide, of love affairs and nervous breakdowns, of political persecution and hair's-breadth escapes all have their place in the rich tapestry of the Vienna Circle, but the tapestry's main thread is the unbroken stream of heated debates among its members. In no way was the Circle the intellectual collective that a few of its members had hoped it would become, nor was it the congregation that its opponents accused it of being. It teemed with vociferous controversies and silent misgivings. How can it be otherwise when philosophers meet?

At the beginning of the tale, near the dawn of the twentieth century, in the lecture hall of the Vienna Academy, physicists Ludwig Boltzmann and Ernst Mach conducted a highly publicized debate on the burning question "Do atoms exist?" At the tale's conclusion, one year after World War II had ended, a fierce clash between Karl Popper and Ludwig Wittgenstein took place in a plush Cambridge sitting room on the burning question "Do philosophical problems exist?" In the fifty-odd years between these two deeply symbolic debates, the role that Vienna played in philosophy was as seminal as its role in music had once been.

The Vienna Circle stood in the center of that extraordinary period of intellectual flourishing: a shining pinnacle of exact thinking, set against

a backdrop of wild fanaticism and maniacal stupidity. Our valiant philosophers were well aware of standing on the perilously listing deck of
a sinking ship, but this only lent greater urgency to their discussions,
which dealt with the limits to knowledge. There seemed to be little time
left. Some of the musicians were already packing up their instruments.

Today, it feels like a long time ago that the ship went down. In
the current era, millions of scientists and hundreds of millions of their
kith and kin take the scientific worldview pretty much for granted. If
pressed, they will admit that it may be threatened in various ways: by
religious fundamentalists of all creeds, by a debilitating flood of trash
culture, or simply by an epidemic lack of public interest. Compared
with all the other threats we face, the danger to science probably does
not seem urgent; however, as the Vienna Circle's story shows, things
can change quickly.

The entire epic of the rise and fall of the Vienna Circle spans less
than half a century. A coffeehouse waiter could have witnessed it all
from a ringside seat, so to speak. As a young *piccolo* he would have
served an *Einspänner mit Schlag* to the portly *Hofrat* Ernst Mach, the
darling of waltz-giddy imperial Vienna; and as an elderly, stooping Herr
Ober he would have commiserated with a grim-faced Wittgenstein on
the undrinkability of the postwar *Ersatzkaffee*.

If I were a Jim Jarmusch, I would convey the waiter's tale through
a series of short episodes forming a movie called *Coffee and Cigars*. But
alas, I'm not an artist—just an elderly, stooping professor who grew up
in the shadow of the Circle. And so I'll simply tell you its story from the
beginning, as best I can.

# A Tale of Two Thinkers

*Vienna, 1895–1906: Famous physicist Ernst Mach hired as philosopher. Mach makes ready to meet philosophy halfway. Analyses shock waves, science's history, dizziness, and other sensations. Rejects "Thing-in-Itself." Rejects atoms. Rejects ego and absolute space. Assaults metaphysics. Though celebrated by waltz-giddy Vienna, Mach bows out after stroke; physicist Boltzmann takes over. Boltzmann claims atoms are needed, claims disorder increases, claims to be his own successor. Likens metaphysics to migraines, suffers from both. Hanging suicide of Boltzmann. "It came as no surprise," writes Mach.*

## STUDENT HIRES PROFESSOR

In 1895, an otherwise unremarkable university administration took the bold step of appointing a physicist to a chair in philosophy. The university was in Vienna. The physicist's name: Ernst Mach.

During the nineteenth century, forbidding walls between the disciplines were starting to grow, and academic hierarchies were becoming more and more rigid. If an elderly scientist took to dabbling in philosophy, well, that was his own affair, but to entrust him with a chair in philosophy, when he had not even studied Kant or the Scholastics, was seen as highly out of order.

The enterprise started well enough: the university chair in Vienna seemed made to measure for Mach. But after only a few years, he had to resign out of the blue when a sudden stroke paralyzed him. His lectures were then entrusted to another physicist, the celebrated Ludwig Boltzmann. However, this, too, lasted but a few years, for Boltzmann committed suicide by hanging himself. It seemed as if a beautiful new tradition—having physicists teach philosophers—had been nipped in the bud. Nonetheless, it was out of that novel tradition that the Vienna Circle was to grow a couple of decades later. The two world-renowned physicists had imbued an entire generation of students with their passion for philosophy.

Mach and Boltzmann were alike not just in their looks but also in their careers. They had similar heavy physiques, bushy beards, and thin-rimmed glasses; in their youths, they learned from the same teachers, and as university students they both enjoyed great success. Most importantly, Mach and Boltzmann were headstrong and opinionated, and they relished it. Neither of them ever shied away from any philosophical controversy—least of all from debating each other. Their fierce debate about the reality of atoms became part of the great lore of the history of science.

Curiously enough, Mach's appointment as a philosophy professor came about largely because of a mere student, and that student later turned things around by getting his PhD under Mach's supervision. This was irregular by any standards! However, that student, Heinrich Gomperz (1873–1942), was not just anybody; he was a young man with connections.

The Gomperz family was one of the wealthiest and most influential families in the city. It stood on a par with the Rothschilds, the Wittgensteins, the Liebens, the Gutmanns, and the Ephrussis—the fabulously wealthy Jewish dynasties of Vienna's liberal *Gründerzeit,* or "Age of the Founders," which referred to those who had founded financial and commercial enterprises spanning Central Europe. The Habsburg dual monarchy (that is, the Austrian Empire and the Kingdom of Hungary) was firmly established, and its new oligarchs were enjoying an unprecedented economic boom. This provided them with luxurious town palaces, most of them on the Ringstrasse, Vienna's new circular boulevard; with castlelike country resorts and private sleeping cars; with glittery balls orchestrated by Johann Strauss, plush boxes in Gustav Mahler's opera house, and mausoleums of the finest marble in Vienna's huge central cemetery. You can say what you like about the belle époque, but back then it really paid to be a millionaire.

Heinrich's father, Theodor Gomperz (1832–1912) had turned down a career all planned out for him. Instead of becoming a banker or an industrialist, he chose to devote himself to his private studies; he never needed to obtain a doctoral degree. He succeeded perfectly well without one and soon was recognized as one of Europe's foremost classical philologists. He was duly elected a member of the Imperial Academy of Science and was appointed full professor at the University of Vienna. His three-volume history of classical philosophy, *Greek Thinkers,* was a standard reference work for many decades.

The interests of Gomperz *père* went far beyond the classics, though. They extended to modern thinkers such as Auguste Comte (1798–1857) and John Stuart Mill (1806–1873). These positivists, as they called themselves, had no truck with ancient dogmas and doctrines and made light of all hallowed gospels of religious or metaphysical creeds. No holy writings, no mystic insights: all knowledge was to be based solely on hard-nosed scientific facts. This radically new approach shocked the stalwart keepers of philosophical traditions such as the natural theology of Saint Thomas Aquinas, the moral metaphysics of Immanuel Kant, and the absolute idealism of Georg Wilhelm Friedrich Hegel, all of which were firmly entrenched in the curricula of German-speaking universities. And so those tradition-keepers returned the fire. Thanks to their efforts, words like *positivistic, materialistic,* and *utilitarian* soon took on a highly negative flavor, connoting a shallow soul and a contemptible impotence, unable to fathom the true depths of idealistic philosophy.

Theodor and Heinrich Gomperz, however, were not scared of tackling bold new intellectual ventures, and both of them admired Ernst

FIGURE 2.1 Whisperings in the Academy: Theodor Gomperz and Ernst Mach.

Mach's original views, which were so refreshingly different from the traditional lore of philosophy. A lecture given by the illustrious experimental physicist captivated them. Years later, Gomperz *fils*—by then himself a lecturer in philosophy—confided to Ernst Mach: "When you gave a talk on causality in the early nineties here in Vienna—I think it was at the meeting of Natural Scientists—my father handed me your manuscript to read. I returned it to him the next morning with the words: 'Why, here is the philosopher you are looking for to fill the third chair in philosophy!' My father took up this suggestion, as you know; and hence I, though but a student at the time, was in some sense instrumental in your appointment."

Prodded by his son, Theodore Gomperz lost no time in sounding out Ernst Mach, whom he knew well from the Imperial Academy of Science: "Most respected colleague, I am broaching you today with a request of a very unusual nature, and I will make so bold as to ask you for a prompt reply. In a spontaneous fashion among myself and some colleagues, the desire has arisen to humbly ask you whether it would be hopeless for us to try and solicit you to accept one of the university chairs here in Vienna, of which some are already vacant and others soon will be."

The polite request was met with a gracious and positive response, and in the end Ernst Mach accepted the new Chair for the History and Theory of Inductive Sciences, specially renamed on his behalf, at the University of Vienna. Taking such a step from physics to philosophy had long been in the cards for Mach. As he himself wrote: "My life's task was to start out in science and later meet philosophy halfway."

## Mach Makes a Name for Himself

Ernst Mach was born near Brno (Brünn, at the time) in Moravia. He grew up in Untersiebenbrunn, a small village near Vienna as profoundly rural as its quaint name ("Amid Seven Founts") would suggest. There, his father, a former schoolteacher, ran a farm and, in his spare time, home-taught his children.

At age ten, Ernst Mach was sent to a boarding school in the Benedictine monastery of Seitenstetten in Lower Austria. It soon became clear, though, that the sickly child was not up to the taxing requirements of the *Gymnasium* (one type of Austrian secondary school), and thus it came about that little Ernst returned to rustic Untersiebenbrunn. His father, after all, was still able to provide instruction. As this left Ernst with plenty of free time, he went to work as an apprentice to a cabinetmaker.

One day, while rummaging through his father's bookshelves, the inquisitive apprentice stumbled upon a curious title: *Prolegomena to Any Future Metaphysics That Will Be Able to Present Itself as a Science*. The author's name was Immanuel Kant. This was a decisive moment, as Mach would often fondly recall. In his words: "The 15-year-old boy eagerly devoured this clearly written and relatively accessible book. It made a tremendous impression on him, destroying the boy's naïve realism, whetting his appetite for the theory of knowledge, and ridding him, thanks to the influence of the metaphysician Kant, of any inclination to do metaphysics himself . . . I soon turned away from Kantian idealism. While still a boy, I recognized the 'Thing-in-Itself' as a needless metaphysical invention, as a metaphysical illusion."

Later on, a spirited opposition to Immanuel Kant would unite all the thinkers of the Vienna Circle. In fact, the ideas of the noted Prussian philosopher had never enjoyed great favor in Vienna. As Otto Neurath quipped, "The Austrians figured out how to avoid the detour through Kant." Only Karl Popper, playing his favorite role of "official opposition" to the Vienna Circle, would agree with Kant—at least now and then. And later on, it came out that Kurt Gödel was a closet Kantian.

Soon after his first encounter with metaphysics, the young Ernst Mach tried again to attend a *Gymnasium*—this time at the Moravian monastery of Kremsier (today Kromeriz), run by the order of the Piarists. His second attempt was more successful: "The only disagreeable moments were provided by the endless religious exercises, which incidentally produced the opposite of their intended effect."

After graduating from this school, Mach enrolled in mathematics and physics at the University of Vienna. The physics institute there was in full bloom, thanks to the high-quality research of Christian Doppler (1803–1853), Johann Loschmidt (1821–1895), and Josef Stefan (1835–1893). This heady period was unprecedented. For centuries, the University of Vienna had been in the grip of the Jesuits, and the Habsburg rulers had tended to encourage music rather than the exact sciences. Thus it was only in 1847 that an Imperial Academy of Science was established in Vienna—a couple of centuries after similar academies had been set up in Florence, London, and Paris. Even an intense lobbying effort by the polymath Gottfried Wilhelm Leibniz (1646–1716), himself a one-man academy, had been to no avail. It was only with the dawn of liberalism that Austrian science was at last able to throw off its shackles. Now it was time to catch up with the rest of Europe.

Young Ernst Mach was one of the talents whose moment had come. His resourcefulness and manual dexterity, due in part to his cabinetmaking stint, were soon appreciated at the physics institute. While still a student, Mach constructed an ingenious apparatus that convincingly demonstrated the Doppler effect: namely, that the perceived pitch of a sound rises when the sound's source approaches the listener. To illustrate this, Mach attached a whistle to a vertical disk. When the disk was set to spinning, the pitch of the whistle would alternately rise and fall for anyone standing in the plane of the disk, while to the ears of an observer standing along or near the axis of rotation, the pitch remained perfectly constant.

At age twenty-two, Mach received his doctorate. The following year, he earned the right to lecture at the university. Barely twenty-six years old, Mach became a professor in Graz, first in mathematics and later in physics. He married in 1867.

That same year, Mach was appointed to a chair in experimental physics at Prague. He was not yet thirty. He remained in Prague for the next thirty years, until his return to Vienna. Prague's German-speaking university had been founded in the Middle Ages, even before Vienna's. When Mach arrived, it was in the throes of a fierce political struggle. Emperor Franz Josef, after being defeated by Bismarck's Prussia in 1866, had been forced to concede far-reaching autonomy to the Hungarians. And now the Czechs were clamoring for the same rights! To the Austrians, such a thing was utterly unthinkable. Ernst Mach, during his years as dean and later as rector, or head, of the University of Prague, found himself caught in the middle of severe nationalistic turmoil, similar to the upheavals in Ireland. He advocated creating a new Czech university from scratch, rather than splitting up Prague's venerable old university, the Alma Mater Carolina, founded all the way back in 1348. In the end, however, his idea failed.

Working on shock waves in his physics lab was much more to Mach's taste. He soon made a name for himself—indeed, quite literally so. To this day, "Mach one" refers to the speed of sound, while "Mach two" means twice the speed of sound, and so forth. Thanks to his experimental work, he became a pioneer of scientific photography. He caught images of bullets in flight—a remarkable achievement in an era when portrait photos were often blurred because the sitter grew restless as the minutes dragged by. Mach's pictures of streamlines and shock waves thrilled his contemporaries and inspired, a few decades later, the Italian Futurists' attempts to capture in images the nature of great speeds.


## A Glimpse Behind the Scenes

Even more than his experiments, Mach's ideas on the foundations of physics brought him worldwide acclaim. As Karl Popper would later write: "Few great men have had an intellectual impact upon the twentieth century comparable to that of Ernst Mach. He influenced physics, physiology, psychology, the philosophy of science, and pure (or speculative) philosophy. He influenced Albert Einstein, Niels Bohr, Werner Heisenberg, William James, Bertrand Russell—to mention just a few names."

There have been scores of scientists who philosophized, and quite a few philosophers who tried their hand at science. But Mach was exceptional. He pioneered a new discipline: the philosophy of science. Science itself became the topic of inquiry. The time was ripe. No longer could science be viewed as the hobbyhorse of just a few isolated thinkers and visionaries. In the nineteenth century, science had turned into a global enterprise spanning the generations. It was universally recognized as the engine driving the Industrial Revolution. The question could no longer wait: Given that human progress is based on science, what is science itself based on?

To understand the underpinnings of knowledge was and still is one of philosophy's principal tasks. How do we know that there is a tree over there? Or that Napoleon once lived? Or that a dog can feel pain? Mach addressed a more practical matter, one that could not be sidestepped or shrugged off: the foundations of *scientific* knowledge, the growing, hard-fought knowledge that belongs to all and that affects everybody. He addressed these issues in three books: *The Science of Mechanics* (1883), *Principles of the Theory of Heat* (1896), and *Principles of Physical Optics* (which appeared posthumously in 1921).

What is the true meaning of physical concepts such as force, heat, and entropy? What is matter? How do we measure acceleration? Mach tackled such questions from the bottom up, starting with the simplest observations, and continuing with a critical analysis of the historical roots. From the start he intuited the intimate link between the philosophy of science and the history of science.

The first paragraph of *The Science of Mechanics* comes right to the point: "The present volume is not a treatise about applications of the principles of mechanics. Its aim is to clear up ideas, expose the real significance of the topic, and get rid of metaphysical obscurities." Then Mach proceeds: "The gist and kernel of mechanical ideas has, in almost every case, been developed through the investigation of very simple special examples of mechanical processes. The historical analysis of the way such examples were first understood will always be the most effective and natural means to reveal this kernel. We may even claim that it is the only way that can lead to a full understanding of the general results of mechanics."

Textbooks, then as now, aim to lead the student as quickly as possible to the state of the art. But for a critical analysis of the tools—the concepts and methods—it helps to know how they evolved. Thus Mach's approach to physics was historical. On the other hand, he had little interest in the history of philosophy, in contrast to traditional philosophers. Modern times had arrived. It was best to start from scratch, building up from the basics.

With the acumen of a psychologist, Mach analyzes concepts such as "physical force," for example—a notion that, although it is familiar to everyone, took a long time to emerge with scientific clarity: "Let us direct our attention to the concept of force. . . . Force is a circumstance leading to movement. . . . The circumstances giving rise to movement that are best known to us are our own acts of volition, the results of nerve impulses. In the movements that we ourselves initiate, we always feel a push or a pull. From this simple fact arose our habit of imagining all circumstances that give rise to movement as akin to volitional acts, and thus as pushes or pulls."

A physicist views the vast universe as filled with all kinds of forces, a concept derived through a long and arduous intellectual process. It seems odd to ground this notion in intimate bodily sensations first experienced as a toddler. But what else can we do? "Whenever we attempt to discount this conception [of force] as subjective, animistic, and unscientific, we invariably fail. Surely it cannot profit us to do violence to our own natural thoughts and to deliberately inhibit our minds in this regard."

In this way, Mach reduced physical concepts to directly experienced sensations such as pushes and pulls—that is, to sensory impressions. Hence his physical interests led him inevitably toward physiology. In this field, too, he struck gold. For instance, he pinpointed the sense of balance in the inner ear, thus adding a sixth sense to Aristotle's famous list of five. This discovery was also made, roughly at the same time, by Josef Breuer (1842–1925), a Viennese physician who later, along with Sigmund Freud, contributed to the founding of psychoanalysis. Still later, Robert Bárány (1876–1936) extended Breuer's and Mach's findings and was rewarded with a Nobel Prize in Medicine, the first one to go to Vienna. Why was Vienna such a fertile ground for the study of dizziness? Might it have been due to the then prevailing craze for the waltz?

## THRIFTY THINKING

Science has to restrict itself to empirical facts, but it is obviously more than a mere stockpiling of them. For Mach, the main aim of science was an *economy of thought:* that is, to describe as much as possible in as concise a manner as possible. Newton's law of gravitation, for instance, covers, in one short equation, countless phenomena, ranging from the fall of an apple to the orbit of the moon. Mach writes: "All of science tries to replace or economize experience by mental models, since models are easier to deal with than experiences, and can even replace them in some situations. . . . By recognizing science's fundamentally economical nature, we rid science of all mysticism."

Mach was radical: in his view, theories serve solely to simplify thought. Natural laws are mere prescriptions guiding our expectations, and causality is nothing but the regular connection of events. In this sense, causal links do not provide an additional "explanation." "Most researchers ascribe a reality beyond the human mind to the basic concepts of physics, such as mass, force, and atom, whereas they have no other purpose than to connect experiences in an economical fashion. Moreover, it is commonly believed that these forces and masses constitute the true field of inquiry, and that if they were specified, everything else would follow directly from the equilibrium and the motion of these masses."

But this view confuses reality with representation, argued Mach. Force, mass, and atom are mere concepts—just intellectual props. "Someone who knew the world only through the theater, and who came across the mechanical contraptions behind the scenes, would likewise come to think that the real world needs a backstage. . . . In that sense, we should not confuse the foundations of the real world with the intellectual props that serve to evoke that world on the stage of our thoughts."

Economic principles govern not only the activity of science but also its teaching: "Instructing individuals in science aims at sparing them the task of acquiring experience on their own, by providing them instead with the experience acquired by others."

When he was young, Mach had had unhappy experiences at school. Hoping to spare others a similar fate, he campaigned tirelessly for school reforms and for improved curricula. He wrote a textbook for secondary schools. Despite the author's renown, it took no small effort to get it approved by the ministry of education. Brilliance is suspect.

A born teacher, Mach wrote splendid essays explaining science to the public, was an advocate for adult education, and never ceased to fight against the "ingeniously contrived barriers which barbarically prevent mature persons of talent, who have missed out on the usual schooling, from attending institutions of higher education and entering into learned professions."

For Mach, education was enlightenment: "I will not meet any opposition when I claim that without at least an elementary instruction in mathematics and science, man will remain a stranger in this world, a stranger in the culture that supports him." Culture was not to be reserved to just one of the two sexes, by the way—Mach used the word *Mensch*, "human being."

Not only in scientific theories, but also in school studies, our thoughts can get tangled up in the backstage clutter of abstract concepts, like a fly in a spiderweb. Science education was still in an embryonic stage: "Without any doubt, far more can be expected from the teaching of science and mathematics once a more natural method of teaching is adopted. This means in particular that young people should not be ruined by being exposed too early to abstraction. . . . The most effective way to disrupt the process of abstraction is to embrace it too early."

And in another passage Mach writes: "I know of nothing more depressing than those poor people who have learnt too much. What they have acquired is a spiderweb of thoughts, too weak to offer support but complicated enough to confuse them." Mach wanted to do away with this spiderweb.

## THE EGO AND ITS SENSATIONS

Mach's most important philosophical work appeared in 1886: *The Analysis of Sensations*. It opens with "Anti-Metaphysical Introductory Remarks"—a clarion call for striking down Immanuel Kant's *Ding an sich*, or "Thing-in-Itself," and for that matter any "thing," or any substance. Mach considered such notions to be useless dead weights, superfluous abstractions lacking any connection to our sense organs. Since science for him was economic thinking, it had no room for such extravagances. Fleeting sensory impressions are all we have to go by.

Mach's empiricism was all-encompassing. For him, all knowledge had to be grounded in experience, and all experience grounded in perception, hence in sense-data, which is to say, in his "sensations": "Colors, sounds, temperatures, pressures, spaces, times, and so forth, are connected with one another in manifold ways; and attached to them are moods, feelings, and desires. In this vast web, only that which is relatively solid and permanent stands out prominently, engraving itself on the memory and expressing itself in language. A relatively greater permanence is exhibited by certain patterns of colors, sounds, pressures, and the like, which are bound together in space and time. Such patterns are recognized as *objects,* and they are assigned names. But in no way are these objects truly permanent."

Within such a pattern, primordial sensory elements can shift, like the colorful pebbles in a kaleidoscope: "A pencil held up before us in the air is perceived by us as straight. Yet dip it at an angle into water, and we see it as bent. We will say that the pencil merely *appears* to be bent but is in reality straight. But what allows us to declare one fact to be reality while demoting the other fact to the status of mere appearance?"

Indeed, why should tactile sensations be privileged, as opposed to visual ones? Why should we trust our fingers more than our eyes? Or should we? "The objects that we perceive consist merely of bundles of sense-data linked together in regular ways. There exists no further object independent of our sensations—no Thing-in-Itself. . . . We thus know only appearances, never a Thing-in-Itself—just the world of our own sensations. Therefore, we can never know whether there exists a Thing-in-Itself. Consequently, it makes no sense to talk about such notions."

And this leads us to the next unsettling thought: *I* do not exist any more than any other thing does: "Among the relatively long-lasting patterns of memories, moods, feelings, etc., there is one pattern that is attached to a special body, and that pattern is called 'I', or Ego . . . Yet this Ego is just as transient as are all other things."

This was a topic that Mach would take up again and again. A striking experience had once left an indelible mark on him: "One bright summer's day in the open air, the world with my Ego in it suddenly appeared to me as nothing but one tightly bundled mass of sensations, just bundled together more tightly in the Ego."

Had Mach been a mystic, he would have taken this as a flash of enlightenment. However, being a hardnosed physicist, he merely drew an ironic sketch back in his study, calling it "The Ego Inspecting Itself." The ego consists of sensations. Behind them, there lurks—well, nothing. Nothing at all. And nothing more remains to be said about

FIGURE 2.2 Mach's Ego
inspecting itself.

it: "'I experience green' means that the element of greenness occurs within a certain pattern of other elements (sensations, memories). When I can no longer experience green—when I die—then these elements will no longer occur in their usual familiar groupings. That is the whole story. . . . No Ego will remain. The Ego cannot be saved."

The notion of the "unsavable Ego" became a catchword with the writers of Young Vienna. Mach's world without objects or substances, consisting entirely of sensory impressions, was impressionistic by definition and thus completely in tune with the tenor of the times—the heady Zeitgeist of the belle époque.

In the nearby Berggasse, a cigar-smoking Sigmund Freud dissected the soul by closely tracking his patients' associations, as well as those of his "chief patient" (namely, himself). The poet Hugo von Hofmannsthal, Young Vienna's wunderkind, attended Mach's lectures. Vienna's foremost writer Arthur Schnitzler adopted Mach's perspective in his "inner monologues" and dissolved the ego into chains of associations and bundles of sensations. Those in the fine arts chimed in and painted not *things,* but *light.* The remarkably versatile historian and cabaret performer Egon Friedell (1878–1938) pithily summarized the work of the impressionists: "In one brief word, they painted Mach."

In the salons of fin-de-siècle Vienna, the venerable physicist/ philosopher with the prophet's head became a lionized celebrity. True, Mach dressed in a somewhat slovenly manner, and his hair was often disheveled—but the waltz-giddy beau monde was struck by this homespun genius and was itching to hear him hold forth on his original views. And Mach came through for them, finding just the right words to thrill the Viennese society of artists and critics, countesses and mistresses, patrons of art and entrepreneurs: "When I say 'The Ego cannot be saved,' I mean that it consists solely of man's way of relating to things and to phenomena; that the Ego totally dissolves into that which can be felt, heard, viewed, or touched. Everything is fleeting: ours is a world without substance, consisting solely of colors, shapes, and sounds. Its reality is in eternal motion, colorful as a chameleon."

The Austrian writer Hermann Bahr (1863–1934) waxed ecstatic: "In this phrase 'The Ego cannot be saved,' I found spelled out what had tormented me for these last three years. The Ego is a mere name; it is a mere illusion. It is a quick fix that we exploit to put some order into our thoughts. Nothing truly exists but combinations of colors, sounds, temperatures, pressures, spaces, times, and their associated moods, feelings and desires. Everything is eternally changing."

The spell cast by these ideas was not restricted to Vienna's *haute bourgeoisie.* Mach also attained a particular eminence among Marxists. More than a few of them hailed his work as a fresh new approach to materialism. The Austro-Marxists proved particularly receptive—so much so that Vladimir Lenin felt obliged to call these unruly dissidents to order. In his 1908 book, *Materialism and Empiriocriticism,* written expressly to oppose their heresy, he thundered, "All our Machists are deeply mired in idealism." Mach must have felt surprised at being accused of idealism, but by claiming to resolve matter into mere bundles of sensations, he certainly was a threat to materialists.

Foremost among the Machists in braving Lenin's ire was a young theoretical physicist named Friedrich Adler (1879–1960). He was the son of Viktor Adler, the highly respected founder of the Austrian Social Democratic Workers' Party, and he looked uncannily like a clone of his father. Ten years after Lenin's attack on Mach and his disciples, Friedrich Adler fired back with a book of his own, titled *Ernst Mach's Victory over Mechanical Materialism.* He wrote it while in a cell on death row—but more on that later. Indeed, although Friedrich Adler would never be a member of the Vienna Circle, the side plot in which he stars will form an important part of the Circle's tale.

Three years after his appointment to the University of Vienna, Mach suffered a paralyzing stroke during a long train trip. He was no longer able to move his right arm or right leg. In 1901, after making valiant attempts to resume his lectures, he finally had no choice but to resign for reasons of ill health. He declined the emperor's offer to make him a baron, as it ran against his democratic convictions. But he could not resist being named a lifelong member of the Austrian House of Lords, or *Herrenhaus,* along with his trusted old friend, Theodor Gomperz.

Despite his infirmity, the aging Mach remained as intellectually agile as ever, ceaselessly sparring with some of the foremost scientists of his time, such as Ludwig Boltzmann and Max Planck. A halo of controversies surrounded him. Indeed, his opinions, while seductively unconventional, raised substantial problems when pursued carefully, such as: If all science rests on sense-data, then what about things that cannot be perceived? Must we discard them as extravagant fictions? And what about other people's sense-data, which we ourselves can never experience? Must we discard them, too? Mach constantly had to defend himself against charges of being a solipsist—he, who had proclaimed the end of the Ego!

## BOLTZMANN'S FORMULA

Ernst Mach was not the first physicist to call into question the Ego's existence. A century before, in a similar vein, Georg Lichtenberg

(1742–1799) had quipped that we should say "It thinks" rather than "I think." And Mach's Viennese colleague Ludwig Boltzmann clearly shared Lichtenberg's outlook when he railed at "the bizarre opinion that we can think as we choose to think." The lives and thoughts of Mach and Boltzmann were closely intertwined.

Ludwig Boltzmann, born in 1844 in Vienna, came from a background as modestly middle-class as Ernst Mach's. Soon after Ludwig's birth, his father, a tax official, was transferred to the finance department of the town of Linz. There, the boy's remarkable talents, especially in mathematics and music, were quickly noticed. And much like Ernst Mach as a boy, little Boltzmann was privately tutored before entering the *Gymnasium*. His young piano teacher, a certain Anton Bruckner, was just starting to make a name for himself as the town organist of Linz.

When Ludwig was fifteen, his father died. His widowed mother spent her entire inheritance on the education of her sons. After Ludwig had finished his *Gymnasium* studies, the family returned to Vienna. There, the young man studied mathematics and physics, got his doctorate in 1866, and obtained a lectureship, just like Mach, at the tender age of twenty-three. But Boltzmann's interests focused more on theoretical than on experimental physics. Later, he would joke: "I disdain experiments the way a banker disdains coins."

His professor Josef Stefan had urged him to read James Clerk Maxwell's treatises on physics, handing him an English grammar book as well, for at the time Boltzmann did not speak one word of the language. He proved to be a quick learner. Already his second paper, *On the Mechanical Interpretation of the Second Law of Thermodynamics,* turned out to be groundbreaking. Soon he was recognized as the physicist most able to grasp and extend Maxwell's work on electromagnetism and thermodynamics.

By the time he was twenty-five, Boltzmann was appointed full professor of mathematical physics in Graz. In 1875, he became professor of mathematics in Vienna, but remained there for just three years; then he returned to Graz, accepting a chair in experimental physics for which Mach had also been under consideration. Of course, Boltzmann did

FIGURE 2.3  Boltzmann proposes.

not really disdain experiments, as he had claimed, and was delighted to accept such a chair—but there was also another reason behind his return to Graz.

During his previous stay in that city, he had met a young lady named Henriette von Aigentler, who was uncommonly fond of mathematics and physics. Boltzmann convinced the authorities to allow her to attend university lectures, something unheard of at the time. His motive was not entirely selfless. In 1875, he asked in a letter for Henriette's hand. He wrote:

> No matter how little I believe that emotions could or should ever be inhibited by the cold and inexorable consequences of the exact sciences, it nevertheless behooves us, as the representatives thereof, to act only after well-considered judgment, rather than to follow fleeting whims.
>
> As a mathematician, you surely do not find numbers, which rule the world, to be unpoetic. And so: my salary is currently 2400 florins per year. My active annual bonus is 800 florins. Last year, my fees from lecturing and examining amounted to about 1000 florins; this latter

revenue, however, is subject to change from year to year. . . . The sum total is not small and will suffice to keep a household going; however, in view of the enormous rise in prices these days, it will not afford you many distractions and amusements.

Boltzmann's well-penned if starchy proposal was accepted, and the marriage resulted in five children—the same number as in Ernst Mach's family.

The next fifteen years in Graz were Boltzmann's most productive period—not just in terms of progeny but also in scientific output. He became one of the founders of the kinetic theory of gases, which provides a mechanistic underpinning to thermodynamics. Not only was this a major breakthrough for physics, but it was also philosophically relevant, as it provided a causal explanation based on a mechanical model, a feature that Mach was slow to accept.

According to Boltzmann, gases consist of particles that are constantly rushing about and colliding like billiard balls—the greater the temperature, the faster they move, although they don't all have the same velocity. As they collide with each other and with the walls (thus exerting measurable pressure on the walls), some speed up, while others slow down. Boltzmann's equations statistically summarizing such particles' behavior soon became central pillars of physics, and today they play a pivotal role in many areas of technology—for instance, in the theory of semiconductors.

Of course, gas particles are not really miniature billiard balls. In light of this, shouldn't we say that the statistical theory of gases, rather than providing an *explanation,* merely provides a *picture*? But then again, aren't the tiny particles in the container far more real than a mere picture? And doesn't their constant whizzing-about in fact *cause* the pressure? Even the mysterious notion of entropy, which always increases with time for any closed system, becomes intuitive and understandable when rephrased in terms of statistical mechanics.

According to Boltzmann, entropy is related to the probability of the state of the particles in the vessel, which is greater when the system is

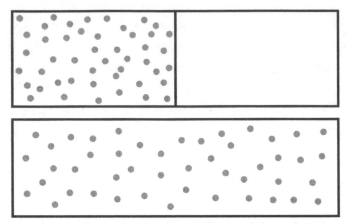

FIGURE 2.4  Gas molecules, first confined and then released.

more random (just as a shuffled deck of cards is more likely to look random than to be in pristine order). In other words, entropy is a measure of the disorder of the system when examined on a microscopic level. And it should come as no surprise to learn that disorder increases with time if things are left to themselves—just look at your desk!

Mach, however, remained unimpressed: "To reconcile the molecular hypothesis with entropy is a bonus for the hypothesis, but not for the law of entropy." In his eyes, the sole duty of a theory was to concisely relate *observables,* such as pressure or temperature. Boltzmann's statistical recasting of thermodynamics had therefore stepped out of bounds.

Moreover, the new theory raised some thorny issues. For instance, if disorder always increases with time, then this fact must define the direction of time's flow. To be concrete, suppose all the molecules in a gas are placed in the left half of the container and are then released. As they bash into each other, they will quickly spread all around the container. If left to themselves, they will never again occupy just the left half of the container. Things will never go back to the simpler, more ordered state in which they started. At least, no such reversal has ever been observed so far. And so this effect of ever-increasing disorder clearly distinguishes past from future, thus creating an unambiguous arrow of time.

Two serious objections were raised to Boltzmann's theory, and to this day neither of them has been settled in a generally accepted way. They are called the paradox of recurrence and the paradox of reversibility.

The *paradox of reversibility* was first raised by Boltzmann's fatherly friend and mentor Josef Loschmidt. The laws of mechanics, which govern the collision of billiard balls and of all objects, do not distinguish future from past. Thus if we watch a film of billiard balls colliding frictionlessly on a table, we cannot tell whether it is running forward or backward. But if we watch a film of a drop of cream dissolving in a cup of coffee, we can easily tell. So how does time acquire its arrow?

The *paradox of recurrence* originated with the German mathematician Ernst Zermelo (1871–1953). According to the laws of probability, every state that has been reached once must be reached again, and again, and again. This is an ironclad theorem. Hence the particles in the container eventually *must* all return to the left half of the container in which they were originally confined. Except that they don't!

Such tricky riddles can be troubling for even the most cool and composed of thinkers—and "cool and composed" was hardly what Boltzmann was.

## Professor Restless

All throughout Boltzmann's life, his temper had oscillated between extremes. Jokingly, he ascribed his mercurial streak to having been born in the wee hours of Ash Wednesday, the night between Carnival and Lent. His psychological agitation increased with age, and this started to worry colleagues and friends.

He accepted a professorship in Berlin only to give it up right away, and yet, a short while later, he reasserted his interest in it. In 1896 he accepted a chair in Munich, and soon thereafter, a different one in Vienna. In the year 1900, after endless vacillations, he accepted an offer from the University of Leipzig. But then in 1902, like a gas molecule bouncing about in a container, he returned to Vienna. On this occasion, he became his own successor, as he gleefully pointed out on resuming his chair: "One usually starts out one's inaugural lecture with a paean praising one's predecessor. Today, however, I can fortunately spare myself this oft-challenging task, for the fact is, I am my own predecessor."

The authorities were not so amused by Boltzmann's fickleness. This time, Boltzmann was required to pledge his word of honor to Emperor Franz Josef himself: never again would he accept an offer from abroad. No more job-hopping! But Boltzmann's love for travel did not abate in the least. He proceeded to visit Constantinople, Smyrna, Algiers, and Lisbon, and he also crossed the Atlantic three times to travel across the United States. In *A German Professor's Trip to El Dorado,* he recounted the third of these voyages, which included a brief stay at the newly founded Stanford University, in a humorous manner. (He called himself German rather than Austrian as he was referring to his cultural background rather than his nationality.)

By that time, Boltzmann enjoyed worldwide fame. Two of his former collaborators in Graz, Walther Nernst (1864–1941) and Svante Arrhenius (1859–1927), would later receive the Nobel Prize. Among his Viennese students were the brilliant and captivating Lise Meitner (1878–1968), who later codiscovered the fission of uranium, as well as the theoreticians Paul Ehrenfest (1880–1933) and Philipp Frank (1884–1966).

## The Great Debate

The paths of Mach and Boltzmann crossed frequently. This couldn't help but lead to a certain degree of rivalry, even though one was an experimentalist and the other was a theoretician. Thus in 1874, Ernst Mach was elected to the Imperial Academy of Science, but not Ludwig Boltzmann, who had also been on the ballot; conversely, in 1894, it was Boltzmann who was selected for the chair in physics at the University of Vienna, but not Mach, who also had expressed interest in it.

The two physicists respected each other highly, but their courtesy could not hide the fact that they tended to hold different views. This tension climaxed in their famous debate on the reality of atoms. Do atoms truly exist, or are they just mental objects, somewhat like the concept of a point?

The controversy polarized the world of physics and chemistry. Nobel Prize winners joined the fray, such as Wilhelm Ostwald (1853–1932) on Mach's side, and Max Planck (1858–1947) on Boltzmann's (although Planck had only recently shifted sides in the controversy). Later, Karl Popper would write: "Both Boltzmann and Mach had large camps of supporters among physicists, and they were engaged in an almost deadly battle—a battle over what kind of research should be done in physics."

"An almost deadly battle" is obviously an exaggeration, but the debate was extremely heated. Boltzmann needed atoms for his thermodynamics, and he wrote a passionate plea titled *On the Indispensability of Atoms in Science.* However, since atoms cannot be directly perceived, Ernst Mach treated them as mere models, constructions of the mind, not all that different from his old bugaboo—Kant's Thing-in-Itself. Whenever atoms were mentioned, Mach would slyly ask, with a clear Viennese lilt in his voice: "Did you ever see one?"

Today, nanotechnology allows us to see atoms, in a sense—and in that sense, the debate has been settled, and in Boltzmann's favor. But in essence, it was a debate about a question of philosophy rather than one of physics, and in *that* sense, it is still far from settled. The debate on whether atoms exist turned less on atoms themselves than on what "to exist" means.

## BOLTZMANN'S *NATURAL FILOSOFI*

When, after Mach's stroke, it became clear that he could no longer give lectures, a search was started for a successor. It promised to be difficult and time-consuming. That is why Mach's lectures—but not his chair—were temporarily entrusted to Boltzmann. This seemed rather ironic, given their public disagreements, but a tradition of sorts was emerging, with Boltzmann becoming one more physicist to teach philosophy at Vienna's University.

FIGURE 2.5 Boltzmann lectures
on philosophy.

Boltzmann's opening lecture in 1903 proved to be a spectacular success. Newspapers reported that there had been a nearly life-threatening throng, with the eager crowd stretching way out into the street. Upon learning of this, even the old emperor became curious and invited Boltzmann for a private audience.

The first words of Boltzmann's first philosophical lecture alluded to his illustrious predecessor Mach—and not just as a perfunctory tip of the hat. After courteously saying, "To praise Mach would be like carrying owls to Athens" (or, to substitute an English idiom, "like carrying coals to Newcastle"), he quickly turned to his true point: "I thus believe that I can best honor Mach by doing my utmost to further develop his ideas with the help of my own."

Of course, in this sentence, the phrase *to further develop* meant, in only a slightly coded way, *to fully destroy.*

Knowing that everyone present must have been wondering why he had been given the honor of taking on Ernst Mach's philosophy course, Boltzmann promptly brought up the atomic debate: "Up till now, I have written just one single essay on philosophy, and I was led to doing so by pure chance. One time, in the meeting room of the Academy of Science, I was hotly debating with a group of academicians, including his Honor the Court Counselor Professor Mach, on the value of atomic theories, a dispute that had once again taken on great intensity among physicists. Quite out of the blue, in that distinguished group, Mach tersely declared, 'I don't believe that atoms exist.' This remark just kept on echoing in my mind."

Although they belonged to opposite camps concerning the existence of atoms, the two physicists were united in their distrust of metaphysics. Boltzmann was not one to mince words: "Whereas I felt some qualms about plunging myself into philosophy, philosophers seemed to have none about intruding into science. . . . I first encountered philosophers a long time ago, and at that time I had no idea what they meant by their utterances, and therefore I tried to become better informed about the basics of philosophy." Boltzmann's weapon of choice was the club rather than the foil: "To head straight into the deepest depths, I first turned to Hegel; but oh! what obscure, vacuous balderdash did I find there! My unlucky star then ushered me from Hegel to Schopenhauer. . . . "

At the time, Schopenhauer was extremely popular in Vienna. Like a bull, Boltzmann charged at him head-on, and in so doing he created a furor. The Viennese public, intoxicated, flocked to his lectures, avidly hoping for more fun and more blood. But when, later in the course, Boltzmann turned to the foundations of mathematics, he rapidly lost much of his audience. "My lectures on philosophy," as he ruefully wrote, "did not have the hoped-for success. I talked about set theory, non-Euclidean geometry, and the like. This proved too mathematical for my public, and many gave up." But Boltzmann saw no alternative. He flatly stated: "What the brain is to man, mathematics is to science."

It is not easy to understand, however, why mathematics turns out to be so central to physics. The truth of mathematical propositions does not hinge on sense-data, after all. How, then, can mathematics agree with Mach's radically empiricist viewpoint? "No equation," said Boltzmann, "ever represents any phenomenon with absolute precision. Each equation is an idealization, stressing commonalities and neglecting differences, and therefore going beyond experience."

Both atoms and differential equations are abstract concepts that "go beyond experience." This was bound to make them suspicious for Ernst Mach. But Boltzmann did not sympathize with such qualms, being far too much a pragmatist—a recent term, by the way, that had become popular in the New World. Mach's view of science as an "economical way of thinking" did not convince Boltzmann either. "We hesitate to ascribe to mere 'thriftiness of thinking' the exploration of the physical and chemical nature of stars, or of stellar motions and interstellar distances, not to mention the invention of the microscope and the discovery of the origins of our diseases."

Boltzmann came out with only a few philosophical writings, found mainly in his *Principles of Natural Filosofi* (with that title, he was making fun of a then popular movement to reform German spelling, which in the end went nowhere). The notes from his "Filosofi" course remained unpublished until eighty years after his death, yet they still make good reading.

Much of Boltzmann's thinking seems remarkably modern—for instance, his interest in the analysis of language and in the theory of evolution. Boltzmann saw traditional philosophy as having been rendered obsolete by Darwin's great insights, and in fact he aspired to become "the Darwin of inanimate matter." More than a decade before Einstein's general theory of relativity, Boltzmann toyed with the idea that space was curved. His students summarized this in a cute though rather clumsy couplet:

> *Tritt der gewöhnliche Mensch auf den Wurm, so wird er sich krümmen;*
> *Ludwig Boltzmann tritt auf; siehe, es krümmt sich der Raum.*
> A worm stepped on by a man will coil;
> Enter Boltzmann, though, and space itself will roil!

Boltzmann's demeanor may have struck his listeners as majestic, but deep inside, he was wrestling with philosophy and wallowing in self-doubt. Despite his many harsh words about philosophers, he did not believe that the problems of metaphysics had all been solved. He suffered from metaphysics—an incurable illness. "Metaphysics appears to exert an irresistible charm on the human mind, and this temptation, despite all our vain attempts to lift the veil, has not lost any of its intensity. It seems impossible to squelch our inborn urge to philosophize."

## Looking for the Final Cure

Boltzmann was well aware that the habit of asking questions, though usually healthy, can lead a person to an obsession with sterile pseudo-problems, "much as a baby grows so accustomed to sucking at the breast that eventually it will contentedly suck on a mere pacifier." For example, the instinctive urge to always ask for a cause may lead us to ask for the cause that lies behind the Law of Cause and Effect. This would surely be carrying things too far; but who will tell us where to stop? Will philosophy? Boltzmann wished that that were the case: "Which definition of philosophy imposes itself with irresistible force upon me? I have always suffered from the frightening feeling, weighing down on me like a nightmare, that great riddles, such as how I can exist, or that a world exists at all, or why the world is exactly this way and not some other way, will remain forever unsolved and unsolvable. Whichever branch of science would succeed in solving such riddles seemed the greatest, to my eyes, and thus the true queen of sciences; and this is what I called philosophy."

Alas, the true queen was in exile and her riddles have no solution. And yet those riddles never cease to haunt us:

> My knowledge in science increased. I took in Darwin's teachings and learned from them that I had been mistaken in asking these questions, since they had no solution; but nonetheless the questions invariably returned, and always with compelling intensity. If such questions are illegitimate, why then can't they be discarded? And to make matters worse, countless others rise in their wake. If there is something else

behind perception, how can we ever find out what it is? Or if, on the other hand, there is none, does this mean that a landscape on Mars fails to exist simply because no conscious being ever gazes down on it? If none of these questions makes sense, then why can't we discard them all, or what can we do to squelch them once and for all?

This last question, more than any other, haunted Boltzmann. Not only is there no sensible answer; there is no sensible *question*! So why can't we stop asking, then?

My current hypothesis is totally different from the doctrine that there are certain questions that lie beyond the realm of human understanding. Indeed, according to that doctrine, this would indicate a lack or a flaw in the human capacity for knowledge, whereas I maintain that the existence of such questions or problems is an illusion of the senses. On first glance, it seems surprising that the urge to answer these pressing questions does not fade away even after they have been recognized as illusions. Apparently our habit of thinking is too ingrained for us to be able to let them go.

It is precisely as with the well-known optical illusions, which persist even after their cause has been made clear. Hence the feeling of insecurity, the lack of satisfaction, which overwhelms a scientist who dares to philosophize.

Insecurity indeed. A thinker who loses control of his thinking is but one step removed from madness. Unable to throw off his haunting obsessions, Boltzmann slept poorly, and his neurasthenia gradually worsened. His nearsightedness became so extreme that he had to don three pairs of glasses, one on top of the other, whenever he wanted to play the piano. Headaches, exhaustion, depression, and a terrible, agitated restlessness made his life a burden. Thinking, for him, was becoming agony.

In his philosophical distress, Boltzmann turned to Franz Brentano (1838–1917), a charismatic thinker who embodied a pre-Raphaelite's idea of a philosopher. Brentano had once been a Catholic priest, and thus a professional comforter of souls, which was just what Boltzmann needed. When Brentano had married, he had been forced to resign his position at the University of Vienna, to the great consternation of his students. He had been extremely popular there. Among his aficionados was an audacious young medical student named Sigmund Freud. In one of his lectures, Brentano argued that it makes no sense to speak of the unconscious. The young medical student took note of this claim and had second thoughts about it, but whatever doubts Freud had did nothing to diminish his admiration for his professor.

It was actually Brentano's vacant philosophy chair that had been renamed and offered to Ernst Mach. However, after stepping down, Brentano kept on lecturing in a private capacity. He was hardly in need of a salary, for his wife belonged to the Lieben family, one of Vienna's foremost banking dynasties. But then she died, and after that he moved out of the glittering Palais Todesco on Vienna's Ringstrasse and settled in the hills outside Florence. Gradually his eyesight failed him.

Boltzmann asked for Brentano's philosophical help and guidance, writing to him: "The irrepressible urge to philosophize is like nausea caused by a migraine, like wanting to throw up something that is not there." But despite the nausea, he could not stop: "The sublime, majestic task of philosophy is to make things clear, to finally cure humankind of this migraine."

Increasingly often, Boltzmann had to cancel his lectures because of ill health. Stays in fancy spas did not help. From a sanatorium he wrote to his wife: "I'm sleeping very poorly and am completely beside myself with sadness. If only somebody would come and fetch me, I would leave at once. They don't allow me to leave by myself. Please come, Mama! Or send someone! Please, have mercy, and don't ask for anyone's advice; just decide on your own. Please, forgive me everything!"

In the spring of 1906, Boltzmann had to cancel all his lectures. And then, on September 5 of that year, he committed suicide while on a holiday on the Adriatic coast, near the castle of Duino, an eerily romantic spot where the poet Rainer Maria Rilke would later write his famous elegies. Boltzmann's daughter, when she returned from some errands, discovered her father's body dangling on a short rope affixed to the cross of a window.

In an obituary in the distinguished German newspaper *Die Zeit,* his former rival Ernst Mach wrote: "In well-informed circles, it was known that Boltzmann was not likely ever to resume his university chair again. There was talk of the need to keep him constantly under close observation, because he had already attempted suicide before."

And Franz Brentano wrote in a letter to Mach, reminiscing about their common successor Boltzmann, who now had preceded them in death: "This extremely talented scientist lacked neither philosophical interests nor the pure love of truth. And yet, into what strange speculations didn't he enter! You yourself certainly know a great deal about him, but not as much as I do, if it is true, as he told me, that I was the first human being with patience enough to hear him out to the end. . . . And very honestly, doing so was not an easy task."


## Eloping with Charon

Old, crippled, and growing ever more deaf, Ernst Mach survived his younger rival by ten years. The last major work of his to appear during his lifetime was *Knowledge and Error,* based on his earlier philosophical lectures at the University of Vienna. Mach did not intend "to introduce a new philosophy, but to get rid of an old and stale one." The same "hygienic" goals were later pursued by Ludwig Wittgenstein and the Vienna Circle: cleansing the stable, airing the brain out.

Mach stressed that he was "no philosopher at all, but a scientist," and that "a scientist can rest content when he recognizes, in the conscious psychic activity of his research, a methodically purified, sharpened, and refined version of the instinctual activities occurring every day in the natural and cultural life of animals and people."

Maybe scientists, knowing when to stop, were wiser than philosophers: "Science has progressed almost more through deciding what to ignore than through deciding what to study."

In his notes, which now he had to scrawl with his left hand or type on a machine specially adapted to his needs, Mach returned again and again to his firmest lifelong convictions: "The aim of science: the fitting of facts into thoughts, and the fitting of thoughts to each other." "The Ego varies. It changes and expands, or shrinks. Sometimes, it vanishes altogether—and not necessarily in the unhappiest moments." "Sensations are the common elements of all possible physical and psychic experiences. Given this, many troubling pseudo-problems simply go away."

These three quotes encapsulate for posterity what Ernest Mach stands for. As a kind of personal motto, he added: "To give up nonsense is not to resign."

Ernst Mach, the feisty philosopher of the unsavable Ego, bore his infirmity with a serene, almost Buddha-like detachment. One of his visitors wrote: "I faced a saint who had overcome the last traces of earthly gravity, and in whose eyes there shone the unflinching kindness reached through universal understanding." And the American psychologist William James (1842–1910), who founded philosophical pragmatism, enthusiastically declared, after an encounter with Mach: "I do not think that anyone has ever left me with such a strong impression of unadulterated genius."

In 1913, Ernst Mach moved into his son's home in Munich. Dutifully informing the Imperial Academy of Science of his change in address, he added, with a literary wink: "Should this letter be my last, I ask you merely to assume that Charon, that old joker, has run off with me to a station so remote that it does not yet belong to the International Postal Union."

In 1916, Ernst Mach died at age seventy-eight. In an obituary, Einstein praised his "grandiose one-sidedness" and claimed that "even those who view themselves as Mach's opponents are hardly aware of how much of his viewpoint they have soaked up, so to speak, with their mother's milk."

# A Trial Run for the Vienna Circle

*Vienna, 1906–1914: Youthful fan club of Mach and Boltzmann meets in science café. One Viennese circle among hundreds, it soon scatters. Hindsight spots it as forerunner of Vienna Circle. Theoretical science hits all-time high: Albert Einstein, Bertrand Russell, David Hilbert present novel paradigms. Old Vienna discovers modernity. Freud, Klimt, Schnitzler, and consorts dig deeper. Eminent author warns: we are afloat in midair. Counterintelligence confirms the warning. Spy colonel commits assisted suicide.*

## The Waltz of the Circles

Around 1910, the salons and coffeehouses of Vienna were teeming with discussion groups. Some of them would wind up exerting a decisive influence on the twentieth century. Art, science, and social reform triggered heated debates. Circles formed around Sigmund Freud, Karl Kraus, Gustav Klimt, Viktor Adler, and Arthur Schnitzler. The topics ranged from the avant-garde to Zionism, from school law to modern drama, from feminism (though not under that name) to psychoanalysis, and from urban planning to the history of art. In this bubbling cauldron, the intellectual and personal exchange between groups was intense.

Circles that met to discuss philosophical questions were particularly common. One of them was the Socratic Circle, centered on Heinrich Gomperz, the young scholar who had had a part in attracting Mach to Vienna. The members of this group reverently acted out the dialogues of Plato. Other circles were devoted to Immanuel Kant, Søren Kierkegaard, or Count Leo Tolstoy.

Many of these circles were offshoots of the tireless Philosophical Society of Vienna, which had been founded in 1888 by faithful followers of Franz Brentano as a reaction against his forced retirement from his university chair. Brentano himself, in the inaugural address of the Philosophical Society, laid down the program for his disciples. It was simple enough: their brief was to steer clear of traditional German philosophy, which he summarily dismissed as "pathological."

A profound distaste for the German idealists who followed in the wake of Kant runs like a red thread through Austrian philosophy. Well before Brentano, the mathematician and philosopher Bernard Bolzano (1781–1848), another defrocked priest who was eventually forced to resign from his university chair, had exclaimed: "Germans! When will you finally give up your aberrant ideas, which make you ridiculous and obnoxious to your neighbors?"

Boltzmann tooted more or less this same horn whenever he was given the chance to address the Philosophical Society. For example, he opened one of his lectures, which had been announced under the rather bland title *On a Thesis of Schopenhauer,* with the offhand comment that he had originally intended to use a somewhat more provocative title, to wit: *Demonstration That Schopenhauer Is an Insipid and Ignorant Philosophaster Who, by Ceaselessly Propagating Hollow Twaddle, Spreads Nonsense Far and Wide and Forever Perverts Brains from Top to Bottom.* Amusingly enough, Boltzmann was actually giving Schopenhauer some of his own medicine, for none other than Schopenhauer had used exactly these same phrases to rail against the arcane writings of George Wilhelm Friedrich Hegel. Not that Boltzmann cared one bit for Hegel either.

FIGURE 3.1 The University of Vienna in 1910.

## THE PROTOCIRCLE, OR *URKREIS*

Among the scholars who gravitated around the Philosophical Society were a number of fresh young PhDs who took great pleasure in meeting each other in various coffeehouses in town. In the years before 1910, this was just one group among many. With hindsight, however, one realizes that this short-lived coterie constituted the key link between Mach and Boltzmann, on the one hand, and the Vienna Circle, on the other. No one would have guessed it at the time. After all, these young people were scientists and did not see themselves as heirs to a philosophical tradition, let alone as philosophers. But they had grown up in Mach's and Boltzmann's town: this was enough to stamp them for life. Although Mach was the epitome of a sage and Boltzmann was the opposite, the two played equal parts in molding that small clique of thinkers.

"Strange to say, but in Vienna all physicists were disciples of both Mach and Boltzmann. It wasn't the case that, as an admirer of Mach, one felt any disinclination towards Boltzmann's atomic theory." Thus wrote Philipp Frank, who had started writing his doctoral thesis under Boltzmann and completed it only after the suicide of his supervisor. Half a century later, Frank painted the following portrait of this first Vienna Circle (today called the Protocircle, or *Urkreis*):

> I belonged to a group of students who met every Thursday evening in one of the old coffeehouses of Vienna. We would stay there till midnight and beyond, discussing questions of science and philosophy. . . .
>
> Our interests were widespread, but we kept returning again and again to our principal problem: how can we avoid the traditional ambiguity and incomprehensibility of philosophy? How can we bring science and philosophy closely back together? By "science" we meant not only natural science, but also social sciences and humanities. The most active members of our group, who attended most regularly, were, in addition to me, the mathematician Hans Hahn and the economist Otto Neurath.

Decades later, Hahn and Neurath would become the founding fathers of the Vienna Circle, and thus they will soon take center stage in this story. But at this point, they were just fresh young PhDs with a weakness for philosophy.

Hans Hahn was the son of a Viennese Court Counselor. He had studied in Vienna, and after his doctorate he spent several semesters in Göttingen, Germany, at that time a mecca for all mathematicians. Philipp Frank had a similar curriculum vitae. As for Otto Neurath, he was the son of a Viennese professor and had spent the better part of his university years in Berlin, studying economics, sociology, and history. It was he who made sure that the *Urkreis* did not neglect social sciences and the humanities.

In Vienna, Neurath and Hahn had attended the same school. Otto, who always had an eye for the fair sex, did not fail to take notice of

Hahn's younger sister Olga (1882–1937), a brainy girl who had set her sights on becoming one of the first female students of mathematics in Vienna. Otto's attachment was deeper than a mere flirtation. When Olga, at the age of just twenty-two, lost her eyesight and sank into a deep well of depression, Otto took it on himself to pull her up out of her gloom, organizing private tutoring, and in the end enabling Olga to obtain her PhD in mathematics.

Little is known about this first Vienna Circle. It probably also included Richard von Mises (1883–1953), who, after studying mechanical engineering at Vienna's Technical University, worked on developing water turbines. Later, von Mises would make up the Berlin outpost of the Viennese group. This young man was brimming with arrogant self-confidence. In a report on his doctoral thesis, one of his readers grumbled: "This treatise, written in the style of a revelation, is an immodest imposition on the reviewer."

Modest or not, the young men of the *Urkreis* were firmly determined to make their way in their respective fields. In addition, they all shared a lively interest in pinpointing the true basis of all scientific knowledge, and so they avidly read anything that philosophy could offer them on that topic.

As it happens, all these young postdoctoral thinkers were of Jewish origin, and they couldn't help but be keenly aware of the growing antisemitism in Vienna, a repulsive scourge that incessantly pursued Sigmund Freud, Stefan Zweig, and Arthur Schnitzler in their thoughts and dreams, and that caused Theodor Herzl (1860–1904) to turn to Zionist action. And it was not only Jews who deplored the rampant racism. "Will things improve with another emperor?" asked the aged Brentano, by now nearly blind, in a letter to the aged Mach. Nothing makes it seem probable, he went on; except that improbable things were always likely to happen in that Austrian Empire of Improbabilities.

All the members of the *Urkreis* belonged also to the Philosophical Society of the University of Vienna: Hans Hahn since 1901, Philipp Frank since 1903, Otto Neurath since 1906, and Olga Hahn since 1908.

The young thinkers of the *Urkreis* eagerly fell in with the Philosophical Society's intense anti-metaphysics crusade. They were not professional philosophers, but that hardly mattered: as coffeehouse philosophers, they were warmly welcomed by the Philosophical Society and were given free rein to deliver lectures or participate in discussions whenever they felt like it. The Society thus became a second home to them—or rather, a third home, as their second home was the coffeehouse.

The small group that revolved around Hahn, Neurath, and Frank would soon scatter to the four winds without leaving any lasting traces. The young men all had careers to attend to, after all. Their *Urkreis* was merely a sideshow among the grand fireworks of Viennese modernism, and by today it would long since have been completely forgotten, had it not been resurrected twenty years later.

## ALBERT EINSTEIN

There is no written record of the discussions that chained the members of the *Urkreis* to their coffeehouse tables late into the night. History does not even record the name of the coffeehouse. But it's not hard to guess the key figures on whom their discussions probably focused: Heinrich Hertz, Henri Poincaré, David Hilbert, Bertrand Russell. . . .

And then there was a young German-born patent clerk in Bern, Switzerland, who was about to take physics on a wild roller-coaster ride: Albert Einstein.

It was Einstein (1879–1955) who in 1905 finally closed the book on the debate that had divided Boltzmann and Mach. From that moment on, whether atoms existed was no longer an issue. True, they were still invisible, but they could hardly be disputed any longer. To carry out this remarkable coup, Einstein had needed no new apparatus. He had merely used the potent weapon of deep thought, pointing it at a well-known phenomenon: Brownian motion.

Robert Brown (1773–1858) was a Scottish botanist who had noticed, back in 1827, while peering through a microscope, that minute particles suspended in liquids were constantly jiggling back and forth

FIGURE 3.2 "Einstein transforms the ultimate basis of our knowledge of nature in a much deeper manner than Copernicus did . . . The guide who showed the way up to these summits is Albert Einstein. Through an amazingly acute analysis, he cleansed the most fundamental scientific concepts from hidden prejudices that had gone unnoticed for centuries." (Moritz Schlick)

in a random fashion, almost as if they were tiny animate creatures. But this they were not. Brown published his observations but could offer no explanation for this puzzling behavior. (As is often the case, Brown's discovery was made independently by another person—in this case, the Dutch biologist Jan Ingenhousz, who had written about it over forty years earlier—but it nonetheless became known as *Brownian motion*.)

Einstein hypothesized that the visible particles' unpredictable tumbling motions were caused by constant collisions with far smaller particles making up the surrounding liquid—in other words, by collisions with invisible atoms or molecules, the existence of which was still, in those years, a matter of fiery debate. But Einstein, by making careful statistical calculations based on numerical data observed in the random motion of the visibly jiggling particles, managed to deduce both the size and speed of their invisible dance partners. This ingenious analysis proved to be the final nail in the coffin of the anti-atomists, and the picture of atoms that Einstein had given was soon afterward fully confirmed in follow-up experiments carried out by the French physicist Jean-Baptiste Perrin (1870–1942).

Although Mach offered some feeble resistance to Einstein's arguments, his view was doomed. From that moment on, the term *atomic hypothesis* was no longer used, except by historians of science.

The year after Einstein's explanation of Brownian motion was published, Ludwig Boltzmann died at his own hand. Like Moses, Boltzmann met his maker close to his Promised Land, but sadly, it's not clear whether he ever realized that his views about atoms had at last been vindicated. Adding to the irony is the fact that a former student in Boltzmann's institute, the Austrian-born physicist Marian Smoluchowski (1872–1917), after leaving Vienna and taking up a professorship in Lwów, now in Poland, had independently derived the same results as Einstein.

Boltzmann had apparently resigned himself to the fact that some people, rather than saying, "Atoms exist," would always prefer to say, "The corresponding mental representations form a simple and useful picture of the observed phenomena," even if the two ways of speaking meant almost the same thing.

In his final years, Boltzmann spent more time in the Philosophical Society than in physics seminars. He sensed that the heyday of classical physics, his home turf, was essentially over. Electrons, X-rays, and above all, that dark horse called radioactivity, were heralding a revolution any day now. And it was to be Einstein's revolution more than anyone else's.

Albert Einstein was about the same age as the budding scientists in the *Urkreis*. He was a serious and resolute young man who at age fifteen had dropped out of his Munich high school and had given up both his German nationality and his Jewish creed. One year later, believing he had managed to teach himself all that he needed, he applied to the Zurich Polytechnic (now the Eidgenössische Technische Hochschule, or Federal Institute of Technology)—but to his shock, he failed the engineering department's entrance exam. This was quite a blow, but young Albert was resilient, and he enrolled in a secondary school in the Swiss town of Aarau. The next year he passed the exam and was accepted by the Zurich Polytechnic; in 1900, he graduated with a teaching diploma.

To make ends meet, he started giving private lessons in mathematics and physics, but with only moderate success. "I want a tutor for my boys, not a Socrates," said one exasperated client, and fired him (or so the story goes). Eventually, though, with the help of a friend, Einstein landed a job as an assistant examiner in the patent office in Bern, with the humblest possible rank of Technical Expert, Third Class.

It was in that most unlikely venue that Einstein's genius exploded like a supernova. In 1905, the year of his PhD, the curly-haired patent clerk published, in rapid succession, four truly groundbreaking papers, not only proving the existence of atoms and founding the theory of relativity but also putting forth the impossibly brazen idea that light, despite being a wave, is made up of particles.

This last-mentioned paper, published in March of Einstein's annus mirabilis, was in fact the first fruit of that year, and it set forth a hypothesis that Einstein himself considered to be the most revolutionary idea he ever had. It was so revolutionary that for a long while, nobody took it seriously—not even his admirers. Not even the *Urkreis,* for that matter. Nevertheless, the tale must be told, for it is a challenge to every theory purporting to describe how science is actually done.

Einstein's paper, titled "On a Heuristic Viewpoint Concerning the Production and Transformation of Light," started out by pointing out a simple but non-obvious analogy between the electromagnetic spectrum of a so-called blackbody (an evacuated cavity containing only light waves, whose walls are being held at a fixed temperature) and the distribution of molecular velocities in a similar container filled with an ideal gas (the Maxwell–Boltzmann distribution). This analogy led Einstein, after several pages of mathematical considerations, to the astonishing conclusion that light must, in some unfathomable sense, consist of particles.

This hypothesis was far more radical than Max Planck's earlier hypothesis, in 1900, about the existence of quanta of energy in a blackbody system. Planck's idea, which restricted the ways that material objects could vibrate, was the first quantum hypothesis ever, and although it was surprising and hard to reconcile with previous laws, it did not seem profoundly threatening to the entire edifice of physics.

But to suggest that light had a particle nature was definitely threatening. Thanks to James Clerk Maxwell's great equations, published in the mid-1860s, and Heinrich Hertz's great experiments roughly twenty years later (and countless other pieces of evidence), anyone who knew anything about light was convinced beyond a shadow of a doubt that light was waves; indeed, that fact was an unshakable pillar on which huge amounts of the physics of the day rested. It was therefore a complete and radical break with virtually all of classical physics when Einstein proposed that light might consist of particles. This heresy really *did* threaten the entire edifice.

Einstein knew he was taking a gigantic leap in the dark, and so, at the end of his paper putting forth his "heuristic viewpoint," he briefly proposed, in a sort of appendix, three ways of testing his audacious light-quantum hypothesis. One of his three possible tests involved the photoelectric effect, which is the fact that when light strikes a metal plate, electrons may come jumping out of it, or may not, depending on various factors that, in 1905, were not yet clear.

This effect was a quite recent discovery. In fact, it had first been noticed in 1887 by the German Heinrich Hertz (1857–1894), while he was carrying out the set of experiments that definitively proved Maxwell's theory that light consisted of electromagnetic waves! This effect was just an annoying nuisance, from Hertz's point of view, mildly interfering with his experiment, but eventually that minor nuisance would come to be seen as the first tiny chink in the armor of Maxwell's great theory. The twist is irresistibly ironic: here we have one and the same experiment simultaneously confirming and undermining the greatest piece of physics of the nineteenth century!

Although it was discussed now and then, Hertz's photoelectric effect was not of great concern to most physicists in 1905. Einstein, however, with his usual intuitive flair, suspected that with careful experiments, one might be able to use this curious effect to confirm, or to refute, his light-quantum hypothesis. Therefore, he used his hypothesis to make a prediction—a simple straight-line graph—about what the results of future experiments on the photoelectric effect would be. Interestingly enough, in his paper, Einstein pointed out that the straight line he was predicting was in conflict with the best experimental data then available for the photoelectric effect. His paper was therefore not *explaining* those data but predicting different data. How much bolder could Einstein possibly have been? Not only did his proposal fly in the face of all accepted theoretical ideas about light, but it contradicted experimental data about the effect that he claimed could confirm his hypothesis.

Well, Max Planck and the rest of the physics world reacted with utter disbelief to the arguments in Einstein's light-quantum paper. In fact, for many years thereafter, no one cited it, no one took up its ideas, and Einstein was the sole person in the world (perhaps with the exception of his wife) to believe that light might have a "corpuscular" nature. Even Planck, who greatly admired Einstein, stated in 1913 about his much younger colleague: "The fact that sometimes, as for instance in his hypothesis on light quanta, he may have gone overboard in his speculations should not be held against him too much, for without occasional venture or risk, no genuine innovation can be accomplished, even in the exact sciences."

One decade after Einstein's paper, the American physicist Robert Millikan described, in a famous book about the photoelectric effect, how he had confirmed Einstein's straight-line graph to an unusually high degree of accuracy, but then he claimed (wrongly) that at this point, Einstein himself had abandoned his hypothesis of light quanta. In order to underline the deep strangeness of this situation, Millikan added: "Experiment has outrun theory, or, better, guided by erroneous theory, it has discovered relationships which seem to be of the greatest interest and importance, but the reasons for them are as yet not at all understood."

In 1922, a curious twist took place. At long last, Einstein was awarded the Nobel Prize in Physics. But what was the prize being given to him for? For having proven that atoms exist? No, that was old hat. For light quanta? Certainly not! No one believed in that foolishness! For relativity, special or general or both? No, not that either, even though his work on relativity had turned physics upside down and had made Einstein a household name all over the world. Einstein was getting the Nobel Prize "for his services to Theoretical Physics, and especially for his discovery of the law of the photoelectric effect." This Nobel citation made precious little sense, for at that time no one but Einstein himself accepted the reasoning leading to his law of the photoelectric effect, even though the law itself had been beautifully confirmed by Robert Millikan in several years of high-precision experiments.

In 1923, however, things changed drastically. In that year, the American experimentalist Arthur Holly Compton discovered an anomaly in the way that light scattered off electrons (namely, the light's wavelength changed) that Maxwell's equations could not explain but that Albert Einstein's "erroneous" ideas about light quanta (soon renamed *photons* by chemist Gilbert Lewis) explained perfectly. All of a sudden, physicists around the world started rethinking their knee-jerk rejection of Einstein's 1905 "heuristic viewpoint," and soon all were rushing to embrace it with open arms.

From that time on, physicists had to use not one, but two conflicting images of what is called light; both Maxwellian electromagnetic waves and Einsteinian photons, seemingly incompatible with each other, were needed for its full description. This kind of dual nature was absolutely unprecedented in the history of physics.

And so, what is light, really? The wave/particle duality could not be explained. But perhaps explanation is not the ultimate aim of a physical theory. Already in 1905, Einstein had been careful to speak only of a "heuristic viewpoint," to be used as a mental tool. He took great pains to avoid speculating on the true nature of light.

## RELATIVITY

But back to 1905, the annus mirabilis. Einstein's paper announcing his special theory of relativity was published only three months after the light quantum paper. Although it, too, was highly revolutionary, it was far easier for large numbers of physicists to understand and to accept. Many were thrilled by its combination of great simplicity and great depth.

Some three centuries earlier, Galileo Galilei had realized that the motion of a body can only be measured with respect to another body—that is, *relative* to the other body. In other words, a measurement of position or speed requires what is called a *frame of reference*. In two frames of reference that are moving with respect to each other, the velocities assigned to moving objects do not agree. After all, what is at rest in one frame is moving in the other. A fly buzzing about inside an airplane seems to have a rather low speed to the passengers but a very great speed to people on the ground. This seems obvious enough. But measurements had shown that the speed of a beam of light does *not* depend on which reference frame you measure it from. Light in vacuum has *only one fixed speed,* no matter where you are or how you are moving. A beam of light inside an airplane has exactly the same speed for the passengers and for people on the ground.

What could rational people possibly make of such hocus-pocus? Ever since Nicolaus Copernicus and his followers, people had gotten used to the idea that our Earth moves at a speed of some 100,000 kilometers per hour around the sun, but the new experiments showed clearly that this motion in no way affected the speed of light.

Another salient anomaly was that the standard theory of electromagnetism seemed to give different explanations, depending on whether a magnet was moving with respect to an electrical circuit or the circuit was moving with respect to the magnet. This was truly odd, because there really was just one single situation being looked at from two different points of view.

These deep anomalies in the known laws of physics led Einstein to take a fresh and deep new look at the concepts of space and time and speed, and this in turn led him to revising the concept of simultaneity. Two observers need to synchronize their clocks if they want to agree on what it means for two things to happen "at the same time." Synchronizing requires an exchange of signals. Signals take time, even if they travel at the speed of light. The upshot is that two events happening at the same time for one observer may occur at different times for another observer who is zooming by. In other words, people in differently moving reference frames will disagree about whether two events are both happening "now."

Einstein rewrote the equations of classical mechanics using his new ideas about space and time, and lo! the "ether" vanished from the theory, being literally replaced by the void. And although one could no longer speak of an "absolute space" (meaning a hypothetical reference frame that is *not* in motion), there now was, thanks to Einstein's insights, an absolute *speed*—namely, that of light in vacuum, no matter what reference frame it was measured in.

Almost as an afterthought to this realization, Einstein derived the most famous equation of all of physics, $E = mc^2$, connecting energy $E$, mass $m$, and the speed of light $c$. This finding was announced in the final paper that Einstein published in his annus mirabilis.

It is comforting to know that Albert Einstein's four earth-shattering contributions to physics during that spectacular year did not go unnoticed by his peers. On April 1, 1906, he received a promotion in the Swiss patent office: from Technical Expert, Third Class, to Technical Expert, Second Class.

Einstein reached all of these earthshaking results without ever touching a single piece of lab apparatus. But revolutionary as his theories were, they were all firmly rooted in the philosophy of science of their day—in the ideas of Mach, Hertz, and Poincaré.

## Two Henrys: Hertz and Poincaré

Heinrich Hertz's book *The Principles of Mechanics* (published posthumously in 1894, the year of his untimely death) became as influential as Ernst Mach's *The Science of Mechanics*. Hertz stressed the role of mathematical models in describing scientific facts. His view was that we do not need an intuitive mechanistic understanding of the phenomena. All we need is to be able to check the model by means of computation and measurement.

Even further went the thoughts developed by Henri Poincaré (1854–1912), the leading French mathematician, in his book *Science and Hypothesis*. According to him, the laws of nature are free creations of the human mind, and their purpose is to relate observed facts in a consistent manner. Several different models can describe the same set of observations; when that happens, settling on one model rather than another is purely a matter of convention, based on whatever appears to be simpler and handier. There is no objective "fact of the matter." Moreover, abstract notions such as force and electric charge are defined only by the ways in which they are used. Asking for what lies "beneath them" or what they are "in reality" is useless metaphysics.

Thus, the idea of an ether at rest in absolute space—an idea that lay at the heart of Poincaré's theory of space and time—is quite compatible with physical observations, as long as we assume that rulers shorten in the direction of their motion and that clocks slow down whenever they are moved. Well before Einstein, Poincaré had understood the role of synchronizing clocks via electromagnetic signals. By introducing a concept he called *local time,* he was even able to explain all the same effects as Einstein could. His theory was thus an equal alternative to Einstein's. However, the theory of relativity proved to be far more elegant and practical than Poincaré's ether theory, and it became the convention that eventually was chosen. Or, if you prefer, Einstein's theory came to be seen as The Truth, while Poincaré's theory came to be seen as a great try but a near miss.

This case is a striking irony in the history of science: Henri Poincaré unwittingly served as a perfect example of his own claims that scientific truth was a matter of convention. Here were two alternative theories—his and Einstein's—that had equal predictive power (at least for experiments that could be done at that time). Poincaré had come tantalizingly close to scooping Einstein on relativity, but instead he developed his ether-based theory. This move was essentially betting on the wrong horse—a decision all the more intriguing, given that in *Science and Hypothesis* he had written that "without any doubt, the ether will one day be rejected as useless."

## David Hilbert

All physical theories rely on mathematics. But what does mathematics rely on? Ever since Euclid, any self-respecting mathematical theory was presumed—at least ideally—to consist of *theorems,* which are the collective progeny, using various pathways of strictly logical reasoning, of a set of axioms. And axioms were propositions that were simply taken as given. But given by what? By whom?

In Euclid's time, the axioms of geometry were considered to be obvious, which is to say, given by the evidence of our spatial intuition. But intuition can be treacherous. Moreover, the Greeks had noticed that one of Euclid's axioms is not quite as obvious as the others. That axiom, the so-called *parallel axiom,* states that in the plane, to every straight line L and every point P not located on L, there exists exactly one straight line containing P and not intersecting L. This line is the unique parallel to the line L through the point P. Now since straight lines go on forever, nobody can inspect them all the way from one end to the other; how, then, can we be sure that the two lines do not meet at some remote spot, far beyond our range of vision?

For two millennia, geometers had attempted to circumvent this problematic axiom by deducing it logically from the others, which were simpler and more intuitive. Their strenuous efforts, however, met with no success, and early in the nineteenth century, it gradually dawned on the mathematical world that these efforts might never succeed. It even occurred to a few mathematicians that it might be rigorously impossible to prove the parallel axiom.

And then, roughly at the same time (in the 1820s), two daring mathematicians (János Bolyai in Hungary and Nikolai Lobachevsky in Russia) realized that if the parallel axiom is replaced by the alternative axiom that given a line L and a point P, there are *many* lines that pass through P but that never meet L, then one simply obtains another geometry—a non-Euclidean geometry—with *many* parallel lines rather than just one. Its theorems seem exotic at first, and they differ

FIGURE 3.3 "Based on a great deal of preparatory work, David Hilbert undertook to erect geometry upon foundations whose security is never threatened by an appeal to intuition." (Moritz Schlick)

in fascinating ways from those of the familiar Euclidean geometry—for instance, in the new geometry, the sum of the three angles of a triangle is always less than 180 degrees—but the key thing is that the set of theorems as a whole is in no way less consistent. From a purely logical point of view, both geometries are equally correct.

What this revealed is that in principle, the axioms and theorems of something called "a geometry" don't need to have anything to do with human intuition. Our human imagery of points and lines, our sense of what they "are"—their nature, so to speak—is strictly our private affair. These intuitions are of course useful in daily life as we navigate through space, but to an abstractly reasoning geometer, all that matters is how certain purely theoretical entities that are given the names *points* and *lines* relate to each other according to the axioms that are arbitrarily given and the set of theorems that flow logically out of them. Geometry, in short, need have nothing to do with the physical world into which we were born.

This viewpoint was advocated most forcefully by David Hilbert (1862–1943), the leading mathematician of his time. Hilbert's birthplace was Königsberg, Prussia, the city of Immanuel Kant.

Hilbert did not start out in life as a wunderkind. As he later confided: "I did not bother much with mathematics at school, because I knew I would do it later." The young Hilbert was not in a hurry. He knew how to take the long view.

He also had a knack for being dead on target. Whether in algebra, analysis, number theory, or applied mathematics, Hilbert never failed to come up with seminal new results. He was gifted with *le coup d'oeil*, as the French might say. In 1895, Hilbert took a post at the University of Göttingen as the successor of the towering geniuses Carl Friedrich Gauss and Bernhard Riemann. Within a brief period, he managed to turn that small university town into a scintillating global hub of mathematics and theoretical physics, unrivaled for the next four decades.

Hilbert's book *The Foundations of Geometry* became a paragon of the modern conception of a mathematical theory. The slim volume provided the axiomatic framework for Euclidean geometry with great rigor, and without any recourse to intuition. The trick was simple. The basic concepts were not defined at all, except through their mutual relations. Hilbert omitted, for instance, Euclid's statement "A point is that which has no part," but kept "Any two points lie on a line."

To ask what points and lines *really are* makes as little sense as asking what chess pieces really are. Who cares? What matters are the underlying rules. The meaning of the basic concepts is utterly irrelevant. Hilbert made it plain: "Instead of calling these things 'points,' 'lines,' and 'planes,' we could just as well call them 'tables,' 'chairs,' and 'beer mugs.'" His quip became a catchphrase in mathematical circles.

In physics, it's the same, but with a difference. Some people believe that the ideal physical theory should emulate the model of geometry. That is, the role of the axioms would thus be played by certain basic laws, as few in number and as simple in structure as possible, which relate the most elementary concepts of physics to each other. Starting from these axioms, we can then logically derive vast numbers of consequences, just as in mathematics. However, physics aims to reveal facts about the real world, and so the concepts of physics have to be related to measurements, and the consequences of the basic laws have to be tested by careful observations.

Thus a *physical* geometry coexists with all the diverse mathematical geometries. The physical geometry describes actual space, and it should apply, for instance, to the corners, edges, and faces of rigid bodies. One could thus construct physical triangles made of metal rods and measure the sum of their angles. If this sum happened to deviate from 180 degrees, we would be faced with two options: either our space is non-Euclidean, or else our rods deviate from straight lines. Which alternative we prefer is a matter of agreement. It is up to us to decide which convention suits us best.

What would the axioms of physics look like? Or the axioms of probability? Is there a mechanical way to inspect a statement of mathematics and tell whether it is true or false? These are a few of the twenty-three deep questions that David Hilbert challenged the mathematical world to tackle, at the International Congress of Mathematicians, held in Paris in 1900. Hilbert hoped to see at least some of his problems solved in the dawning century. Some of them remain unsolved. All of them had a lasting influence on the field.

BERTRAND RUSSELL

During the nineteenth century, mathematics had been relentlessly brought to heel, disciplined with ironclad logic. Not only was its intuitive appeal sacrificed to the new rigor; logical reasoning itself was strictly regimented and put through its paces. Indeed, it had become obvious that mathematicians needed more than old-fashioned Aristotelian logic. In response to their needs, England's George Boole (1815–1864), Germany's Richard Dedekind (1831–1916), and Italy's Giuseppe Peano (1858–1932) developed their own versions of purely symbolic logic in order to make it possible to formalize even the most complex of mathematical proofs. This tendency was pushed to its limits by the *Begriffsschrift,* or "concept script," concocted by the German logician Gottlob Frege (1848–1925).

FIGURE 3.4 "It is virtually impossible to overstate the value of Bertrand Russell's way of philosophizing. I am firmly convinced that it is the method of the future—the only method that can realize Leibniz's dream of bringing the rigor of mathematics to the treatment of philosophical questions." (Moritz Schlick)

The young Bertrand Russell (1872–1970) had likewise set his sights on the twin challenges of founding mathematics on logic and of turning logic into a mathematical discipline.

Russell was born into the British aristocracy; his grandfather had been prime minister on two occasions. Little Bertie grew up as an orphan under the strict regime of his devoutly religious grandmother. He was privately tutored before enrolling in mathematics at Cambridge. For many years, the dread of mental illness haunted him. There had been such cases in his family. What gave him great relief, though, and turned his thoughts away from suicide were the cold certainties of mathematics.

But in 1902, Bertrand Russell discovered a paradox that cast grave doubt on these cold certainties. What made matters even more worrisome was that this was a paradox in set theory, the theory that in those days was starting to be used as the firm bedrock on which to erect all the rest of mathematics. What a disaster!

Sets are collections of elements. These elements can, in turn, be sets, just as folders can contain other folders. One can easily imagine a set that contains itself as an element (for instance, the set of all sets; that is a set). But of course many sets do *not* contain themselves (for example, the set of all cats; it is not a cat).

And so, what about *the set X of all sets that do not contain themselves?* Does X contain itself? If it does, then it doesn't; and if it doesn't, then it does. To spell this out more carefully, if X does not contain itself, then X must be—by the definition we gave of it— one of the elements of X, and hence it contains itself; and conversely, if X *does* contain itself, then it is—again by its definition—not an element of X, and therefore it does not contain itself. This unstoppable flip-flopping between *yes* and *no* is certainly a troubling state of affairs.

A related paradox was devised by the German philosopher Kurt Grelling (1886–1942), who worked with Kurt Gödel for a while, and who was a member of the so-called Berlin Circle, a group of philosophers closely allied with the Vienna Circle. Grelling was Jewish and perished in Auschwitz.

His paradox runs as follows. A word is said to be *self-descriptive* if it accurately describes itself. Thus the adjective *wee* is wee, and hence is self-descriptive. The adjective *huge,* by contrast, is not huge, and hence is non-self-descriptive. Here are a few more examples. The word *pentasyllabic* has exactly five syllables, and hence is self-descriptive; on the other hand, the word *bisyllabic* does not have two syllables, but four, and hence is non-self-descriptive. The adjective *recherché* is recherché, and hence is self-descriptive, while the adjective *unpronounceable* is perfectly pronounceable, and hence is non-self-descriptive.

Now then: Is our newly coined adjective *non-self-descriptive* self-descriptive? If it is, then it is not; and if it is not, then it is. This, too, is a troubling state of affairs.

Russell himself illustrated his paradox with the example of a barber who shaves all the men in his village who do not shave themselves. Does the barber shave himself? If he does, then he does not; if he does not, then he does. Once again, we are faced with a most troubling state of affairs.

The logician Frege was thunderstruck when Russell apprised him of the new paradox. He immediately realized that his whole theory had been totally undermined. The second volume of Frege's *Basic Laws of Arithmetic* was just then due to be published; it was too late to modify the manuscript. All that Frege could do was add an after-word. To this day what he wrote remains a monument to intellectual honesty: "Few things can be less welcome to a scientific writer than to have one of the foundations of his edifice shaken after the work is finished."

In a desperate effort to escape from the deadly clutches of his paradox, Russell invented a *theory of types,* which forbade any set from ever containing itself (or two sets from containing each other, etc. etc.). This more cautious approach, and other people's carefully devised approaches to set theory, some of which are more popular today, made it possible to overcome the Russellian paradoxes.

With his book *The Principles of Mathematics,* which appeared in 1903, the thirty-year-old Russell became the best-known logician of his time. His book's message was a program: mathematics ought to be based on logic, and on nothing else. Together with the older mathematician Alfred North Whitehead (1861–1947), Russell undertook to carry this grandiose project out in extreme detail; their joint work *Principia Mathematica* appeared in three volumes between 1910 and 1913. It became the bible of mathematical logic. The proof of the theorem "1 + 1 = 2" appears on page 362 of Volume Two, but it is written in such a custom-made and prickly notation that it would be unrecognizable as such to most readers, even to most mathematicians.

In *Principia Mathematica,* Russell's paradox seemed to have been tamed, but the disquiet it had caused remained. Can we be sure that there are no other still-undiscovered contradictions unsuspectedly lurking here or there? Of what good are the most elaborate logical proofs if logic itself is unreliable?

Henri Poincaré described this in a parable: The mathematician is like a shepherd trying to protect his herd from wolves by surrounding it with a high fence. No beast can cross it. But what if a wolf is hidden *within* the enclosure?

Thus another one of David Hilbert's twenty-three problems for the new century was: How can we make sure that no hidden contradictions exist within mathematics?

## CAREER STEPS

Russell, Poincaré, Planck, Hilbert, and Einstein: the philosophical revolutions connected with these lofty names animated the coffeehouse discussions of the *Urkreis* in Vienna. But soon the group dissolved: Hans Hahn took a position in Czernowitz in 1911, and Philipp Frank took one in Prague in 1912. Richard von Mises had become a professor in Strassburg in 1909 already. Young scientists have to be ever ready to follow calls from afar, if they have aims of rising in the academic world.

This principle applied to Albert Einstein as well, of course. The first step in his career was short: it led from the patent office in Bern to the University of Zurich, where Einstein became an associate professor. The Viennese physicist Friedrich Adler, of the same age, could also have applied for the same job, and there is no doubt that the Social Democrats in Zurich would have supported him; however, Adler decided not to apply, fully aware of the outstanding merits of Einstein, whom he knew from their student days together.

Soon the two physicists became good friends. They had apartments in the same building, and they often spent their evenings together. "We developed pretty much in parallel," Friedrich Adler wrote to his father. Indeed, the similarities were uncanny: they had married at about the same time, each of them to an intensely intellectual student coming from Eastern Europe; they had children of about the same age, who now played together frequently; they had written their PhD theses under the same professor, who had driven them both to exasperation; they had similar bohemian lifestyles; and their scientific and political outlooks were close.

Friedrich Adler, this doppelgänger of Einstein, ought by all rights to have been a member of the first Vienna Circle, the *Urkreis;* everything fit, except that he was not living in Vienna at the time. Indeed, his father, Viktor Adler, the founder of the Austrian Social Democratic Party, had insisted on sending him abroad for his studies, fearing that at home he would get sidetracked by politics. Young Adler graduated in mathematics and physics in Zurich, just as Einstein had. He then instantly started earning his spurs as a physicist, with his head crammed chock-full of Machian ideas. And just as Ernst Mach refused to believe in atoms, so Friedrich Adler was skeptical about the recently discovered electrons. After all, these newfangled hypothetical particles were even smaller and harder to see than atoms!

In 1911, Albert Einstein was lured away from Zurich by the offer of a full professorship in theoretical physics in Prague, at the former institute of Ernst Mach. Einstein recommended Friedrich Adler as his successor in Zurich. Adler declined, however, having by then become more interested in politics than in physics. His father's forebodings were being confirmed, alas.

Friedrich Adler returned to Vienna and devoted himself to Social Democratic causes. He put physics on the back burner for a while, sourly noting that "my ideas in that field had proved inaccessible to other physicists." But soon his new colleagues complained—jokingly at first—that his thinking was far too mathematical and that he had fallen prey to "the disease of logic."

Einstein was political as well, but unlike Adler, he could not possibly give up his first love, which was physics. However, before he could take up his appointment as a professor in Austria (for Prague at the time was still in Austria), he was required to declare his religion. Which religion it was, was left entirely up to him, but he had to belong to *some* religion or other; Franz Josef, the old monarch, insisted on such things. And so Einstein readily obliged and signed in as a member of the "Israelitic" creed.

By that time, he was already working on a dramatic extension of his theory of relativity. The *special* theory of relativity is called "special" because it is *limited*, in that it deals solely with observers who are moving at uniform speeds with respect to each other; the *general* theory of relativity was meant to be far more comprehensive, covering observers moving in arbitrary ways relative to each other. (It is not commonly known that the adjective *special* was applied to the earlier theory of relativity only after the general theory had been formulated, to indicate that the earlier theory was "special" in the sense of being restricted to a much smaller class of situations.)

Einstein's generalization from frames moving at uniform speeds relative to each other to frames moving arbitrarily meant that he was turning his focus to frames that were *accelerating* with respect to each other. And in his usual uncannily spot-on fashion, Einstein intuited that gravitation and acceleration were somehow deeply interrelated. This insight sprang out of his noticing a peculiarity about Newton's law of gravitation. Anyone since Newton could have made the extraordinarily simple observation that Einstein made, but somehow nobody else had picked up on it. A fantastically important idea had sat there right in front of everyone's eyes, yet had gone unnoticed for three hundred years. Einstein spotted this beautiful shell sitting all by itself on the beach, and he plucked it right up.

What Einstein noticed is that any object is pulled by gravity in proportion to its mass (this comes from Newton's law of gravitation), and yet its resistance to a gravitational pull (indeed, to any pull) is *also* proportional to its mass (this is the meaning of Newton's third law of motion). This coincidence, if one can call it that, means that the two effects cancel each other out. The upshot is that if various objects are all subjected to the same gravitational field, they will all move along identical trajectories, independently of their masses.

So far so good, but what does that matter? Well, Einstein recalled from his old student days in Zurich that the same statement also applies to objects in an *accelerating* frame of reference, such as, say, a box falling in space toward the Earth, its velocity ever increasing from moment to moment. Strangely enough, an observer trapped inside such a perilously hurtling box would be unable to tell, solely from watching the objects inside the box, that the box was not simply at rest in empty space, because all the objects would drift about as if there were no gravity whatsoever, as if no forces were acting on them at all. (This is today's well-known phenomenon of "weightlessness," which we have all seen in videos of astronauts in space stations in orbit, but back in those days that phenomenon was of course unheard of, and almost unimaginable—except to Einstein.)

Conversely, if a box were being pulled upward by a magical angel who was constantly accelerating, then to an observer inside, objects would appear to fall to the floor, just as they would if the box were sitting on the ground. In short, the scene inside the angel-pulled box sailing ever faster through space would look indistinguishable from the scene in a box sitting stock-still on the ground, but in the gravitational field of the Earth (or of the moon, or of Jupiter, etc.).

With these two thought experiments based on the coincidence he had noticed, Einstein had thus come to a crucial realization: that acceleration and gravitation were profoundly analogous. Indeed, he went further, boldly jumping to the conclusion that they were not just *analogous* but actually *indistinguishable*. With this flash of insight, he had come up with an astonishing out-of-the-blue connection between two totally familiar phenomena that on their surface have nothing to do with one another. He called this brand-new idea the *principle of equivalence*, and later in life he looked back at the moment of discovering it as "the happiest thought of my life."

Acceleration was a phenomenon that had to do with motion, and thus with space and time, and thus with four-dimensional geometry. So Einstein had put his finger on a profound yet never-suspected link between geometry and physics (specifically, the force of gravity). But what was the detailed nature of this deep conceptual link? Einstein struggled for many years with this riddle, trying to express the link in the form of precise equations. At one point, he learned that David Hilbert was also struggling with the same questions, and Hilbert was certainly the more skilled mathematician. Yet Einstein was not fazed in the least by the news, and did not lose a beat. Instead, he doggedly pursued a hunch coming from the ideas of Ernst Mach, a figure whom he had long venerated.

Mach had claimed that a body's inertia is determined by the fixed stars in the far reaches of the universe, and hence it depends on the distribution of mass in all of space. This sounded vague, but to Einstein it seemed to point in the right direction. Einstein named it *Mach's principle.* To see how it works, imagine you are at the North Pole and would like to measure the rotation of the Earth. You can do it in two different ways: either watch the pinprick of light emanating from any star (other than the North Star) as it describes a full circle above your head, or set a pendulum in motion and watch the plane of its swing slowly rotate, like a Foucault pendulum in a science museum, until it comes back again to where it started. Intriguingly, the two measurements will agree.

In September 1910, when Einstein showed up at the ministry in Vienna in order to declare his religious status and thereby settle his appointment in Prague, he took advantage of the opportunity to visit Ernst Mach in his modest suburban apartment. This was six years before the latter's death. The old scholar, crippled by his stroke and now growing deaf, received the young renegade eagerly. For years, Mach had hoped to meet the discoverer of the newfangled theory of relativity. He had already asked Philipp Frank to explain space-time to him.

The discussion between Einstein and Mach inevitably touched on the atomic hypothesis and the philosophy of science. Were the general laws of physics really nothing but some kind of economical ordering of observations? Einstein was greatly relieved when he heard that Mach meant "economic" in a *logical* sense rather than a psychological sense. This came much closer to his own views than anything he had found in Mach's writings. As for the idea of atoms really existing, Mach conceded that this idea would indeed be economical, and hence scientifically valid, as long as it could connect a set of observations that would otherwise remain isolated from each other. Einstein realized that this, coming from Mach, was quite a concession, and so he graciously left it at that. On the same afternoon, he also paid a visit to the famous Viktor Adler, the father of his good friend Friedrich.

Einstein found Prague beautiful, but the city could not hold him for long, although he pleasantly noted that the faculty meetings there were more entertaining than any play in town. On leaving that great city, Einstein made it publicly known that, contrary to rumors, he had not encountered anti-Semitism there. This was his first step as a spokesman for what he liked to call his "tribe."

In 1912, Einstein returned to Zurich, this time as a full professor; however, soon thereafter, the peripatetic relativist moved once again, this time to the Kaiser Wilhelm Institute in Berlin, the forerunner of today's Max Planck Institutes. He thereby automatically became a German citizen again. The German Reich offered ideal conditions for the undisputed shooting star of physics. Viennese journalists sourly noted that no institution in Austria could compete with the opportunities that the Kaiser Wilhelm Institute offered to its researchers.

Boltzmann's former disciple Philipp Frank, that pillar of the *Urkreis,* became Einstein's successor in Prague, on the latter's recommendation.

Frank's friend Hans Hahn, the mathematician, had left Vienna already in 1911 to take up a professorship at the University of Czernowitz, which had been founded in 1875 at the farthest confines of the Habsburg empire. In later years, the question whether this melting pot of nationalities belonged to the Ukraine, to Poland, or to Romania was hotly contended. But in 1911, Czernowitz quite simply belonged to the Habsburgs. It was the capital of one of their crown-lands, the Dukedom of Bukowina. That was as plain as day! The newly appointed professor Hahn could therefore marry without qualms. His chosen lady fair was Lilly Minor, who, together with Hahn's sister Olga, had been among the first Viennese women to acquire a doctorate in mathematics.

Hahn's appointment at the Imperial and Royal Franz Josef University in Czernowitz was a typical first step in the usual *cursum honorum* through the rustic provinces, also known as the "oxen's tour." Ideally, such a roundabout provincial circuit should finally lead a rising scientist, if all went well, via stopovers in Prague or Graz or other such places, to a chair at the University of Vienna. In a similar manner, stage actors in Vienna had to work their way up the ladder before reaching the stage of the coveted *Burgtheater,* the city's foremost playhouse. Needless to say, many became stuck on their way, professors and actors alike.

Young Hans Hahn, however, never seemed to entertain any doubts that he would one day return to Vienna. Indeed, on the eve of his departure for his own oxen's tour, he announced to his coffeehouse friends that as soon as he returned, they would all resume their regular

Thursday night discussions, but this time "with the support of a university philosopher."

And curiously enough, this is just what came to pass, some fifteen years later.


## The Viennese Modern Age and Other Sensations

Unsurprisingly, the philosophical questions that attracted the *Urkreis* in the wake of Russell, Hilbert, and Einstein remained arcane to most of their contemporaries. The wider public noticed, at best, lurid headlines such as one that graced the *Neue Freie Presse*, Vienna's most prominent newspaper: "The Minute Endangered: A Mathematical Sensation."

This enigmatic phrase, of course, alluded to the strange notions coming out of relativity theory. But in other fields, too, scientists were coming up with "sensational" new results that profoundly affected the radically modern generation that was growing up on the brink of World War I.

Technical progress had led to breathtaking innovations. Radio signals bridged the continents; X-rays allowed one to peer into living organisms; machines heavier than air carried daredevil pilots into the sky.

And yet, at the same time, the foundations of science had turned increasingly abstract and inaccessible, and their potential uses looked more and more ominous. Something of that deeply felt unease came to light in the famous Faculty Paintings produced by Gustav Klimt for the University of Vienna between 1900 and 1907. For each of the *Fakultäten* (schools) of Philosophy, Medicine, and Law, one was commissioned.

Klimt's eerie images let loose a storm of indignation; they showed naked men and women drifting in forlorn trances through an uncanny void. He had been asked to depict *The Triumph of Light over Darkness*. After all, the Faculty of Philosophy was home to the Science Institutes. It had thus been expected that the artist would deliver an optimistic glorification of progress. What he came up with instead was his own personal shell-shocked reaction to the terrible lesson of all scientific rev-

FIGURE 3.5 *Philosophy* (mural by Gustav Klimt).

olutions: Man is *not* the measure of all things—far from it. Humanity is just a freakish fluke in a huge and utterly alien world.

And thus Vienna was rocked by scandal. Tempers flared, and the newspapers eagerly joined the fray. At the Philosophical Society, the art historian Franz Wickhoff (1853–1909) gave a lecture on Klimt's paintings titled "What Is Ugly?" In it, he defended Klimt; however, most of his professorial colleagues could not recognize their beloved science in Klimt's demonic murals. Many of them, Boltzmann included, signed petitions in protest.

In the end, a thoroughly frustrated Gustav Klimt withdrew from the contract and paid back his advance. By then, he had spent it all, of course, but he was helped out financially by private patrons, among whom was Ludwig Wittgenstein's father. Never again would Klimt accept a commission coming from the state. Today, the only way we know Klimt's notorious Faculty Paintings is from reproductions; the originals were all destroyed in the closing days of World War II by retreating SS units.

In the years leading up to World War I, modern art was not any less exciting than modern science, and Vienna's intellectual youth were able to enjoy both to the hilt. The most skilled at provoking avant-garde scandals were the students who congregated periodically in the Academic Union for Literature and Music. This group found itself in the thick of every controversy, no matter whether it involved the austere, smooth architecture of Adolf Loos, the twelve-tone music of Arnold Schönberg, or the vibrantly colorful paintings of Oskar Kokoschka.

The lively discussions of the *Urkreis* have to be seen against this culturally supercharged backdrop. After all, the artists were sitting just one coffeehouse table away. Thus, Hans Hahn's younger sister Luise was a painter, and Philipp Frank's brother Josef was an architect. The young scientists were fully in tune with the galvanized mood of the budding century.

The unadorned functionality of the modern worldview pervaded literature and architecture much as it did science. Ornament and sentimentalism were seen as suspicious leftovers from the older generation. The new prevailing mood was factual, practical, businesslike, and close-shaven. Full beards and corsets were discarded as relics of a bygone era.

## The Science with the Evil Eye

The young writer Robert Musil (1880–1942) belonged to this generation that was so spellbound by science. He had no truck with sentimental postures. He loved the inexorability of mathematics, which he characterized as "the science with the evil eye." He proposed that in between reading two German novels, one should always compute at least one integral, so as to lose weight.

Musil had studied engineering in Brünn, where his father was a professor. The bright young man was not only well-versed in mathematics but also attached to it through a romantic interest. He had indeed, as he later wrote, "elected to fall in love with Elsa von Czuber," the daughter of a mathematician. His deep smittenness, however, led nowhere. Elsa (whose name was actually Bertha! and what would Dr. Freud have made of this?), the captivating Fräulein Czuber turned to another. She married an archduke who renounced his title and his income in order to gain her hand. The times were modern, but operettas were never too far away.

In 1902 Robert Musil left "Kakania." This was his name for the dual monarchy (the Austrian Empire and the Kingdom of Hungary), whose central institutions were all signed with the letters *k.k.,* pronounced "ka-ka," and standing for *kaiserlich-königlich,* "imperial and royal." Of course it was clear to all that Musil's "Kakania," despite its aristocratic façade, really meant "Dunglandia."

Robert Musil moved to Berlin, where he studied mathematics, physics, psychology, and philosophy, and eventually earned his PhD in 1908. His doctoral thesis, titled "Appreciation of the Teachings of Mach," starts out with these lines: "The word of a scientist carries great weight, wherever today an exact philosophy probes questions of metaphysics or the theory of knowledge. The times are long gone when a picture of the world could emerge full-blown from a philosopher's head."

In spite of his admiration for science, Dr. Musil became no scientist. Even while still a student, he had finished writing a novel, *The Confusions of Young Törless.* One of those confusions was due to "the riddle of the imaginary unit, a quantity with which one can perform calculations, although demonstrably it does not exist."

This first novel met with considerable acclaim. Thereupon Musil declined the offer to become an assistant professor at the University of Graz, and resolved to devote himself entirely to writing. Did he come to regret this decision? The unfortunate fact is that his literary career soon stalled. Eventually Musil had to accept a job as librarian at the

Technical University in Vienna. This compromise, however, left him plenty of time for writing.

But soon he felt profoundly agitated. A psychiatrist named Otto Pötzl (1877–1962), whom we will encounter again, and again, diagnosed him as suffering from severe neurasthenia. Musil resigned from his perhaps overly cushy job and returned to Berlin. There, while working for a famous publishing house, he tried—in vain, as it turned out—to convince an unknown young author from Prague to modify a bizarre story that the latter had submitted. The strange tale was titled *The Metamorphosis,* and its author's name was Franz Kafka.

In 1913 Musil published his essay "The Mathematical Man," a reflection on the crisis that had been triggered by Bertrand Russell. Musil began his text by citing the essential role of mathematical thinking in creating all the machines that dominate daily life. He stressed, however, that all this mathematical thinking was driven not by practical engineering needs but by unadulterated mathematical curiosity. So far, the argument is familiar. But then came Musil's turnaround:

> Suddenly, the mathematicians—those who brood in the innermost reaches—came upon something deeply flawed at the very crux of the whole structure, something that simply could not be fixed; they actually looked all the way to the bottom and found that the whole edifice [of mathematics] was floating in midair. And yet the machines still worked! We must therefore assume that our existence is a pale ghost; we live it, but only on the basis of an error without which it would never have arisen. Today, there is no other way to experience such astonishing sensations as those of mathematicians.

"Floating in midair": Musil seems to be describing what Klimt's figures are experiencing in their trance while sailing through a viscous void. But some strong minds are still able to keep their head on their shoulders. Indeed: "This intellectual scandal," as Musil wrote, "is endured by the mathematicians in exemplary fashion—that is to say, with confidence and pride in the daredevil boldness of their reason. Let no one object that outside their field mathematicians have banal or silly brains, apt to be left in the lurch by their very logic. There, they are not professionals, while in their own field they are doing just what we should be doing in ours. Therein lies the exemplary lesson of their existence; they are ideal role models for the intellectuals of the future."

The Viennese author Hugo von Hofmannsthal (1874–1929), while less versed in scientific crises than Robert Musil, was likewise aware of "something deeply flawed at the very crux of the whole structure, something that simply could not be fixed." He wrote: "Our era is fated to rest on what is slipping, and we are aware that what prior generations had taken to be solid is slippery."

In the same year that Musil wrote about "The Mathematical Man," the *k.k.* monarchy was stunned by an espionage affair that could not be suppressed by the censors, try as they might. It turned out that one Colonel Alfred Redl, the figure in charge of counterespionage, had been a spy in Russia's service all along. His job had thus consisted in fighting himself. Or, to reuse Henri Poincaré's metaphor, the shepherd's dog had turned out to be a wolf.

When this lurid affair came to light, a well-meaning colleague of Redl's discreetly placed a gun in his hand. Thus Redl's job was now to execute himself. This he did, leaving many unanswered questions behind.

One question that was never even asked, however, was this: Did the Austrian Secret Service, officially known as the Bureau for Evidence, by any chance offer "another way to experience such astonishing sensations as those of mathematicians"? Indeed, if one looked "all the way to the bottom," in Musil's words, one likewise observed that "the whole edifice was floating in midair."

The scandal of Colonel Redl must have convinced the Viennese, if they needed convincing, that absolute certainty was a chimera.

Indeed, the foundations of Kakania had already started to crumble.

# The Circle Starts Rolling

*Vienna, 1914–1922: Eastern front crumbles. Erstwhile Einstein doppelgänger Friedrich Adler shoots prime minister after lunch. Einstein asks judge to pardon Adler. Adler thinks Einstein mistaken. War-injured mathematician Hans Hahn, discharged from army, accepts chair in Vienna, renews youthful affair with philosophy. Munich court condemns war economist Otto Neurath for abetting treason. Neurath, deported to Vienna, claims ships can be rebuilt on high seas. Moritz Schlick, Berlin-born Einstein protégé, takes over Vienna Circle, faces gloomy future in Austria. Claps hands to start Thursday evening meetings.*

## New Dimensions

If the ground beneath one's feet starts to crumble, one has to find something to cling to. The *k.k.* generals clung to their war plans, although they knew that the Russians were privy to them. They may have reasoned that since the Russians knew that the Austrians knew that the Russians knew the Austrian plans, the Czarist army would surely anticipate some changes and would be caught by surprise if there weren't any.

If so, this reasoning misfired. Before the *k.k.* troops were properly assembled, the Eastern Front had gone to pieces. Within a few weeks,

Czernowitz fell to the Russians. Thus in the late summer of 1914, Hans Hahn lost his home as well as his job at the Franz Josef University, both of which had seemed so secure so recently. His wife, Lilly, and their little daughter, Nora, had to seek shelter in Vienna.

Hahn was drafted into the *k.k.* army. In 1915, on the Italian front, he was struck by a bullet. It lodged in one of his vertebrae, so close to the spinal cord that the surgeons did not dare to remove it. After a few months in the hospital, Hahn was discharged from the army, and that bullet would remain in his body for the rest of his life.

Despite this bleak outlook, Hans Hahn resumed his mathematical investigations into spaces having infinitely many dimensions, and a promising new field of research opened up. Hahn became one of the founders of functional analysis, together with his Polish rival Stefan Banach.

To the uninitiated, the idea of four dimensions seems abstruse enough. But mathematicians were no longer tied to intuition. In two-dimensional space, a point is represented by two numbers, its co-ordinates; in three-dimensional space, by three. Replacing visualizable points in space by sets of coordinates robs space of its depth, so to speak, but depth is not needed for computations. One can compute whether or not one has an image of what is going on.

Instead of two coordinates or three, why not four, five, or a hundred? Mathematicians love to generalize, and so they couldn't resist studying spaces of arbitrarily many dimensions, even infinitely many: a point in such a space corresponds to a sequence of coordinates that goes on forever, like the decimal expansion of pi. In such spaces, although they're abstract, one can generalize the familiar formulas for distances and angles, and this allows one to do geometry. Calculations replace intuition—or to be more precise, they assist it; after all, mathematicians cannot help forming mental pictures of the entities that engage their thinking day after day. But they do this privately, on the sly.

Oddly enough, functional analysis turned out to be immensely useful for physics. In Boltzmann's statistical mechanics, for instance, the state of a gas containing $10^{23}$ molecules is nothing but a point in a space having $6 \times 10^{23}$ dimensions. After all, each molecule is defined by its position in three-dimensional space (that means three coordinates), and by its velocity—three more coordinates per molecule. Even more importantly, the newfangled type of analysis, in which infinite sequences and even functions were treated as points, turned out to be indispensable in the booming field of quantum physics.

For the war-injured Hans Hahn, the outlook was growing brighter. In 1917, he was appointed professor in Bonn. There, by chance, he ran into a former colleague from his Czernowitz days, the flamboyant economist Josef Schumpeter (1883–1950). Unlike most economists of his time, Schumpeter was convinced of the value of exact quantitative methods, and as a fresh young PhD he had even written an essay called "On Mathematical Methods in Theoretical Economics."

Despite the unquestionable excellence of the University of Bonn, Hans Hahn was out of his element on the banks of the Rhine and never could feel truly settled there. Moreover, the pacifist leaflets that he distributed did not endear him to the German authorities. By that time, however, Hahn had already set his sights on a further change of address. At the University of Vienna, a chair in mathematics was about to be vacated. Suddenly he saw a golden opportunity to revive his philosophical circle, the old *Urkreis*, just as he had promised to do many years earlier.

Hans Hahn was ready to return home.

ALMOST UNFAITHFUL

Hans Hahn was born into Vienna's so-called second society. He grew up in the refined milieu of the fin-de-siècle described so aptly by Arthur Schnitzler, featuring wickedly witty conversations in literary salons, meaningful glances exchanged during musical soirees, dimly lit coffeehouses oozing with megalomaniacs, and invigorating weekends on the Semmering, a fashionable mountain resort within commuting distance from Vienna.

Hahn's father had begun his career as a journalist and music critic, and had risen to become head of the Telegraphen-Correspondenz-

FIGURE 4.1 Hans Hahn seduced by philosophy.

Bureau. Thus he belonged to the highest stratum of the civil-servant bureaucracy; after all, the telegraph, being by far the fastest means of communication, was now the monarchy's nervous system, and this monarchy was second only to that of Russia in size. Whoever opened the *Amtskalender*—the almanac that meticulously listed all the civil servants under the emperor—would find that the first name in the list was that of Hans Hahn's father.

The father had his hopes set on his son studying law, but after one year the lad switched to mathematics. He enrolled for several semesters in Strassburg and Munich before returning to the University of Vienna. Both for his doctorate and for his *Habilitation*—the right to lecture at a university—he was examined by Professor Ludwig Boltzmann, already a legend by that time.

Hahn had a remarkable talent for attracting talent. His first circle of friends at the university acquired the nickname "the inseparable foursome." Aside from Hahn himself, this quartet of students included Gustav Herglotz (1881–1953), who was studying mathematics and astronomy, and later became professor in Leipzig and Göttingen; Heinrich Tietze (1880–1964), an exceptionally versatile mathematician, later professor in Munich; and Paul Ehrenfest, a former schoolmate of Herglotz who, under Boltzmann, wrote his doctoral thesis on the mechanics of Heinrich Hertz and subsequently made important contributions to quantum mechanics and general relativity.

After Boltzmann's suicide, the brilliant young Ehrenfest was asked in his stead to write the article on statistical mechanics for the newly created *Encyclopedia of Mathematical Sciences.*

Ehrenfest's article, a masterpiece of clarity, became a classic of physics. However, since its author, in contrast to Einstein, stubbornly refused to declare belief in any religious denomination whatsoever, he could not apply for a professorship in Franz Josef's empire. Thus Paul and his gifted wife, Tatyana, lived and worked together for five years in St. Petersburg, but without securing any permanent position. Then, on Einstein's recommendation, Ehrenfest was appointed to a prestigious chair in theoretical physics in Leiden. In the Netherlands, a professor could be as much of a freethinker as he wished to be.

Hans Hahn, too, was asked to author an article for the *Encyclopedia*—a mark of high professional esteem. He wrote this piece jointly with Ernst Zermelo, a former student of David Hilbert's. Zermelo, who once had caused endless headaches for Boltzmann with his troublesome "paradox of recurrence," had also managed to discover Russell's set-theory paradox, quite independently of Russell, and even slightly before Russell found it. (In science, it is common practice for a result to be named after the person who discovered it last.) The Göttingen-based Zermelo was ideally placed to bring Hahn up to date on the foundations of mathematics, and some twenty years later, this investment paid off very well.

The inseparable foursome scattered while they were still students. Hahn became the center of another clique, this one consisting of young PhDs. This was the *Urkreis.* But even after the careers of the inseparable foursome had separated, they kept in touch. In 1909, Hahn wrote to his friend Paul Ehrenfest (now called Pavel) in St. Petersburg: "In the last year I have been almost unfaithful to mathematics, seduced by the charms of—philosophy. It began splendidly with Poincaré, Mach, and Hertz, but then along came Kant, and inexorably that led to Aristotle and company. The contempt with which our colleagues speak about these thinkers today strikes me as utterly absurd; many seem to seriously believe that a man whose name still is, after 2000 years, as powerfully resonant as it was back then, wrote nothing but silly nonsense." Hahn confided to Ehrenfest: "I am not easily affected by emotions. But to a friend who is as far away as you are, I must confess: at times, in my fleeting attempts to dip into the metaphysics of Aristotle, I have felt awestruck—and I deeply regret the lack of any opportunity to ponder these things in depth."

In the Vienna of the 1920s, Hahn then managed to carve out for himself just such an "opportunity to ponder these things in depth." The competition for the vacant professorship of mathematics had been intense, but in the end, Hahn wound up at the top of the list—and he did not wait to be asked twice.

## A New Form of Romanticism

Vienna 1920: the town had lost much of its glamour. There was no longer an emperor, and the multiethnic empire had crumbled. The glorious projects of a subway system and a canal to the Baltic, both scheduled to be launched in 1914, had had to be shelved because of the war. What remained of the Austrian Empire was a hydrocephalic republic, a country far too small with a capital far too big, and economically doomed, by general agreement. The great hunger had descended on Vienna in the third year of the war, and had not gone away during the war's bitter aftermath. The population had been decimated by epidemics, in particular the so-called Spanish flu. The painters Gustav Klimt and Egon Schiele now were dead, as was the architect Otto Wagner. The state's finances had been torn to tatters, and even Hahn's former

FIGURE 4.2 Hans Hahn had his office in the
New Physics building, close to the Strudlhof staircase.

colleague Josef Schumpeter, who also had returned from Bonn and for a time had served as secretary of state, found he could not help matters, no matter how hard he tried. Nothing worked. For many, it felt like the end of the world. Karl Kraus wrote a bitter satirical play, *The Last Days of Mankind,* intended, he said, to be performed only in a theater on Mars.

Hahn, on his return, had to sell the beautiful villa in the resort town of Semmering, which he had inherited from his father. On the other hand, he was able to hold onto the house in Neuwaldegg, one of Vienna's most elegant residential areas, only a stone's throw from the *Wienerwald,* or Vienna Woods. The coffeehouses were busier than ever, and the Vienna Philharmonic had not forgotten how to play. Hahn used to attend their concerts holding musical scores in his lap. After all, his father, the late Court Counselor, had once been a music critic.

The institute building for mathematics, physics, and chemistry had been finished shortly before the war started. It was located on the newly named Boltzmanngasse, close to the elegant Strudlhof staircase. The novelist Heimito von Doderer (1896–1966), whose characters frequently roamed the district, described the building as "slick and inscrutable in content." For him, it breathed "something of that new kind of romanticism that emanates only from the most exact of sciences." Doderer belonged to the "lost generation" that was now flooding the university's lecture rooms.

Hahn's two colleagues in the mathematics department, Wilhelm Wirtinger (1865–1945) and Philipp Furtwängler (1869–1940), enjoyed outstanding reputations. But Furtwängler was paralyzed from the neck down, and Wirtinger was turning stone deaf and was highly irritable. Both were about twenty years older than Hahn, and they were only too glad to share some of their working load with the vigorous newcomer.

The lecture rooms were hopelessly packed. There was no coal with which to heat them during the winter, and no paper on which to print the *Monatshefte für Mathematik und Physik,* Vienna's highly regarded mathematics journal. It took a donation by the rich Wittgenstein family to resume publication, with a backlog of several years. Scientific contacts with other countries had been severed. An academic salary could cover only half of one month's bills, at best. The new state was tottering on the verge of bankruptcy, and the currency was in free fall.

All this did not augur well for the resurrection of a philosophical circle. But Hans Hahn, a tall man with a booming voice, was not to be deterred from his great ambition.

And at his side there stood a comrade even taller and louder than he was—a veritable mountain of a man: Hahn's old friend from their school days, Otto Neurath, who had also returned to Vienna. Or to be more precise, he had been returned.

## THE PROBLEM OF MAXIMAL PLEASURE

The two friends, both around forty at the time, came back from Germany under very different circumstances: Hahn by accepting the offer of a prestigious chair at his Viennese alma mater, and Neurath by crossing the border sometime in the wee hours as a deported prisoner, unceremoniously sent back to where he had come from. As founder and president of a short-lived agency for Central Economic Planning in Munich, Neurath had attempted to introduce "full socialization," meaning nationalization, under two ill-fated regimes headed by Bavarian Soviets. The experiment had failed spectacularly, with Germany's army, the Reichswehr, putting a bloody end to it.

Otto Neurath had championed socialization for as long as he could remember. His father, Wilhelm, had risen from what used to be called the lower classes to become professor of national economy and statistics at the Vienna Agricultural Institute (once the Royal College of Agriculture). Relentlessly, the elder Neurath heaped one biting critique after another on the liberal economic system. Capitalism was doomed. In his view, private competition was bound to cause overproduction, and thus to end in waste, crisis, and misery.

The same subversive views were trumpeted by a close confidant of Otto Neurath's father, Josef Popper (1838–1921), a notorious Viennese eccentric who by chance was also a close friend of Ernst Mach's. Popper, who used the pen name Lynkeus ("Lynx-eyed"), proclaimed, with missionary zeal, the state's duty to feed its citizens. This was a highly unconventional opinion at the time, being cautiously explored only by the Fabian Society in late-Victorian England, and a few other dreamers. Popper-Lynkeus wrote treatises about social utopias and published them under catchy titles such as "The Right to Live and the Duty to Die." Otto Neurath, having grown up surrounded by such influences, had ardently endorsed a planned economy and the abolition of money from childhood on.

After studying mathematics and philosophy for two semesters in Vienna, Neurath opted in 1900 to move to Berlin, an incomparably dynamic city bustling with confidence and strength. You did not have to be German to be convinced that the dawning century was to be Germany's. In that teeming metropolis, Neurath studied national economics, sociology, and history. He earned his PhD writing a thesis on the

FIGURE 4.3  The red-maned Otto Neurath.

economics of the ancient world, especially of societies based on barter rather than money.

When his father died back in Vienna, Otto learned the realities of a moneyless economy in a most direct and painful manner. To make ends meet, he had to sell many of the thirteen thousand books he had inherited and fill out a form applying for state support.

In 1907, Otto Neurath accepted a position as instructor in Vienna's New Academy of Commerce. It was he who, in the coffeehouse debates of the *Urkreis,* made sure that the humanities were never given short shrift. Moreover, he was eager to expound his views during the meetings of the Philosophical Society.

In his lecture called *"Das Problem des Lustmaximums"* ("The Problem of Maximizing Pleasure"), the young firebrand proposed a radical new vision of utilitarianism. This was at just about the time when Sigmund Freud had announced his celebrated "pleasure principle" (or *Lustprinzip*—the instinctive drive for pleasure) in the meetings of his Wednesday Society. But Otto Neurath, unlike Freud, was not concerned with the innermost recesses of the human soul. Not individuals but societies were what interested him; in his eyes, a society was to be judged by how well it succeeded in optimizing the sum total of pleasure within it. Should there be no consensus on this topic (as seemed likely), then "the diverse views about the best way to organize society would have to resort to combat, to see which of them would prevail."

This towering young man with the flaming red mane readily attracted attention—that of ladies in particular. Already as a cocky schoolboy he had started a romantic relationship with the famous Swedish writer Ellen Key (1849–1926). In 1907, he married Anna Schapire (1877–1911), a Marxist feminist six years older than he was. The writer Arthur Schnitzler pithily portrayed her in his diary as "a philosophizing Russian." There was more than a breath of *La Bohème* in her affair with Otto, and it ended in tragedy: Anna died while giving birth to their son, Paul.

Half a year later, Otto married again. His second wife was an old flame of his: Hans Hahn's younger sister Olga, she who had lost her eyesight in her early twenties. Together, the two of them even published a few papers on her specialty, mathematical logic.

It proved too hard, however, for Olga, with her inability to see, to raise Otto's little son, and so the boy was sent to a children's home and grew up mostly there. During the first ten years of his life, he hardly set eyes on his hyperactive father.

In a series of articles dating from 1909 onward, Otto Neurath founded the field of war economics. Wars were too frequent, too momentous to be treated merely as vexing disturbances of the usual market equilibrium. Here Neurath found clear harbingers of a centrally planned economy. Moreover, here were examples where money lost its relevance as a medium of exchange, and where people turned back to barter. Neurath's articles put war economics on the map of scholarly disciplines. Material for his studies abounded—indeed, right in Austria's own backyard: the Balkan wars followed each other in quick succession—first in 1912, then in 1913.

Neurath viewed his discipline as perfectly value-free—"a science just like ballistics, which similarly does not depend on whether one is for or against the use of cannons." Through his novel ideas, the unconventional young sociologist met with professional success. It seemed that war economics was a promising new academic niche.

When World War I erupted, once again in Austria's own backyard, and this time as a result of Austrian blunders, the interest in Neurath's newly invented field of study underwent a tremendous upsurge. But Neurath had to spend two years in humdrum military service before he was ordered to show up at the Ministry of War. There, he became a section head in the committee for war economics.

The German ally, too, began to show interest in him, and appointed Neurath as the founding director of the Museum for War Economics in Leipzig. The museum's mission was to document the importance of a planned economy. During the final year of World War I, Neurath commuted between Vienna and Leipzig, and en route, so to speak, he also acquired the right to lecture (a *Habilitation*) at the University of Heidelberg. The famous Max Weber, a sociologist and an economist, among other things, was said to set great store by him.

In August 1918, the new museum in Leipzig proudly opened its first exhibition. It was also its last. The topic of the exhibition was the economic blockade imposed by the enemy on Germany and Austria. And indeed, the blockade took its toll. The empires of Germany and Austria collapsed, and the museum was shut down.

## THE PLAN OF ALL PLANS

In Neurath's view, the end of the war and the subsequent November Revolution in Germany offered a splendid opportunity to get full socialization underway. The main job had already been done, he claimed: now all that was needed was to turn the centrally planned war economy, which was already in place, into a system that was geared to peacetime needs.

Neurath presented his plans first in Saxony, where no one took them up. However, success beckoned in Bavaria. This came as a surprise, since the revolutionary movement had fared poorly in the Bavarian elections of March 1919, and its leader had been assassinated;

FIGURE 4.4 Otto Neurath featured on a hate poster.

despite this setback, the soviets (councils) of soldiers and workers did not back off. A socialist-led government came legally to power in Munich. It entrusted Neurath with setting up a central economic agency to make full socialization really take place.

In the midst of enormous political turmoil, Neurath turned all his energies to the task of elaborating "the plan of all plans"—the master plan of a planned economy. This, after all, was his dream. He pursued the task even after, in April, a Soviet government had putsched itself to power in Munich, soon to be overthrown by another. These short-lived governments were but the briefest of flashes in the political pan. When order was restored, a wave of arrests swept through Bavaria.

Neurath's pretrial detention lasted for about as long as his job for the central economic agency had lasted: six weeks.

His line of defense was clear: he claimed to be merely an apolitical civil servant and "social technician." Numerous witnesses, including Neurath's blind wife, Olga, who had lived through the pandemonium in Munich, vouched for the purity of his intentions. Aside from Olga, the industrialist and writer Walther Rathenau (1867–1922), who had run Germany's war economy; the sociologist Max Weber (1864–1920), who had discovered a Protestant work ethic lurking in the corridors of capitalism; and Otto Bauer (1881–1938), the leader of Austrian Social Democracy—all tried to exonerate Neurath through letters of support. Bauer, who had worked cheek by jowl with Neurath in the Austrian committee for war economy, wrote to the court of law: "With the best of conscience, he was able to serve, in succession, the k.k. War Ministry, then a center-left coalition government, and finally the Bavarian Soviet republic, for he had convinced himself that any of these governments would be as good as the other for realizing his socio-technical projects."

## A WAR HORSE RETURNS

Despite the favorable testimony of Otto Bauer, Otto Neurath wound up being convicted. On July 25, 1919, he was sentenced to eighteen months in prison for having been an accessory to high treason. Still, he could thank his lucky stars that he had been accused only of aiding treason rather than of treason itself, for had it gone the other way, he could well have lost his head.

Shortly after Neurath's sentencing, Otto Bauer, who had accepted the position of Minister for External Affairs of the Austrian Republic, undertook measures with the Bavarian government to free the political prisoner. This led to an intergovernmental wrangle that lasted several months. The ever resourceful Neurath used his incarceration profitably to write a book: *Against Spengler*. The German historian Oswald Spengler (1880–1936) had written a cultural history called *The Decline of the West*, which appeared in 1918 and immediately exerted a tremendous influence on the German-speaking world. Spengler's vision of the organic blossoming and inevitable fading of high civilizations hit a sensitive nerve, for at the time, Europe had just witnessed the downfall of three empires. Having tasted Armageddon, many people had lost all faith in progress; they could already see Asiatic hordes angrily banging at the gates and were expecting a bloody fight to the finish for world domination.

However, Neurath had no patience for such a Twilight of the Gods. He vehemently argued against Spengler's apocalyptic vision of decline and collapse. Neurath was convinced that steps could be taken to prevent any such downfall. In his anti-Spengler treatise, he wrote (as he also did on many later occasions, with different words): "We are like mariners who must rebuild their leaky ship while sailing on the open sea, and who are never able to start from scratch. Whenever a beam is removed, it must be replaced immediately by a new one, with the remainder of the ship serving as a means of support. It is in this way that, thanks to old beams and random pieces of wood drifting by, the ship can be completely built again—but only through step-by-step reconstruction."

The metaphor was not new. Already the ancient Greeks had used it to ask: If all the planks of a ship are replaced, one by one, is the vessel that results still the same ship, or is it a new one? Neurath freed the simile from its status as a mere riddle and turned it instead into a valiant image of persistent human struggle.

A "step-by-step reconstruction," as in the mariner's parable, was also what the Austrian Social Democrats had in mind for the scraps they had inherited after the Danube monarchy had been carved up. However, their coalition government made little headway. Their intense drive for socialization slowly faded. Otto Bauer stepped down as head of the committee entrusted with the task; he also resigned from his ministerial position for foreign affairs, which had made it possible for him to intervene with Bavaria on Otto Neurath's behalf. His efforts to achieve that goal had met with no success.

In the end, it fell to Chancellor Karl Renner (1870–1950) to arrange for Neurath's repatriation to Austria. The Austrian government guaranteed that the detainee would henceforth refrain from any form of agitation whatsoever against Bavarian authorities, and Neurath himself solemnly pledged never again to set foot on German soil. The Bavarian government wanted to make sure, once and for all, that it had rid itself of the obnoxious Viennese. The German historian Karl von Müller spoke for many when he decried Neurath as "a demagogue called in from Austria."

Let it be noted that the same historian soon thereafter became a disciple of another Austrian demagogue, but one of an altogether different stamp. In 1919, however, ex-corporal Adolf Hitler was still serving in an obscure capacity with the Reichswehr in Munich. Right about that time, he was discovering his unusual gift for political rhetoric, or to put it perhaps more accurately, his flair for rabble-rousing. Five years later, after his abortive "Beer Hall Putsch" against the Bavarian government, he would be sentenced by a court in Munich to *Festungshaft* (imprisonment in harsh conditions in a fortress), just as Otto Neurath had been. Also like Neurath, the future *Führer* used his incarceration profitably to write a book (*Mein Kampf*). However, unlike Neurath, Hitler was not repatriated to Austria: the Austrian government would not entertain the idea.

It seems probable that it did not cost Otto Neurath too much to abstain from any future political activity in Germany. There was plenty to do in Austria, after all. And so he jumped into the fray with alacrity.

Writer Robert Musil, who met Otto Neurath during the first weeks after the latter's homecoming in Vienna, noted in his diary: "Has a notebook with very many entries. Whatever has already been taken care of gets neatly crossed out. Appears to be continually somewhere else in his thoughts, but then, out of thin air, he offers some nice words: "Please convey my regards to your wife"—this despite the fact that we were with her just fifteen minutes earlier. Is constantly on the go, making contacts here, there, and everywhere." Musil summarized Neurath with the words: "Some kind of a professorial war horse. But with explosive energy."

Now as always, the favorite target of Neurath's explosive assaults was metaphysics. He viewed it as a reactionary smokescreen, whether it came in the guise of philosophical idealism or of theological revelation. Neurath never tired of fighting this tricky tool of the bourgeoisie, and from early 1921 on, his brother-in-law Hans Hahn was also his brother-in-arms in the pitched battle.

## A Philosopher à la Carte

The two friends were certainly not about to run out of new topics for philosophical inquiry. Since the prewar days of the *Urkreis,* much had occurred to provide rich grist for the mill. In particular, Hilbert, Russell, and Einstein had not been idle. This alone would have been reason enough to relaunch the coffeehouse meetings of yore and to return to the philosophy of science.

But someone was still missing: a philosophy professor from the university whom Hans Hahn had long hoped would join them. He probably had in mind his Viennese colleague Adolf Stöhr (1855–1921), the official successor to Ernst Mach's chair. Stöhr had started out as a physicist, just as had Boltzmann and Mach, but his thinking leaned toward the analysis of language. He wrote: "If there were no words, there would be no nonsense, or at worst there would be errors. . . . Nonsense cannot be thought, it can only be spoken." Today, this sounds like vintage Wittgenstein. Thus Stöhr would doubtless have been a suitable addition to the Circle, but he had fallen incurably ill and was no longer able to teach. Alas, he would not join their ranks.

Two full professorships in the School of Philosophy had recently opened up. The hiring committee was already in place for them, and now it received the additional charge of finding a successor to Adolf Stöhr. The newly arrived Hahn managed to wangle his way onto the committee, latecomer though he was. Lady Luck was winking at him.

There remained the question of whom to appoint. On this, though, Hahn was in the enviable position of getting incomparable advice—from Einstein himself.

Albert Einstein had become a celebrity of the first order. In January 1921, the great theoretical physicist, who to many looked like more a dreamy violin virtuoso, was scheduled to give two lectures to an academic audience in Vienna, and a third one to a general public. The frenzied crush at the doors of the popular talk was unimaginable. No university in town had a lecture hall anywhere near large enough to accommodate such a mass of people. And thus Einstein was booked for the great hall of Vienna's Concert House.

Tickets were sold on the black market. The mega-event was as new an experience for Einstein as it was for the Viennese public. And luckily, all went well; the physicist cast a spell on his rapt audience. Indeed, the crowd was so mesmerized that Einstein had almost walked offstage before anyone realized that his lecture was over. "The applause set in late, but was all the more frenetic," as the newspapers reported.

Philipp Frank, Einstein's successor in Prague, had come to Vienna for the great event. The two physicists knew each other well. Thus it came to pass that Frank, Neurath, and Hahn, who had been the three pillars of the *Urkreis* before the Great War, invited Albert Einstein for a wintry walk on Vienna's favorite excursion spot, the Kahlenberg hill. From up there, the glorious panorama of the city on the Danube offered no visible trace of decay. The tall spire of the City Hall could be made out, contrasting with the ponderous roof of the nearby university building.

The stage was all set for taking the long view. How to connect this dawning new age with the great old days of Ernst Mach? It seemed almost inevitable that the quartet of scientists should come to speak about the man who, at that time already, was known as Einstein's *Hausphilosoph,* or "pet philosopher"—namely, Moritz Schlick. Schlick was a Berliner, but he quickly proved to be the ideal center for what was to become the Vienna Circle.

## Oedipus in the Headlines

Before turning to Schlick and tracing the steps that led him to Vienna, it is time to conclude the story of another link between Einstein and Vienna. It concerns Friedrich Adler—the Viennese physicist who, because he was off studying in Zurich, had missed out on the *Urkreis* but later became a friend of Albert Einstein's, and for a while was almost his double. In the Vienna Circle story, this is merely a side plot, but it is one that sheds a lurid light on the nature of exact thinking in demented times.

Friedrich Adler, having given up on an academic career, had returned to Vienna as secretary of the Social Democratic party. His comrades soon called him "the logician," because he always followed his convictions absolutely rigorously, without pity, wherever they might lead. And so, at the height of World War I, in the autumn of 1916, "the logician" followed his convictions without pity and killed the *k.k.* prime minister, Count Karl von Stürgkh.

The count had decided to declare a state of emergency, in order to divest the parliament of its power. In Adler's eyes, there seemed to be no legal way left to prevent absolutist rule from taking over. Thus he resolved to commit a political act of homicide. He had seen Schiller's play *William Tell* as a child, and his years in Switzerland had not dimmed his enthusiasm for the legendary tyrant-killer.

On the morning of October 21, 1916, Friedrich phoned his mother and told her not to expect him for lunch. Then he headed directly to the restaurant Meissl und Schadn, renowned equally for its boiled beef and its aristocratic clientele. As everyone in town knew, this was the spot where the prime minister always took his midday meals. Having ascertained that his quarry had already arrived, Friedrich Adler chose a nearby table and ordered a three-course menu, to calm his nerves. Then he paid and discreetly removed the catch from the gun in his pocket.

A slight contretemps arose, due to the presence of a lady at a neighboring table. Adler waited patiently over coffee for nearly one hour, until she finally left the room. Then he rose, drew his gun, walked up to the prime minister, and pumped several bullets into the count's head. Officers dining nearby did not even have time to reach for their swords. After a short commotion, Adler handed over his gun and waited for the police. He had lost his spectacles in the scuffle, but not his sangfroid. When the police officer asked him why he had shot the count, he coolly replied that it was none of the officer's business. He would offer explanations in the proper place.

FIGURE 4.5  Friedrich Adler commits his crime.

From the moment he was arrested, Friedrich Adler insisted that he was fully responsible for his act. "I have not committed this assassination with an unsound mind," as he said to the prosecutor and later the judges. His goal had been, so he declared, "to make people start thinking." Had he merely shouted "Down with absolutism! We want peace!" he would not have been heard. Censorship would have made sure of that. But his *shots* had been heard, and therefore his words at the trial would be heard. If it came to a trial, that is. Adler's nightmare was that he would be sent off to a mental asylum instead. If that were to happen, his sacrifice would have been made in vain. Adler's family, on the other hand, saw in a plea for insanity the only way to save him from the gallows.

To prove that he was in complete possession of his mental faculties, Friedrich Adler resumed his physics research during pretrial custody. He wrote a book on Ernst Mach, sometimes working on it throughout the night. His father, Viktor Adler, tried to convince the authorities that it was precisely this manic hyperactivity that proved that his son was non compos mentis. A psychiatric examination confirmed the father's diagnosis: "hypomania," it said, "and circular neurosis." There had been plenty of nervous disorders in the family. The panel of experts listed eleven cases spread out over five generations. But they added that Friedrich Adler was of sound mind—merely a fanatic.

One of the experts in charge was the psychiatrist Julius Wagner-Jauregg (1857–1940), who ten years later would receive a Nobel Prize for his method of fighting syphilis with malaria. But at the time of the murder, Wagner-Jauregg was helping the Austrian war efforts by trying to fight shell shock with electric shocks. His rival Sigmund Freud grudgingly admitted that the method sometimes worked.

Freud himself refrained from getting involved with the Friedrich Adler case, possibly because he felt too close to it. Indeed, his by-now world-famous address—Berggasse 19—had once been Viktor Adler's. And the latter, a former psychiatrist too, even played a cameo role in Freud's *The Interpretation of Dreams*. It's a small world—and particularly so in Vienna.

But under the present conditions, Freud left it to his epigones to try to decipher the father–son relation in the Adler family—and they had a field day doing so. It was obvious enough: Friedrich Adler's subconscious wish to kill his father had been diverted towards the pater patriae instead, which is to say, the emperor; and since the emperor was not easily accessible, the next best thing to do was clearly to assassinate the emperor's prime minister. It was a classic case of an *Ersatzhandlung* (or "displacement" in Freudian terms)—what else? And anyone looking for an ideological or ethical motive was just a naïve simpleton. Albert Einstein, for instance.

## An Excess of the Mathematical

Indeed, Einstein spontaneously offered to testify in favor of his old friend Friedrich, "whose selflessness had made him land in the soup."

He even drafted a letter to the emperor, asking His Majesty to pardon the murderer of the prime minister. This was a new emperor, since Franz Josef, well in his eighties, had died a few weeks after the assassination of Count Stürgkh. Thirty-year-old Karl the First, fated to also be Karl the Last, was an unlikely candidate for a pater patriae.

In the end, Einstein did not send off his letter; he added some doodles to the draft and then covered the back of the sheet with formulas of cosmology. But he was ready to act as a character witness, and he even asked Adler's former Zurich colleagues to do so as well. Moreover, Einstein gave a newspaper interview praising the profound selflessness of his former colleague, and illustrating this quality with the story of young Adler in 1909 withdrawing his application for a professorship in Zurich in favor of another candidate "who was by far superior." (Einstein tactfully skirted around the fact that this other candidate happened to be Albert Einstein.)

Friedrich Adler, for his part, spent his days and nights in jail zealously seeking ways to refute his former colleague's special theory of relativity. Constantly in a psychic state of great exuberance, Adler finally convinced himself that he had found a better approach, based on a set of special frames of reference. He was sure that he had therewith reached "the highest pinnacle of physics."

When Friedrich Adler wrote, while still awaiting trial, that he had made a discovery "which may be said to be the greatest possible one, given the current state of physics," his father quickly forwarded the letter to psychiatrist Wagner-Jauregg as new evidence for his son's insanity. Wagner-Jauregg summoned his assistants right away—including Otto Pötzl, the young psychiatrist who had diagnosed Robert Musil's neurasthenia—and returned to Friedrich Adler's prison cell for another round of examination.

The prisoner found this hard to stomach. "I confided in you as a friend," he wrote bitterly to his father, "and in return you send me psychiatrists!" He raged that by trying to save his son's life, Viktor Adler was merely furthering his own political interests. In order to calm Friedrich down, both his father and his lawyer had to renounce all thoughts of a plea of insanity. And anyway, the psychiatrists had found no reason to change their opinion. The accused was a fanatic, but he was not insane.

At his trial, Friedrich Adler was at the top of his form. Straight off, he declared the trial to be unconstitutional. He therefore saw no need to defend his deed. And yet he would do his best to explain its motivation: this, actually, had been the very purpose of the act. *He had had to commit the assassination because that would enable him to publicly explain why he had had to commit it.* This may sound like a vicious circle, but it actually had its own bizarre logic.

Adler's self-defense took six long hours. Ever the physicist, he began by describing the change in perspective required to switch from the Ptolemaic to the Copernican system. The judge groaned: "Must we hear this?" But Dr. Adler coolly replied that we live today "in the age of relativity" and kept pressing on, relentlessly. You, he said to the judge, view humanity as divided into *nations;* I, the accused, see it as divided into *classes.* This requires a change in perspective—just as had recently occurred in Russia's revolution of February 1917, while Friedrich Adler had been awaiting his trial.

Adler had raised a touchy issue. The bug that had bitten the Russian workers and soldiers might well soon spread across the no-man's-land and invade Austria. "Do not speak out of the window!" ordered the judge. But Adler was not to be stopped. Step by step, he walked the court through his "proof" that Count von Stürgkh's suspension of the constitution had left him with no other resort than assassination.

It also emerged during Adler's plea that his action had been a protest less against the *k.k.* government than against the all-too-tame role of the Social Democratic opposition. He had felt ashamed for Austria, he confessed, and ashamed for his father. It was obvious that he deeply loved both. The Oedipal perspective was not so preposterous after all. And indeed, after the death of the old emperor, it had become clear to the whole country that Viktor Adler was the only father figure left.

At his son's trial, Viktor Adler stated right away that Friedrich was and always had been the person dearest to him. And he added that if Austrian politics did not make you lose your mind, this just meant that you didn't have one. The feeble joke was clearly an attempt to ask Friedrich to forgive him for having ever dared to call in the psychiatrists.

Viktor Adler went on to explain that his kindhearted son had committed the assassination "due to an excess of the mathematical," which he, the father, had never guessed would be possible. He also mentioned that Albert Einstein had reported that his friend Friedrich was occasionally prone to imagine that he could uproot trees. And lastly, Viktor Adler went on record with the statement that "if a mathematician draws a line, he believes in the line" and would never let himself be deflected from his chosen course.

Indeed, when the judge asked the accused why he had never thought of the effect on his parents and his children, Friedrich Adler haughtily replied that political assassination was admittedly problematic, but "to reserve it for childless orphans is not worth discussing." And from his trial's first moment to its last, Adler made it clear that he had always taken the death sentence for granted.

He got it. Friedrich Adler was condemned to death by hanging.

The young monarch, however, commuted the sentence to eighteen years in jail. From his prison cell in the Stein Penitentiary on the Danube—the "upriver" of the Viennese—the prisoner pursued his correspondence with Einstein, using complex arguments based on the ideas of Ernst Mach. He even devised an experiment to prove special relativity theory wrong. Einstein patiently explained why it could not work. In a letter to another friend, Michele Besso, Einstein confidentially described Friedrich Adler as "a fairly sterile rabbinistic dunderhead, riding Mach's old nag to death." Yes, replied Besso, "but Mach's old nag served you rather well" . . . And while Einstein was unimpressed by Adler's physics, he kept holding the latter's selflessness in high esteem. "I am curious to know," he wrote encouragingly to his jailed friend, "who of us will first visit the other." The war to end all wars was ending.

In the fall of 1918, Emperor Karl ordered the release of Friedrich Adler. It was a gesture toward the Social Democrats. The story goes that Karl even sent his official car, the Royal Automobile, to pick up Friedrich

Adler on his way home. Friedrich's father sat in the back of the luxurious limousine; he was the first to welcome his "boy" back from jail.

It was a hero's return. Friedrich Adler's alternative to relativity theory may have been dead wrong, but his political calculation had proven dead right. In the first days after the assassination, the political left had condemned the murderer as a stark raving monster (how could he do such a thing to his father!), but it did not take long before he started being hailed as a martyr for his democratic convictions. His trial had been reported almost in its entirety. Censorship had lost its teeth, and indeed the Austrian parliament was reconvened a few days after Adler's conviction.

Even while Friedrich Adler was in jail, the communists of German-Austria had offered him the leadership of their party. His old friend Leon Trotsky had asked him to become honorary commander in chief of the Red Army. And no less than V. I. Lenin suggested that he accept the presidency of the Petrograd Soviet: apparently Lenin was willing to forgive Adler for his notorious weakness for Machian ideas, which had raised his hackles ten years earlier. But Friedrich Adler disapproved of Bolshevik politics and thus remained faithful to his father's Social Democrats.

Viktor Adler expired on the eve of the proclamation of the German-Austrian Republic. The triumphant trip home in the Court Automobile had probably been his last outing. As for that republic, born out of defeat, it consisted essentially in the German-speaking people of the former Habsburg monarchy—about one eighth of the population. These Austrians almost unanimously wanted to join Germany, but the Allied Powers were not going to allow such a move. The German-Austrian Republic thus had to drop the term *German* and promise never to ally itself with its larger neighbor to the north.

The two books that Friedrich Adler had written while in prison—*Ernst Mach's Victory over Materialism* and *System Time, Zone Time, Local Time*—were hastily printed within a few weeks, and then were never heard of again. On the other hand, the minutes of Adler's trial became a best seller and were even translated into dozens of languages. They still make for good reading—a kind of *William Tell 2.0*. Playwright Arthur Schnitzler could not have done better.

The first general elections in Austria after the war led the Social Democrats to victory, if only by a narrow margin. The head of the party was now Otto Bauer. Friedrich Adler, suddenly the undisputed moral guide of the left, was elected to parliament. When he learned about communist plans for an uprising, he convinced the worker delegates, in a passionate plea, to desist.

And he never turned back to physics again. But it so happened that the last gift that Albert Einstein sent him to his prison cell in Stein was a book called *Space and Time,* which had been written by Moritz Schlick—the man who now came to Vienna as Ernst Mach's successor.

## BETWEEN EPICURUS AND EINSTEIN

Moritz Schlick had been born in Berlin in 1882. His father, a well-to-do merchant, ran a business selling combs and ivory. Schlick's family traced itself back to old Bohemian nobility on his father's side, while his mother took pride in being related to Ernst Moritz von Arndt, a Prussian poet from the time of the War of Liberation against Napoleon.

When he was six years old, Moritz contracted scarlet fever and diphtheria; the aftereffects plagued him for many years. Though a sickly child, he did well in school. Early on, he read Kant, arriving at conclusions quite similar to those of the young Mach: Kant's metaphysics did not convince him at all. Some years later, young Schlick wrote: "After thus having sentenced theoretical philosophy to death, life itself urged me to delve into the most important part of practical wisdom, the study of man and the human condition, something I had always maintained belonged to science rather than to philosophy."

On leaving school, Schlick was presented with a copy of Ernst Mach's *Principles of Mechanics.* In hindsight, he liked to call this "a hint of fate." Schlick went on to study physics, mostly in Berlin. Later, he would declare, "I turned to physics in a philosophical spirit and in response to a philosophical urge."

Under the advisership of the illustrious Max Planck, Schlick wrote his doctoral thesis called "On the Reflection of Light in an Inhomogeneous

Layer." In 1904, he obtained his PhD summa cum laude. Moritz Schlick and his friend Max von Laue, who was one year older, were Planck's favorite students. And this *Max und Moritz* duo (the title of a famous nineteenth-century German children's tale) went on to do quite well for themselves. Max von Laue, by the time he was thirty-five, had won the Nobel Prize in Physics for his work on crystalline structures, five years before his doctoral supervisor Max Planck was similarly honored.

As for Moritz Schlick, he did not set his sights on a career in physics. "It did not agree with my nature," he declared. While still at school, he had started working on a philosophical book that he finished in 1907, at the ripe old age of twenty-five. Its title: *Wisdom of Life.* Its subtitle: *An Essay on the Theory of Bliss.*

This work was far more than a youthful peccadillo. Although Schlick would soon come to regret its turgid, stilted style, which tried too hard to emulate the style of his lifelong hero Nietzsche, he remained forever faithful to the ideas that he had developed so precociously. Again and again, in lectures and writings, Schlick propagated an Epicurean ethic of kindness, in contrast to a Kantian ethic of duty.

The young philosopher was still working out his "theory of bliss" when he met an American pastor's daughter named Blanche Guy Hardy, in a boarding school in Heidelberg. A few weeks later, the young girl wrote to him in innocent candor: "My dear Doctor Schlick, you have probably long since forgotten me, but please try to recall me to your memory, so that I do not need to feel all too self-conscious." Blanche closed with the words: "Should this letter seem too unconventional, then please forgive me with the comment: 'She is an American.'"

Schlick did not take long to reply. He had not forgotten the American girl. Courtship ensued, and one year later—Blanche had in the meantime returned to the United States—Moritz Schlick proposed, by mail. Their wedding took place in Massachusetts in 1907. Schlick's *Wisdom of Life* came out at about the same time.

In the ensuing years, Schlick did his best to enhance his profile as a philosopher. As he considered his knowledge of psychology to be quite deficient, he enrolled as a guest student at the University of Zurich.

However, when he submitted a *Habilitation* thesis called "On the Concept of Truth," he failed to obtain the coveted right to lecture. Apparently one of the Zurich professors was allergic to the very mention of Ernst Mach's name.

Schlick learned the bad news a few weeks after his son was born. Next he tried his luck in Kiel, and afterward in Giessen, again without success. His father began to lose patience. At long last, in the summer of 1911, Moritz Schlick became a lecturer at the University of Rostock, a charming old Hanseatic town on the Baltic Sea. Though it had gotten off to a rocky start, his academic career was now under way.

In the winter of that same year, events took a decisive turn: Schlick's friend from his student days, Max von Laue, suggested that he take a philosopher's look at Einstein's theory of relativity: "You are well-versed in physics, perhaps more so than any of your colleagues from philosophy. Would the problem not suit you particularly well?"

Schlick temporarily set aside his planned work, *The New Epicurus.* He would never complete it. And yes, Max von Laue had been right: Einstein's theory offered the possibility to explore Kant's thoughts from a radically new perspective.

Are space and time really aspects of intuition that are given to us a priori? Why then should the four-dimensional space-time of relativity theory lead to baffling conclusions that are deeply at odds with our intuition? An example is the so-called twin paradox, which had been studied by Max von Laue: If a twin is sent at high speed to a distant star and back, then on his return, he will be younger than his twin brother, who remained at home. What on earth is so intuitive about that? Inspired by such riddles, Schlick felt that Einstein's ideas promised to be a gold mine for the theory of knowledge. Let physics guide philosophy, rather than the other way round.

Because of his medical history, Schlick had been classified in peacetime as "permanently unfit for duty" by the German Army. During the first years of World War I, he was left alone by the military and was thus able to pursue his philosophical work. He lectured on the foundations of mathematics to the students of Rostock—at least to the few of them who were left. The majority of Schlick's time, however, was devoted to his forth-coming essay "The Philosophical Relevance of the Relativity Principle."

When in 1915 he sent his manuscript to Albert Einstein, he received an enthusiastic reply. Einstein immediately seized on Schlick's observation that even if only one of Kant's a priori judgments starts to crack, the whole doctrine is shaken. He congratulated Schlick: "This work is among the best that have been written on relativity."

Within the next few years, Moritz Schlick became the philosophical mouthpiece of Albert Einstein—and this at a time when the latter's ideas were still hotly contended. From a ringside seat, Schlick was able to watch the dramatic events that led to the triumph of the general theory of relativity.

## "To the Pleasure of Two Fine Fellows"

The field equations relating mass and motion were discovered almost simultaneously by Albert Einstein and David Hilbert. To this day, historians of science argue over the details of this mathematical race.

In the summer of 1915, Einstein had been a houseguest of Hilbert's in Göttingen, while he gave several lectures at the university there. This was followed up by an intensive exchange of scientific letters. Each must have felt the other's breath on his neck. By the end of November, events came to a head: both scientists separately submitted to journals their results on the foundational equations of physics.

Einstein's paper was published first, although it had been submitted later. It makes no mention whatsoever of Hilbert. As for Hilbert, he withdrew his manuscript and replaced it with a modified version. The original galley proof has nonetheless survived—but with half a page missing, neatly snipped out by an unknown hand. This has fueled conspiracy theories: What was written on the missing slip of paper?

Even though the two scientists were too dignified to allow the priority dispute to become a public spat, their friendship cooled down for a time. Soon, however, Einstein was able to write to Hilbert that he had overcome his disgruntlement: "I have been fighting against the feeling of bitterness, and this with complete success." It would be a pity, he added, "if two fine fellows ('*rechte Kerle*') could not find pleasure in each other's company."

The theory with which he had struggled so mightily for so many years was at last completed—gravitation had been conclusively combined with geometry. Later, the physicist John A. Wheeler would put it this way: "Spacetime tells matter how to move; matter tells spacetime how to curve."

All of this was taking place during a war of unprecedented ferocity, when thousands of brave youths were falling every day—maimed, gassed, or shot to pieces. Schlick wrote that he sometimes wondered whether the historians in some far-off and more enlightened future, when asked in what period the Great War had raged, would reply: "The Great War? Ah yes, it took place in the period when Albert Einstein was completing his theory of relativity."

The tragic and pointless "war to end all wars" was still in full swing when Schlick published a slim book titled *Space and Time,* a lucid introduction to the theory of relativity. At the time, Schlick was serving in the physics department at a military airfield close to Berlin. In 1917, he had been reclassified from "permanently unfit for duty" to "fit for garrison duty." After all, able-bodied soldiers were growing ever scarcer in the *Kaiserreich.*

Einstein congratulated Schlick effusively on his book: "Your presentation is of unsurpassable clarity and transparence. You have not shied away from any difficulty, but have gone straight to the point, spelling out everything that is important while leaving out all irrelevancies. Whoever does not understand your presentation must be completely unable to grasp thoughts of this sort."

In successive editions of *Space and Time,* Schlick was able to follow the latest triumphs of Einstein's theory. The decisive breakthrough occurred in 1919, when two British expeditions confirmed that light rays from distant stars underwent a tiny amount of bending when they passed close by the sun.

Einstein had already predicted this gravitational effect in 1912. However, the curving of light rays can be observed only during a total eclipse of the sun. Germany had prepared an expedition to take place during an eclipse in 1914, but the world war had ruled that out. Now that the war was over, it was the former enemy—Britain in particular, with an expedition in May 1919, led by Sir Arthur Eddington—that had brilliantly clinched Einstein's theory, and many internationally minded people felt this was a wonderful symbolic event, and they rejoiced in the way that science united people across cultural gaps.

Max von Laue, a friend of Einstein's, was deeply surprised by the bending of light rays. One of the world's experts on the special theory of relativity, he had nonetheless long been suspicious of the "purely philosophical motivation" behind Einstein's general theory. There was "too much Mach" in it, at least to his taste. The equivalence of all frames of reference, no matter how they were moving with respect to each other, was no more than a hypothesis, after all. In 1913 von Laue had written to Schlick: "Fortunately, one of the most immediate consequences [of the general theory]—the bending of light rays in the vicinity of the sun—can already be verified at the next eclipse. At that time, the theory will likely meet a peaceful death."

But the theory did not die. Instead, it electrified the world. And Max von Laue made gracious amends: "In the meantime," as he wrote to Schlick, "I have become friends with the general theory, and this in particular through your little booklet."


## A Few Degrees of Latitude Farther South

Schlick's reputation was growing by leaps and bounds, among physicists and philosophers alike. He became, in the words of Walther Rathenau, "the evangelist of relativity theory." Hilbert invited Schlick to lecture in Göttingen. The future Nobel Prize winner Max Born wrote to Schlick: "We have become a community which has found its prophet—I hope that you accept this honorable position."

Quaint old Rostock had become too small for Schlick. His *General Theory of Knowledge,* which appeared in 1918, was an impressive volume; any philosopher with such a magnum opus on his list of publications could confidently aspire to a university chair. Schlick asked Einstein to help him "to escape from the sleepiness of Rostock," and to draw the attention of the academic world to the fact that "up there in the North there sits a lecturer in philosophy who is endowed, to a reasonable extent, with common sense, and who asks for nothing more than the chance to shift his activities a few degrees of latitude farther south."

Einstein obliged with pleasure. A hoped-for appointment in Zurich failed, however, and Schlick first had to content himself with a professorship in Kiel, again on Einstein's recommendation. Kiel was no improvement in latitude, compared to Rostock. But even before Schlick had settled in Kiel, he received the offer from Vienna, and in the autumn of 1922, he was able, at last, to move "farther south."

Hans Hahn had played an indispensable role in the hiring committee in Vienna, lobbying on Schlick's behalf within the faculty. The committee reached a Solomonic package deal for the three vacant chairs in philosophy. One physicist, one psychologist, and one "genuine" philosopher were appointed. Schlick was elected to the Chair for Natural Philosophy, as the successor of Ernst Mach (although the naming of the chairs had changed).

His election to the faculty was by no means unanimous: the ten votes against him, plus the three abstentions, out of forty-seven votes in all, clearly indicated certain reservations. But then again, even Albert Einstein had his opponents in the academic world.

The chair in psychology—at that time considered a branch of philosophy—was awarded to Karl Bühler (1879–1963), who soon created an outstanding Viennese Institute for Psychology, together with his wife Charlotte (1893–1974), and their colleague Egon Brunswik (1903–1955). The chair in the history of philosophy was given to the neo-Kantian Robert Reininger (1869–1955).

It was no small venture for Moritz Schlick, with his young family, to settle in the crisis-ridden Vienna of the postwar years, which were beset by hyperinflation and great political tension, both overt and covert. Max von Laue, who was on holiday in Austria, wrote to him: "Our experience has been that one cannot talk about prices. Before one's sentence has ended, they have increased again."

Moritz Schlick confessed to Albert Einstein: "It was hard for me, in the end, to move to Vienna, and not only because the future of Austria looks so gloomy. . . . But the Viennese climate is better, and the tasks for a philosophical teacher are greater."

The housing situation had played an important role in his deliberations. Eventually, the Schlick family, with their two young children, found a large apartment superbly located on Prinz-Eugen-Strasse. Schlick's route to work was delightful: the D tram line glided by the Belvedere and Schwarzenberg Palaces, with their baroque parks, before turning into the Ring Boulevard with its shade trees and its showcase buildings. From there, the whole tourist menu unfolded as one continued the tram ride: first the Opera, then the Hofburg, the museums, the Parliament, and, on the next-to-last stop before descending at the university, the Gothic-style City Hall and the Renaissance-era *Burgtheater*, facing each other. Altogether, this was a ten-minute lesson in the history of architecture, translated into marble and granite a mere fifty years earlier. Of course it was just scratching the surface of the past, but it did so with such elegance and class!

And should Schlick decide to leave town for a while, it was but a few steps from his home to the South Railway Station, where trains left both frequently and punctually for Carinthia (the southernmost region of Austria), Italy, and the Dalmatian coast, all favorite destinations for the sun-hungry philosopher. As he later wrote to one of his former school teachers: "We particularly enjoy the beautiful location of Vienna: the Easter holidays I spend mostly on the Adriatic or in southern Italy, during summertime we are usually in the mountains of Carinthia, and the autumn finds me nearly always in northern Italy."

Only weeks before commencing his duties in Vienna, Schlick had been granted a special distinction. The illustrious Society of German Scientists and Physicians, in honor of its centennial meeting in 1922, had invited Albert Einstein, fresh laureate of the Nobel Prize, to give the keynote address. This was to be immediately followed by a philosophical lecture—a talk by Schlick. To be sure, Einstein was quite an act to follow, but what a compliment to be invited to do so!

But then, out of the blue, a glitch. After Einstein and, of course, Schlick had both accepted, the German minister for foreign affairs, Walther Rathenau, was assassinated by a secret organization named Consul. It was rumored that Einstein's name was also on the murderers' black list: as a pacifist and a Jew, he had long been the target of fanatics from the extreme right.

A lecture by this "non-German" at the Society of German Scientists might well provoke a hate killing. It was thus decided that it was better to play it safe. Einstein canceled his lecture and went on a trip abroad. In his stead, Max von Laue spoke on "The Theory of Relativity in Physics," followed by Moritz Schlick on "The Theory of Relativity in Philosophy." In this fashion, Max Planck, who had organized the meeting, had the pleasure of seeing his two favorite ex-students as the two main speakers, back to back. Now, he quipped, nobody could say that it was all mere Jewish propaganda. The event aroused enormous interest, with a record number of seven thousand scientists attending.

## SCHLICK'S CIRCLE

Schlick's lectures in Vienna proved an instant hit. He was basking in Einstein's glory. Students, impressed by the renown of their new professor, thronged to hear him hold forth. A visitor wrote: "Professor Schlick's lectures were delivered in a huge auditorium packed with students of both sexes, and in his seminar a stray visitor was lucky when he did not have to sit on a windowsill."

Schlick's popularity, however, did not go to his head. "He was very sincere and unassuming, almost to the point of diffidence," as mathematician Karl Menger would later write in his reminiscences:

When I listened to some of Schlick's lectures as a student in 1923 and then took part in one of his seminars, he gave me the impression of being an extremely refined, somewhat introverted man. . . . My admiration for his sincerity deepened as I came to know him more closely in later years. Empty phrases from his lips or the slightest trace of pompousness were unthinkable. What caused the appearance of extreme modesty in his dealings with students, however, was Schlick's sometimes exaggerated politeness. While always prepared to correct his views and to learn, he was actually perfectly self-assured. If, after close acquaintance with Schlick, a trace of one's initial doubt about his self-assuredness remained, this would most likely be due to his inclination to idolize certain figures.

As Menger took pains to explain, Schlick's "inclination to idolize" was never wasted on the undeserving: "He had first studied physics under Max Planck, and then came to venerate Einstein. There followed a period of deepest admiration for David Hilbert. Then Schlick became fascinated with Russell."

Soon, highly talented students gathered around him, such as Friedrich Waismann (1896–1959), a Viennese with Russian roots, a little older than some; the Pole Marcel Natkin (1904–1963), an artsy type who was highly entertaining; the hardworking Rose Rand (1903–1980), poor as a churchmouse; and the eager Herbert Feigl (1902–1988), a gangling young man from Bohemia. It was Feigl who claimed that Schlick's lectures had finally made him grasp what scientific philosophy was all about. Schlick earned yet more gratitude by arranging for Feigl to meet with Albert Einstein.

Feigl and Waismann pressed Schlick to organize a private seminar. This suggestion coincided perfectly with Hahn's intention to resuscitate the old discussion group on the philosophy of science. Thus was the Schlick-Zirkel born.

"In Vienna a great deal of philosophy is presently being done," Schlick reported to Einstein. "I hope soon to be able to send you some samples, which surely will interest you."

Regular meetings were arranged. They were held every second Thursday evening at six o'clock, just as with the *Urkreis*. But this time, they were not confined to a coffeehouse. The mathematics institute was

FIGURE 4.6 Moritz Schlick (with shoes) and Herbert Feigl (without).

able to spare a small lecture room right next to Hahn's office, on the ground level of the new university building, the intellectual goings-on inside of which were, in the words of Heimito von Doderer, "slick and inscrutable." The windows looked out on the Boltzmanngasse. At most meetings, a member of the group would read a paper or deliver a report. Sometimes a session was devoted to the discussion of a particular topic. Sometimes guests delivered talks. When the meetings ended, there was always enough time left for a postmortem in the Café Josephinum, a coffeehouse just around the corner.

Karl Menger wrote: "Over the years, the size of the Circle varied from ten to twenty. Each academic year, the list of those attending remained largely the same, except for foreign guests."

The Circle's old guard dated back to before World War I, and consisted of Hans Hahn, Otto and Olga Neurath, and Viktor Kraft, a quiet and attentive philosopher working at the university library. Philipp

Frank often visited from Prague, a town that newly was located in a foreign country with a hard-to-spell name—Czechoslovakia.

Karl Menger once again: "The room was filled with rows of chairs and long tables, facing a blackboard. When we were not in session it was a reading room, occasionally used for lectures. Those who arrived first would shove some tables and chairs away from the blackboard, which most speakers used. In the space thus gained, they arranged the chairs informally in a semicircle in front of the blackboard, leaving one long table for those who brought books along or who wished to smoke or to take notes."

The students came in roughly equal numbers from mathematics and philosophy, many of them still unsure of which discipline they wanted to settle in. Slightly older than most were Felix Kaufmann (1895–1949) and Edgar Zilsel (1891–1944), both of whom were from the tidal zone between mathematics and philosophy. Zilsel had taken a leave of absence from his work as a schoolteacher to prepare for becoming a university lecturer in philosophy. Kaufmann, already a lecturer, was giving university courses on the philosophy of law and was earning an excellent living as a local representative of the Anglo-Persian Oil Company. He was greatly envied for this cushy job.

Karl Menger, who joined the Circle a little later, described its meetings: "People would mill around in informal groups until Schlick clapped his hands. Then all conversations would stop, everyone would take a seat, and Schlick, who usually sat at one end of the table near the blackboard, would announce the topic of the paper or the report or the discussion of the evening."

At first, the topics mostly revolved around the usual suspects: the trio consisting of Einstein, Hilbert, and Russell. But it wasn't too long before they encountered an unexpected turn in the road; from that moment on, the agenda was mostly set by a slender brochure penned by an obscure village schoolmaster, a completely unknown figure—one whose name, though, had a special ring to it.

# The Turn of the Circle

*Vienna, 1923–1928: Two German rookies, Carnap and Reidemeister, join Vienna Circle. Reidemeister jumps ship, carrying knot theory back to its cradle. Schlick has high hopes, expects philosophy will soon be obsolete. Ludwig Wittgenstein envisions limits to thinking. Mysterious heir scolds Russell, buys mortar, spurns fortune. Wittgenstein's enigmatic* Tractatus *becomes showstopper in the Circle. Hermit author claims logic is vacuous. Wittgenstein breaks ten-year silence, polarizes Circle, is hailed as genius.*

## THE TURNING POINT

The slim volume under whose spell the Vienna Circle fell was the *Tractatus Logico-Philosophicus* by Ludwig Wittgenstein.

"The interests of the Vienna Circle," Karl Menger recorded, "shifted from the analysis of sensations to the analysis of language, from Mach to Wittgenstein."

Moritz Schlick hailed a veritable "turning point in philosophy." Wittgenstein's oracular prose fascinated the Circle. Not that all members agreed with the theses that were proposed in the brief essay. Thus Otto Neurath spotted metaphysics—a deadly sin—lurking behind the scenes of most of Wittgenstein's sentences. But Hahn and Schlick were profoundly impressed. In 1927, Schlick wrote to Albert Einstein that he expected

FIGURE 5.1 Teacher Wittgenstein (right) with his pupils.

from the new turning point "nothing less than a complete reform of philosophy, in the sense of overcoming it, of making it superfluous."

For a long while Wittgenstein avoided any direct contact with the circle. But the mysterious figure lived close by: hidden away in the backwoods of Lower Austria, he was teaching spelling and arithmetic to the children of farmers. It took years before he finally deigned to meet with select members of the Vienna Circle. In the meantime, he had been forced out of his school job, having lashed out once too often at his small charges. Patience was not among his virtues.

"No transcendental twaddle," he once wrote, "when everything is as clear as a slap in the face." He used the Viennese slang for "slap in the face," *Watschn,* which might sound more cordial but no less vigorous. In philosophy, a slap in the face may go over just fine, but in education it could get out of hand.

What does logic tell us about the world? How does language affect our thoughts? What is philosophy all about? The terse sentences in Wittgenstein's *Tractatus,* at once enigmatic and crystal clear, electrified the Schlick Circle. The author had numbered them in an original manner, so as to indicate their place in the idiosyncratic trellis of his thoughts. Two times in a row, the Circle went through the text line by line; this took them several semesters.

The first reading had been suggested by the mathematician Kurt Reidemeister, the second by the philosopher Rudolf Carnap. These two young Germans had joined the Circle early on, the first introduced by Hans Hahn, the second by Moritz Schlick. Reidemeister remained only briefly in Vienna, but within a few years Carnap became one of the standard-bearers of the Circle.


## Reidemeister Returns Knot Theory from Its Crib to Its Cradle

Kurt Reidemeister (1893–1971) was born in Braunschweig. His studies of mathematics and philosophy in Freiburg, Munich, Marburg, and Göttingen were interrupted by a sudden call to arms. As luck would have it, he survived the Great War, and in 1920, the young veteran earned his PhD at the newly founded University of Hamburg. Soon his talent was recognized by Hans Hahn, and in 1922 Hahn succeeded in getting Reidemeister appointed to Vienna as associate professor of geometry—no mean feat, given that Reidemeister had published only three papers at that point and had not even reached the minor status of lecturer yet.

In Vienna, both colleagues and students were instantly taken with the jaunty young German geometer. Shortly after a meeting of the Mathematical Society, a nineteen-year-old student friend wrote to Karl Menger that at no other meeting in living memory had there ever been as much lighthearted banter as took place during Reidemeister's talk.

In the course of his three years in Vienna, Reidemeister laid the foundations of mathematical knot theory. For thousands of years, knots had fascinated artists and thinkers alike; they were the very emblem of, well, knotty problems. How to unravel a knot? How to classify knots? Today, knot theory has become a major field in mathematics, and the so-called Reidemeister moves, which superficially alter a knot yet preserve its underlying identity, marked the beginning of the systematic development of knot theory.

While still in Vienna, Reidemeister had been the first to stumble upon Ludwig Wittgenstein's booklet; he gave a lecture on it and suggested that it should be discussed more thoroughly in Schlick's seminar. Soon after that, however, the inventive knotmeister accepted a chair as full professor in Königsberg, the Prussian town where David Hilbert and Immanuel Kant had been born.

As it turns out, knot theory had been born in Königsberg as well. In 1736, the Swiss mathematical genius Leonhard Euler was living in St. Petersburg, Russia, not too far from Königsberg, where, as it happened, seven bridges spanned the River Pregel, linking the mainland to an island in an interesting pattern. A puzzle arose among the citizenry about whether, in making a stroll through the city, one could cross each bridge exactly once. Euler got wind of this puzzle and solved it (and a large family of cousin puzzles), and two hundred years later his pioneering ideas were recognized as having been the first contributions to knot theory. Thus in 1925, Kurt Reidemeister wound up carrying the toddler that was knot theory from Vienna, its crib, all the way back to its earliest cradle, where it had been a baby.

Kurt Reidemeister had forsaken Vienna and the Vienna Circle, leaving behind only warm memories. However, his younger sister Marie, who had visited him in Vienna in 1924, soon returned to stay. She had been bewitched by the bigger-than-life charm of Otto Neurath.

## The Logical Aufbau of Rudolf Carnap

Just like Reidemeister, Carnap too had served in the Kaiser's army throughout the war—a circumstance that tended to disrupt the usual academic career. Carnap turned thirty before he obtained his PhD, but when, soon afterward, he moved to Vienna, he was already carrying in his suitcase the draft of his *Habilitation* thesis, a work that would subsequently become a classic of twentieth-century philosophy.

Rudolf Carnap was born near Wuppertal in 1891. He lost his father at an early age. His uncle was the eminent archaeologist Wilhelm Dörpfeld, a collaborator of Heinrich Schliemann, who was said by some (especially himself) to be the discoverer of Troy. During school vacations, young Carnap did surveying work in excavations in Greece. He came to love measuring things.

Carnap studied mathematics, physics, and philosophy in Freiburg and Jena. The latter town was where old Gottlob Frege was teaching, he whose work on mathematical logic had set new standards of rigor but had also made him the primary victim of Russell's paradox. Frege's lecture courses, generally considered as far too abstruse, attracted but a scanty audience. Since university regulations required that a minimum of three students should attend all lectures, it fell to Carnap to drum up a quorum before each class meeting.

Young Rudolf was not only a budding logician, he was also a fervent member of the so-called Sera-Circle. This group, his first circle, epitomized a particularly romantic branch of the soulful *Jugendbewegung*, or Youth Movement. Like many others of his generation, Carnap hailed the movement's excursions, its musical evenings, and its solstice festivals as portents of a new society based on love and a communal spirit rather than on the tedium of an established order.

When the world war broke out, Carnap volunteered. Before 1914, the zealous young *Wandervogel* ("migrating bird") had naïvely yearned for an armed conflict to break out—once again, like many other confused idealists of his generation. But now, in the midst of one, he was ordered to the front, where he was wounded. During the last year of the war, Carnap worked as a physicist in Berlin, all of his intense patriotic fire having gone up in smoke. Now reborn as a pacifist, he was the author of appeals for peace, which circulated in the underground press. Moreover, the idealistic young soldier became a father.

In 1917, he had married Elisabeth Schöndube, the daughter of a German emigrant to Mexico. After the war, Carnap spent half a year overseas with his in-laws. Further offspring followed.

FIGURE 5.2 Carnap turns to logic.

It was high time for Rudolf to get some solid ground under his feet. His doctoral thesis, titled simply *"Der Raum"* ("Space"), had aroused Schlick's interest, and Schlick was even more curious to see Carnap's planned *Habilitation* thesis. The idea of the latter was to show how our image of reality can be constructed from sense-data by means of purely logical operations—that is, equivalence relations and logical connectives, such as *and, or, if . . . then . . . , not,* and so forth. This ambitious project cleverly combined Ernst Mach's empiricism with Bertrand Russell's formal mathematical logic.

While still a student, Carnap had written to Russell to inquire about an affordable copy of the monumental *Principia Mathematica,* as the German currency had collapsed. However, Russell had no free copy left; he replied instead by sending to the unknown young German student thirty handwritten pages filled with the most important formulas of the three-volume set.

Carnap viewed the logic of Frege and Russell as a tool for the whole of philosophy, helping to sharpen its concepts and arguments. He wrote: "It is historically understandable that at first, the new logic attracted attention only within the narrow circle of mathematicians and logicians. Its outstanding importance for the whole of philosophy has so far been recognized by only a few; its application to this wide field has hardly yet begun. But if philosophy is willing to follow the path of science (in the strictest sense), then it will not be able to do so without this thoroughly efficient instrument for clarifying concepts and cleaning up problematic situations."

Carnap's own "problematic situation" was the difficulty of securing a university position. After a friend suggested that he might look into the idea of becoming a *Privatdozent* (outside lecturer) at the University of Vienna, Carnap put out his feelers. Moritz Schlick lost no time in assuring him of his support. He wrote in reply: "I had the opportunity to speak about the question with a mathematical colleague who is very well versed in the psychology of the faculty. He is convinced that we will not have any problems at all with you, since the well-known obstacles that usually cause the ill will of the majority are altogether absent in your case. We may thus entertain high hopes."

This "mathematical colleague" was Hans Hahn, of course, and the "well-known obstacles" were Jewish origins and Marxist leanings. After Austria's defeat in the war, many professors with Pan-Germanistic convictions, who by that time constituted most of the faculty, considered it their sacred duty to safeguard all institutions of higher learning from "undesirables." The most likely targets of dismissal were young scholars who had not yet obtained their *Habilitation*. Female scholars, by the way, had no place in the system at all; nothing was clearer than that. Female *students* were bad enough!

Just at about the time of Carnap's move to Vienna, the "well-known obstacles" sabotaged the *Habilitation* of Edgar Zilsel, a member of the Vienna Circle. As for Otto Neurath, he didn't even try to gain a foothold at the University of Vienna. His lectureship at the University of Heidelberg had been canceled, on the bogus grounds that he had neglected his duties as a lecturer there. This had taken place in 1919, right after the inglorious end of the Bavarian Soviet republic. Neurath knew perfectly well that in Vienna, an application for the right to lecture would be doomed from the start. He might just as well dash headfirst against a wall.

But Carnap was different. His pacifist leaflets dating from 1918 had not reached Austria, and his racial background gave no grounds for complaint. Thus he had no trouble being accepted as a *Privatdozent* at the University in Vienna, just as Hahn had predicted.

Carnap's *Habilitation* thesis ranks today as one of the major works of analytical philosophy. On Schlick's suggestion, it was titled "*Der logische Aufbau der Welt,*" or in English, "The Logical Structure of the World." However, it was actually only about the logical structure of *science*—that is, any empirical science, no matter whether it dealt with physical or mental phenomena. Carnap had devised a method to reduce all propositions to the immediate experiences of an observer: "The concepts of the scientific domain are stepwise derived from suitably chosen fundamental concepts, and thereby arranged in a genealogy of concepts; and the propositions of the scientific domain have to be stepwise deduced from suitably chosen basic propositions, and thereby arranged in a genealogy of propositions."

Carnap's unique cocktail, blending Russell and Mach together in a subtle mix, was entirely to the Schlick Circle's taste, and the attraction proved mutual. Carnap later reminisced, in his philosophical autobiography: "My interests and my basic philosophical views were more in tune with those of the Circle than with any other group that I ever found."

Carnap had finished the bulk of *The Logical Structure of the World,* or his *Aufbau,* as it soon came to be known, even before he moved to Vienna. But the last sentence of his book shows how quickly he had imbibed the local spirit. Indeed, Carnap quoted what by then was already a password for the initiated, the shibboleth of the Schlick Circle: "Whereof one cannot speak, thereof one must be silent."

This was the sentence Ludwig Wittgenstein had placed at the end of his *Tractatus.*

## A Legend Takes Off

Legends swarmed about Wittgenstein like moths about a light. Some ten years after the publication of the *Tractatus,* a German mathematician stood up at a conference to ask whether a real person existed behind the Wittgenstein legend, a legend that members of the Vienna Circle were inclined to invoke at the merest drop of a hat. The question was of course a joke, but the fact is that much of Wittgenstein's life sounded almost like a fairy tale.

Wittgenstein's father, Karl, was one of the wealthiest men in Europe; as a steel baron, he played the same role in the Habsburg monarchy as Alfred Krupp in Germany and Andrew Carnegie in the United States. He had come from a well-to-do family himself, but at age eighteen he ran away from home, carrying with him nothing but a forged passport. Once he had made his way to America, he struggled there as a waiter, bartender, and violinist for two years.

Then Karl Wittgenstein returned home, by no means contrite about his adventure. For a year he studied at the Polytechnic in Vienna, then quickly advanced from the humble position of technical draftsman in a Bohemian steel mill to being a mighty captain of industry. It sounds like any old rags-to-riches story, but with the twist that Karl Wittgenstein's rags had washed dishes in New York while his riches had materialized back in Vienna.

In 1898, aged barely fifty, Karl Wittgenstein retired from all his commercial ventures and undertook a yearlong trip around the world. Upon his return, he mutated into a shining patron of the arts, becoming the primary financial sponsor of the Secession, a new art movement in vogue in Vienna at the time. He supported the architect and designer Josef Hoffmann, the composers Johannes Brahms and Gustav Mahler, and the painter Gustav Klimt, whom he invited over to his home for cultural soirees. In addition, the big spender became a bête noire of the harshly satirical journalist Karl Kraus, who saw in him the epitome of capitalism.

Ludwig Wittgenstein was the youngest of the steel baron's eight children. Little Luki, as he was called, had private tutors as well as a workbench and a horse of his own. The imperious father had a violent aversion to schools and was filled with headstrong ideas about how to bring up his five sons (three of whom eventually killed themselves). Ludwig was fourteen before he set foot in a public school. He enrolled in the *Realgymnasium* in Linz where, for a brief time, he became a schoolmate, though not a classmate, of Adolf Hitler, a most refractory youth who was older than Ludwig by a mere six days, and who soon dropped out of school with the intention of becoming an artist.

In Linz, Wittgenstein took his final school exams in 1906, receiving uniformly mediocre grades. His ambition was to study under Ludwig Boltzmann, as he had been enthralled by a popular article on heavier-than-air machines that the latter had published in 1895, long before the Wright brothers' first flight. Boltzmann had little faith in dirigibles; they were feasible but extremely clumsy, in his view. Instead he envisioned "dynamical engines of flight" having either vertical or horizontal "screws" (the word *propeller* had not yet been invented)—thus either "helicopters" or "aero-planes" (two words that did exist at that point).

Why, asked Boltzmann, should such dynamical engines not be invented in Vienna? And with ironclad analogic he added: "After all, the Magic Flute and the Ninth Symphony were composed here! Let the rest of the world trump that, if it can!" Boltzmann concluded his call to action with the reminder that in addition to conviction and passion, just one last key ingredient was needed: money.

Whether the young Wittgenstein thought that money would not be an obstacle in his case is not known. Conviction and passion he certainly had, and thus he decided, early on, to construct a kite equipped with a screw. Unfortunately, he was not to be the first to do so: indeed, by the time he finished school, airplanes had already taken off. And Boltzmann's hunch proved to be exactly on the mark: the conquest of the air fascinated the world. A new dimension was opening up, one with limitless challenges.

Shortly after Wittgenstein finished school in the summer of 1906, Boltzmann took his own life, and so the older and younger Ludwig never met. Wittgenstein enrolled instead at the Technical University of Berlin-Charlottenburg. He received his diploma in 1908 and went on to pursue aeronautical studies in Manchester. There he designed huge kites and was granted patents for various types of propellers.

Increasingly, however, the young engineer became intrigued by the mathematics underlying aerodynamics, and then by the logic underlying the mathematics. He made a pilgrimage to Jena in order to meet Gottlob Frege, and Professor Frege advised him to go to Cambridge to study with Bertrand Russell, little suspecting what he was thereby inflicting on his English colleague.

Wittgenstein first met Russell in the autumn of 1911. The encounter proved decisive for both men.

## "The Next Big Step in Philosophy"

The relationship between the two thinkers got off to a rather bumpy start, as is clearly documented by a famous series of letters by Russell to his lover of the moment, Lady Ottoline Morrell:

> My German friend threatens to be an infliction—obstinate and perverse, but I think not stupid. (October 19, 1911)
> My German engineer, I think, is a fool. (November 2, 1911)
> My ferocious German—he is armor-plated against all assaults of reasoning. It is really rather a waste of time talking with him. (November 16, 1911)

But shortly thereafter, the winds began to change:

> My German is hesitating between philosophy and aviation; he asked me today whether I thought he was utterly hopeless at philosophy, and I told him I didn't know but I thought not. (November 27, 1911)

Finally, Russell somehow discovered that Wittgenstein was Austrian, not German (it was about time!), and his reaction was what all Austrians would hope for:

I am getting to like him; he is literary, very musical, pleasant-mannered (being Austrian) and, I think, really intelligent.(November 29, 1911)

Again, the young Austrian asked Russell whether he considered him to be a total fool. If so, he would become an aeronaut; if not, then a philosopher. In response, Russell asked him to submit an essay as a kind of test. "As soon as I had read the first sentence," Russell later reported, "I was convinced that Wittgenstein was a genius, and assured him that he should by no means become an aeronaut. And he didn't."

This was why the twenty-three-year-old engineer enrolled as a student at Trinity College in Cambridge in early 1912. Not long after that, another stellar philosopher in Cambridge, Russell's friend G. E. Moore (1873–1958), also became convinced of Wittgenstein's immense talent. Moore's reason: "Because Wittgenstein always looks frightfully puzzled during my lectures, but nobody else does."

When Wittgenstein's sister Hermine came to Cambridge for a visit, Russell confided to her: "We expect the next big step in philosophy from your brother."

Soon Wittgenstein became a member of the Moral Science Club, which was Cambridge's philosophical society. Before the year was over, he held his first lecture there, with the modest title "What Is Philosophy?" That lecture was the shortest in the club's history. Four minutes sufficed for Wittgenstein to answer his own question: philosophy is the discipline that deals with all those propositions that are assumed to be true without proof by the various sciences.

As the minutes of the meeting dryly reported, "There was no general disposition to adopt this definition." But it was a pretty good definition—and not just beginner's luck.

Half a century later, Russell noted in his memoirs: "Wittgenstein was perhaps the most perfect example I have ever known of genius as traditionally conceived: passionate, profound, intense, and dominating."

"Dominating" indeed. Before long, Wittgenstein launched a devastating attack on Russell's philosophy. The latter confessed to his beloved

FIGURE 5.3 Wittgenstein's hut on a Norwegian fjord.

Ottoline that after this fateful encounter with Wittgenstein, he could never again hope to do any fundamental work in philosophy: "[His critique] was an event of first-rate importance in my life. . . . My impulse was shattered, like a wave dashed to pieces against a breakwater. I became filled with utter despair."

Wittgenstein had convinced the most noted logician of the day to keep his hands off logic, and to give up his planned *Theory of Knowledge.* Eventually, Russell and Moore wound up taking dictation from their student Wittgenstein. The tables had surely turned.

While touring Norway in the summer of 1913, together with his intimate friend David Pinsent, Ludwig Wittgenstein decided to write down his thoughts on logic all by himself. He spent the dark winter months alone in the village of Skjolden at the inland tip of an extremely deep fjord, and there he arranged for an isolated hut to be built for him. Never had a thinker found a more dramatic backdrop after Moses had returned from having scaled Mount Sinai's heights.

During the Christmas holidays of 1913, which Wittgenstein spent in Vienna, his father died, leaving behind him a huge fortune, mostly invested abroad, because the oligarch had no faith that a lasting peace would reign in Austria.

When, eight months later, war broke out, just as his father had feared, Wittgenstein immediately volunteered to serve in the Austrian Army, even though he had been exempted from military service. Before joining the ranks, though, he spent one hundred thousand crowns of his inheritance on an anonymous donation for artists and writers, asking an expert to do with it as he pleased. That expert, Ludwig von Ficker, made wise choices: the money was divided up among such notables as painter Oscar Kokoschka; architect Adolf Loos; poets Rainer Maria Rilke, Georg Trakl, and Else Lasker-Schüler; and the composer Josef Hauer.

In the army, Wittgenstein was at first assigned to the artillery. And again, he donated some of his fortune: this time, one million crowns for the army, to buy a mortar gun—the largest there was, a monster made of steel. This reminded his sister Hermine of the old joke about the smart-aleck recruit who ended up being told by his corporal: "Hell, buy yourself a gun and get independent!"

On garrison duty in Cracow, Poland, Wittgenstein one day stumbled across Tolstoy's booklet *The Gospel in Brief* in a small dusty bookshop. It triggered a sort of religious awakening in him. Later, Wittgenstein was to claim: "The war saved my life." He prayed frequently during the years of the Great War. But he also persevered with his book on logic, discovering that he did his best thinking while peeling potatoes.

He served on the Eastern and Southern Fronts, had himself transferred to the infantry, and became a highly decorated officer. In the intervals between his periods of service on the front line, Lieutenant Wittgenstein finished writing his *Logico-Philosophical Treatise*. He summed up the meaning of this work in his preface: "What can be said at all can be said clearly; and whereof one cannot speak, thereof one must be silent.

"The book will, therefore, draw a limit to thinking, or rather—not to thinking, but to the expression of thoughts; for, in order to draw a limit to thinking we should have to be able to think on both sides of

FIGURE 5.4 The identity card of Lieutenant Wittgenstein.

this limit (we should therefore have to be able to think what cannot be thought). The limit can, therefore, only be drawn in language, and what lies on the other side will simply be nonsense."

## CLEAR AS CRYSTAL?

After Austria's crushing defeat in 1918, *k.k.* Lieutenant Wittgenstein ended up in an Italian prisoner-of-war camp, or *campo di concentramento,* near Monte Cassino. From there, he wrote to Russell: "I think I have solved the problems once and for all." He repeated this bold claim in his preface: the absolute truth of the thoughts expressed in his small work seemed to him unassailable and definitive. He had solved the central problems of philosophy: they all arise from a misunderstanding of how language works.

During World War I, Russell had relentlessly championed his pacifist convictions. In their name, he sacrificed his position at Trinity College and then his freedom. In jail, he wrote an *Introduction to Mathematical Philosophy,* at about the same time as the pacifism-professing murderer

Friedrich Adler, who also was in jail, was writing his book on Ernst Mach, euphorically believing "to have found everything Ernst Mach had been looking for."

Russell, in a footnote on the last page of his new book, mentioned Wittgenstein, stating that he did not know the latter's whereabouts, nor even whether he was still alive. But now he relievedly wrote to the Austrian in the *campo di concentramento:* "Most thankful that you are still alive." He also sent him his *Introduction to Mathematical Philosophy.* However, Wittgenstein, in perusing its pages, found galling proof that Russell had never understood him, and never would.

The *prigioniere* eventually received permission to send copies of his own slim manuscript to Frege and Russell. Frege proved to be of no help at all: he obviously didn't have a clue as to what to do with Wittgenstein's treatise. Russell, too, was somewhat at a loss. After all, Wittgenstein's brusque claims, in his manuscript, that set theory and the theory of types were superfluous made short shrift of Russell's great achievement. Even so, Russell, as he had done before, went out of his way to be helpful. "I am sure you are right in thinking the book of first-class importance," he wrote. "Don't be discouraged. You will be understood in the end."

It was only after his release from prison, in the summer of 1919, that Wittgenstein was able to do something to expedite the publication of his slim manuscript, which he had been carting around for so long in captivity. He finally met up again with Russell in the Netherlands, which had remained neutral during the war. The weather was cold, the discussions protracted and sobering. Nevertheless, Russell promised to write an introduction to Wittgenstein's treatise, with the intention of clearing up some of the more obscure points.

Wittgenstein returned to Vienna, but not to philosophy. Why should he? After all, he had solved all of its problems. This showed, as he wrote, how little had been actually achieved thereby, and this was precisely what made his treatise worthwhile. Wittgenstein dedicated the work to the memory of David Pinsent. His small, delicate friend, flying as a test pilot, had perished in a crash on the military airfield of Farnborough.

The fortune that Ludwig had inherited from his father had further grown, having been wisely invested in the United States. But Wittgenstein chose to give it all away to his siblings. Only one of his brothers was still alive: Paul, the virtuoso pianist who had lost his right arm in the war. Paul now commissioned works for the left hand alone, and both Maurice Ravel and Sergei Prokofiev, among many others, composed piano concertos for him.

"In order to croak with a good conscience" ("*um anständig zu krepieren*"), Ludwig Wittgenstein decided to become a teacher in an elementary school. At the time, he was living with his sisters. One of them was a friend of the mother of Heinz von Foerster (1911–2002), the future systems analyst. Ten-year-old Heinz had just passed his secondary-school entrance exam. This great event of course had to be duly celebrated, with coffee and cakes. Entered Wittgenstein, clad in his usual leather jacket. "So what are you going to do when you grow up, Heinz?" Wittgenstein asked the boy. "Research!" eagerly cried little Heinz. "Well, you'll need to know a lot to do research," said Wittgenstein gently. "I already know a lot!" said Heinz. "Yes," retorted Wittgenstein, "but what you don't yet know is how right you are."

After studying for one year at the teachers' college in Vienna, Wittgenstein began his school service in Trattenbach, a small hamlet blighted by a tall smokestack, somewhere in the mountains of Lower Austria.

Now that it had been completed, his *Logico-Philosophical Treatise* gave him no pleasure. Several publishers had rejected it, and Russell's lengthy introduction did not meet Wittgenstein's expectations. In the end, though, it was Russell's introduction that was the key reason that the treatise was eventually published, in 1921, in the series *Ostwalds Annalen der Naturphilosophie*. By random chance, it turned out to be the last volume ever published in that prestigious series, and because of the time pressures on the publisher, Wittgenstein never got to see the galley proofs. To his horror, he found that the logical formulas had been garbled, and the text teemed with misprints. But worst of all, it contained Bertrand Russell's introduction.

The English translation of the *Treatise* fared better. It was produced in Cambridge by the linguist Charles K. Ogden (1889–1957), who was helped immeasurably by an eighteen-year-old student named Frank P. Ramsey (1903–1930), a remarkable mathematical prodigy. The bilingual edition appeared in 1922, under the weighty Latin title proposed by G. E. Moore: *Tractatus Logico-Philosophicus.*

Wittgenstein confided to Russell that no one would ever understand the book, although it was, as he put it, "clear as crystal." Elsewhere, however, he noted: "I am aware that all these sentences are unclear." As he seemed to realize at least to some extent, his style struck an odd balance between moments of dazzling lucidity and moments of total opacity, reflecting the tension between his yearning for clear expression and his awareness that some things simply cannot be expressed. The writer Ingeborg Bachmann would later declare that Wittgenstein's style was at once cryptic and crystalline.

Thus it says in the *Tractatus:* "Philosophy aims at the logical clarification of thoughts. Philosophy is not a body of doctrine but an activity. A philosophical work consists essentially of elucidations. Philosophy does not result in 'philosophical propositions', but rather in the clarification of propositions. Without philosophy, thoughts are, as it were, cloudy and indistinct: its task is to make them clear and to give them sharp boundaries." (4.112)

And in the same vein: "Everything that can be thought at all can be thought clearly. Everything that can be put into words can be put clearly." (4.116)

But the same author also wrote: "The inexpressible is contained—unexpressed—in the expressed." Throughout his life, the distinction between what can be *said* and what can only be *shown* was a recurring theme in Wittgenstein's thinking. "There are, indeed, things that cannot be put into words. They make themselves manifest. They are what is mystical." (6.522)

The *Tractatus* takes a firm stance on this: a proposition makes sense only if it is the "picture" of a fact. Unfortunately, this desirable trait can be said to be possessed by precious few of the propositions in the

*Tractatus.* Wittgenstein acknowledges this problem with a beau geste: "My propositions are elucidatory in this way: he who understands me finally recognizes them as senseless, when he has climbed out through them, on them, over them. (He must, so to speak, throw away the ladder, after he has climbed up on it.)" (6.54)

## At the End of the Ladder

Wittgenstein had indeed thrown the ladder away. He, who had never seen an elementary school from the inside, became a teacher at an elementary school and dedicated himself to his chosen new task with missionary zeal. He wrote a *Dictionary for Elementary Schools* and prepared squirrels' skeletons for his classroom. He took his pupils on excursions through Vienna or into the mountains, and he force-fed them with mathematics.

Many years later, one girl who had been a student of his reminisced: "We would often start class doing arithmetic assignments right away, and we would keep on doing nothing but arithmetic for hours and hours on end. He did not stick to the timetable as he should have. He was a zealous mathematician and wanted to drum it all into us."

Wittgenstein's zeal was not shared by every student. One of them recalled: "and then he got frightfully angry and pulled our hair, and that, of course, was the end."

His relations with the people of the village soured. The Cambridge-educated country schoolteacher wrote from Lower Austria to Bertrand Russell, who at the time was a visiting professor in China: "I am still at Trattenbach, surrounded as ever by odiousness and baseness. Here people are much more good-for-nothing and irresponsible than elsewhere. Trattenbach is a particularly insignificant place in Austria, and the Austrians have sunk so miserably low since the war that it is too dismal to talk about."

Transfers of the tyrannical zero-tolerance teacher to other schools, located in the nearby villages of Puchberg and Otterthal, did not improve the situation.

In 1923, the young mathematician Frank Ramsey came to visit him. Together, they labored through Wittgenstein's treatise page by page. Wittgenstein told Ramsey that no one could ever do philosophical work for more than five or ten years. He declared himself to be finished with it—not because there was nothing left to do in philosophy, but because his mind had lost its agility. It turned out, however, that it was not philosophizing but schoolteaching that would be merely an interlude in Wittgenstein's life—an interlude that came to an end after six years.

Despite constant pleas by his colleagues to treat his pupils less violently, Wittgenstein had never learned to control his temper. Eventually, an orphan boy named Haidbauer collapsed, knocked unconscious by one of Wittgenstein's "slaps." In a panic, Wittgenstein carried the boy to the office of the headmaster and then hightailed it away.

The official school inspector tried to soothe the devastated teacher: "Nothing much is going to happen," he said. But Wittgenstein quit his job. The enigmatic Herr Lehrer would never again be seen in that quaint village. "The people here are so narrow-minded that nothing can be achieved."

In 1926, after a few desperately unhappy months as an assistant gardener in a monastery in the outskirts of Vienna, Wittgenstein set his mind to a completely different kind of challenge: helping to design the town house that his sister Margaret Stonborough (whose portrait had once been painted by Gustav Klimt) had commissioned for the Kundmanngasse.

The chief architect whom they selected was Paul Engelmann (1891–1965), who had been a friend of Wittgenstein's for many years. Before the war, Engelmann had collaborated with the famous journalist Karl Kraus and the famous architect Adolf Loos. Later on, he gave a short overview of what he had learned "from the three best teachers of my generation: Kraus taught me not to write; Wittgenstein, not to speak; and Loos, not to build."

Wittgenstein did not substantially alter Engelmann's floor plan for the Kundmanngasse house, but he greatly improved many details of its design, infusing everything from the heating system all the way down to the door keys with his radical precision and rigor.

It was only as the construction work was nearing its end that Wittgenstein finally agreed to have contact with selected members of the Vienna Circle. The time had come.

## M or Non-M? (That Is the Question)

In its multiple readings of the *Tractatus,* the Vienna Circle had discovered many points of agreement. Like most of the Circle's senior members, Wittgenstein was a Viennese born in the prewar years, raised on the ideas of Boltzmann and Mach. Moreover, Wittgenstein's thoughts about logic fell on fertile ground: the Circle was well-versed in Bertrand Russell's work. In their seminars, Hahn and Carnap had explored the *Principia Mathematica* in depth.

Already before the war, Wittgenstein had developed the method of truth-tables, which today is the standard way to introduce logical connectives, such as *and, or, not.* Thus, for instance, if A and B are two propositions, then the compound proposition "A and B" is true as long as both A and B are true, and it is false in every other case.

A tautology is a combination of propositions A, B, C, . . . that is *always* true, independently of the truth or falsehood of its building blocks A, B, C,. . . . Therefore, a tautology arguably expresses nothing at all. For instance, the proposition "Either it is raining or it is not raining" is always true, but it tells us nothing at all about the weather. The same holds for "If grass is green, then either grass is green or cows are blue." Tautologies, it turns out, are exactly those propositions that can be derived from Frege's axioms of logic.

In his characteristically oracular style, Wittgenstein had declared:

The propositions of logic are tautologies. (6.1)

Therefore the propositions of logic say nothing. (They are the analytic propositions.) (6.11)

The propositions of logic describe the scaffolding of the world, or rather, they represent it. They have no "subject-matter." (6.124)

For the mathematician Hans Hahn, this view of logic made eminent sense. "To me," he wrote, "the *Tractatus* has explained the role of logic." And as he confided to his favorite disciple Karl Menger: "[At first] I did not have the impression that the book was to be taken seriously. Only after hearing Reidemeister give an excellent report about it in the Circle and then carefully reading the entire work myself did I realize that it probably represented the most important contribution to philosophy since the publication of Russell's basic writings."

And later, Hahn wrote: "Logic is not a theory about the behavior of the world—on the contrary, a logical proposition says nothing at all about the world—it is a set of directives for making certain transformations within the symbolism that we employ.

"Logic therefore does not say anything about the world; it only has to do with the way in which I talk about the world."

Moritz Schlick seconded this opinion: "Logical conclusions express nothing about real facts. They are merely rules for using our signs."

And in Schlick's seminal essay "The Turning Point in Philosophy," we read: " . . . but Ludwig Wittgenstein (in his *Tractatus Logico-Philosophicus*, 1922) was the first to push upwards and onwards toward the decisive turning point."

Rudolf Carnap, too, was deeply impressed: "For me, Wittgenstein was the philosopher who had the greatest influence on my thinking—except, perhaps, for Russell and Frege."

But not everyone in the Circle succumbed to the charm of Wittgenstein's oracular incantations. In particular, Neurath proved completely immune. What do such solemn, imposing pronouncements as the following actually *mean*?

The facts in logical space are the world. (1.13)

The substance subsists independently of what is the case. (2.024)

It is not how things are in the world that is mystical, but that it exists. (6.44)

Neurath smelled nothing but meaningless metaphysics behind most of the pronouncements delivered by the remote, reclusive prophet. As the other members of the Circle were reverently poring over the cryptic sentences in the *Tractatus,* one by one, Neurath gleefully seized every opportunity to point out their sycophancy. In the end, when an exasperated Schlick "told him that his interruptions were disturbing, Neurath offered simply to say the letter 'M' whenever the discussions turned metaphysical. But soon, he came up with an improvement: 'I think,' he said, 'that it would save us all time if instead I were to say 'non-M' in those rare cases where, for once, we are *not* engaging in metaphysics.'"

Student Rose Rand used a questionnaire to document the opinions held by the key members of the Circle on some of the book's most important philosophical theses, before, during, and after their painstaking read-through of the *Tractatus.* Blue meant agreement, red meant dissent, and green that the thesis was meaningless. The outcome was a colorful chart.

## CLOSE ENCOUNTER OF THE THIRD KIND

From 1924 on, Moritz Schlick tried and tried again to meet Wittgenstein, but it took three full years before such an encounter actually took place.

Despite his persistence, Schlick's politeness never faltered. Thus on Christmas Day of 1925, he wrote: "It would give me particular pleasure to meet you in person, and I would take the liberty of visiting you on some occasion in Puchberg, unless you were to let me know that you do not wish to be disturbed in your rural quietness."

Wittgenstein, from on high, told his sister Margaret that Schlick's letter had "pleased" him. Nonetheless, just as he had already done on a few occasions, he put Schlick off in vague terms. In April 1926, Schlick traveled with a few students to the village of Otterthal, having prepared, in his wife's words, "almost with awesome reverence . . . to go on a holy pilgrimage." But once again the attempt led nowhere, for by then Wittgenstein had already quit his job and taken off in great haste.

In the following year, however, Margaret Stonborough was at last able to transmit a positive reply from her brother to Schlick: "He asks me to convey his warmest regards to you, as well as his excuses for being unable to concentrate on logical problems while doing his present work, which demands all his energies. Although he is not willing to have a meeting with a group of people, if the meeting were to be held with you alone, dear professor Schlick, he feels that he might be able to discuss certain matters."

And thus, at long last, the ban was lifted. Margaret got permission to arrange a meeting. Somehow, Karl Bühler, the professor of psychology, also wound up taking part in it. Schlick's wife, Blanche, later recalled that Schlick returned from the meeting "in an ecstatic state." Wittgenstein, for his part, described the meeting to his architect friend Paul Engelmann in more sober terms: "Each of us considered the other one to be mad." (History did not record what psychologist Bühler made of their encounter.)

In a letter to Albert Einstein, Schlick reported that he now was probing "the depths of logic" with utmost enthusiasm, incited by "the Viennese Ludwig Wittgenstein, who has written a *Tractatus Logico-Philosophicus* (published by Bertrand Russell in English and German), which I consider to be the deepest and truest book in recent philosophy altogether. Its author, who does not have the intention to write anything ever again, has an artistic nature and is an entrancing genius, and the intellectual exchanges with him have been among the most overwhelming experiences of my life." They offer, Schlick added, "not so much an extension of knowledge as an intellectual safety zone."

From 1927 on, Schlick and Wittgenstein would meet regularly. Soon, a few carefully selected members of the Circle were also allowed to participate: the trustworthy students Friedrich Waismann and Herbert Feigl, the latter's consort Maria Kasper, and also Rudolf Carnap, who by that time had taken on a central role in the Circle.

"Before the first meeting," Carnap wrote in his autobiography, "Schlick urgently admonished us not to start a discussion of the kind to which we were accustomed in the Circle, because Wittgenstein did not want such a thing under any circumstances. We were even instructed to be very cautious in asking questions, because Wittgenstein was notoriously sensitive and could easily be upset by a direct question.

"When at last I met Wittgenstein, I saw that Schlick's warnings were fully justified. The impression he made on us was as if insight came to him through divine inspiration, so that we could not help feeling that any sober rational comment or analysis of it would be a profanation."

These initial encounters with the Vienna Circle, despite all their strange and arbitrary constraints, gradually led Wittgenstein, along a circuitous pathway, back to philosophy. In the same year that he finished the design work on his sister's house, he returned to his true calling.

The final impetus came from a lecture called "Mathematics, Science, and Language," given by the famous Dutch mathematician Luitzen Egbertus Jan Brouwer (1881–1966). Hans Hahn had organized the talk, and Wittgenstein was invited. Much later, Herbert Feigl described how, after the lecture, he and Friedrich Waismann spent several hours with Ludwig Wittgenstein in one of the nearby coffeehouses. "It was fascinating to behold the change that had come over Wittgenstein that evening. He became extremely voluble, and started sketching out ideas that were the germs of his later writings. That evening marked the return of Wittgenstein to strong philosophical interests and activities."

Schlick, too, reported on that event in a letter to Carnap, who happened to be abroad: "Recently, Brouwer held two lectures in Vienna. But they were less interesting than what Wittgenstein, who attended them both, later told us about them in the coffeehouse."

The champion had reentered the ring. Apparently, there still was something worth doing in philosophy! Now forty years old, Wittgenstein returned for a short while to Cambridge and at long last earned his doctorate in philosophy. For his PhD thesis, he submitted the now world-famous *Tractatus*. His examiners were Professors Russell and Moore, his old friends from the prewar years.

The oral defense seemed like a farce to all three, and it was brief. As the story goes, Wittgenstein got up after a few minutes, patronizingly patted the shoulders of his examiners, and said: "Don't worry, you'll never understand it."


## The Worldview of a Scientist

Wittgenstein had rapidly come to the conclusion that Carnap, too, was among those who would never understand him. After only a few meetings, he refused to ever meet with the sober German again. Two points of dispute had raised Wittgenstein's ire.

The first was Carnap's interest in the artificial language Esperanto, a well-meaning if naïve attempt to bring about international communication (and hopefully also world peace) at a time when English was not everybody's second language. Wittgenstein had no patience with Esperanto.

The other was Carnap's readiness, in principle, to examine, with the methods of science, the claims of parapsychologists. And indeed, why shouldn't science be used to reject poltergeists and to disprove the possibility of communication with the dead? But when Wittgenstein chanced across a book on Carnap's bookshelves that dealt with extrasensory perception, he wrathfully threw it on the floor and forthwith cut off all contact with Carnap forever.

Even without this outburst, the differences between Carnap and Wittgenstein would sooner or later have come to a head. Their personalities were simply incompatible: the inspired artist on the one hand, and the meticulous scientist on the other.

Carnap insisted that "the rigorous and responsible attitude of scientific research should also become the basic attitude of the philosophical worker." Thus, in the preface to his *Aufbau,* Carnap had described his vision of a future philosophy: "By assigning to each individual only a small task in philosophical work, just as is standard in scientific practice, we believe that we may look into the future all the more confidently. Through slow, careful construction, insight after insight is obtained, with each individual contributing to the collective effort only that which he can account for and justify. In this methodical fashion, one stone gets added after another, and thus is gradually constructed a stable edifice, which can be further extended by each following generation."

"A stable edifice" for the coming generations was just what Wittgenstein most loathed. He did not even wish to add "one stone after another" to his own *Tractatus*—quite the contrary! Over the next ten years, he would raze it to the ground. Nor had he any desire to be involved in "a collective effort." Later, he would say: "I cannot found a school, because I do not want to be imitated."

Carnap badly underestimated the vastness of the gulf between himself and Wittgenstein when he wrote: "When Wittgenstein talked about philosophical problems, about knowledge, language, and the world, I was usually in agreement with his views, and certainly his remarks were always illuminating and stimulating. Therefore I regretted it when he broke off contact. He told Schlick that he could talk only with somebody who 'would hold his hand.'"

Giving comforting support and allegiance, however, was not part of Carnap's agenda. On the contrary, he never shied away from questioning convictions—including his own—for this is the crux of a scientist's worldview, and Carnap was committed to it. Indeed, *The Scientific Worldview*—this was the title of the pamphlet with which the Vienna Circle would soon step onto the public stage.

# 6

# *The Circle Makes a Name for Itself*

*Vienna, 1928–1930: Schlick Circle enters new public phase, surprises leader with manifesto. Scientific Worldview announced. Wittgenstein warns of boastful clique. Schlick hails turning point, describes philosophy as an art. Freethinker Hahn confesses creed: God never does mathematics. Carnap slams metaphysics, asserts Heidegger's Nothing means nothing. Neurath places faith in proletarian class, calls intellectuals grease blobs.*

### "A Thunderstorm at Long Last"

The writer Heimito von Doderer, author of *The Strudlhof Stairs,* once compared fame to a battleship: slow to get under way, but then hard to stop.

Moritz Schlick's name had become familiar to philosophers well beyond the German-speaking countries. For the summer term of 1929, he received an invitation to visit Stanford, the young university that had once so delighted Ludwig Boltzmann, who effusively called it his beloved "El Dorado."

Schlick gladly accepted the invitation. Since marrying Blanche in 1907, he had not once set foot in the United States, and at this point their silver wedding anniversary was approaching. The notion of sunny California appealed greatly to the sun-hungry professor. He planned to be back in Vienna by the autumn.

A few weeks before embarking on his journey, Schlick was offered a chair in philosophy by the University of Bonn—a significant token of recognition. In such cases, the home university standardly tries to retain its professor by making a counteroffer. Such auctions, as it were, can go back and forth for a while, and can turn out quite profitable for the prestigious professor. But the Viennese higher-ups gave Schlick the cold shoulder. All that the Austrian ministry was prepared to offer was a trifling salary for Schlick's assistant, the auxiliary librarian Friedrich Waismann, who up till then had worked entirely gratis. Such a feeble counteroffer could hardly be seen as enticing; indeed, it was a snub. Small wonder that the Bonn offer seemed more and more tempting to Schlick. His farewell to Vienna became increasingly likely.

It was only now that the members of the Schlick Circle realized how essential Schlick had been to them. The group collectively wrote a letter to him, addressing him indirectly: "Were Professor Schlick to leave Vienna, the supporters of the strictly scientific worldview would be deprived of their spiritual leader and their renowned representative at the University, and there would be no way to compensate for the loss to Vienna's intellectual life."

When Rudolf Carnap, who at the time was in Davos, undergoing a cure for a lung ailment, heard of the imminent departure, he anxiously wrote to his friend: "This would be a severe blow to the University of Vienna, and to each one of us."

In a postcard to Carnap, Schlick replied that he was still undecided. But on the flip side of these handwritten lines, there was a photo of a charming Viennese girl named Lisl Goldarbeiter, who had just been voted "Europe's prettiest woman." One didn't need to have attended Professor Freud's classes to guess in which direction Schlick would wind up being pulled more strongly.

On the day before embarking on his transatlantic journey, Schlick wrote from Bremen to inform the Austrian ministry that he would have

FIGURE 6.1  Two sides of a postcard from Schlick to Carnap.

accepted the flattering offer from Bonn, had it not been for his colleagues and students who, "at the very last moment," convinced him that important philosophical tasks still required his presence in Vienna—now more than ever.

That said, Moritz Schlick set sail for the New World, planning to be back in autumn.

Karl Menger recalled: "How great was our joy when we heard that Schlick had decided to stay in Vienna. 'This has to be celebrated!' exclaimed Neurath, and we all agreed. 'We have to write a book that will present our views—a manifesto of the circle—and dedicate it to Schlick when he returns in the fall.' And then, with his usual expeditiousness, Neurath instantly set about working on it."

A conference of physicists and mathematicians was scheduled to take place in Prague in September 1929. The manifesto-writing subcircle of the full Schlick Circle decided to shoot for that occasion, setting their sights on completing their small opus by then, so that it could serve as a calling card, so to speak, for the Vienna group.

In the meantime, Carnap had returned to Vienna from Davos, fully recovered and eager to get to work. It was he who wrote the manifesto's first draft, assisted by Feigl and Waismann. For the title, he suggested "Key Ideas of the Viennese Philosophical School." But his outline met with opposition.

"Neurath advises us not to sound so dreary," Carnap noted in his diary. Neurath's objections began with the title. *Philosophical* was a term he wanted to avoid at all costs, having in fact placed it on his "list of taboo words"; as to *School,* this word was prone to triggering unpleasant associations. Moreover, all the members of the Circle were united in holding the notion of "school philosophy" in contempt, and *Philosophical School* came far too close to that bugbear.

Philipp Frank sided with Otto Neurath: "Several of us ardently wished to avoid terms such as 'philosophy' and 'positivism,' particularly on the title page. Also, some of us disliked 'isms' of all sorts, whether foreign or homegrown. So finally we decided on the title 'The Scientific Worldview.'"

Neurath still felt that this title sounded too dry. He therefore suggested adding "The Vienna Circle" as a subtitle, as this phrase would tend to evoke associations with the Vienna Woods, Viennese waltzes, "and other things on the pleasant side of life."

Thus did the Vienna Circle acquire its name. In Vienna, however, people continued to call it merely "the Schlick Circle"; the dozens, nay hundreds, of other circles in town surely felt that it must have taken quite some chutzpah for *that* circle to single itself out as *the* Vienna Circle.

The manifesto was a collective work. Its official authors were Neurath, Carnap, and Hahn, but the student members Waismann and Feigl had also contributed, as did Maria Kasper, another philosophy student who was Feigl's romantic partner. Rudolf Carnap reserved for himself "the sour duty and the sweet right of carrying out the final revision." Since the document was being composed as a surprise for the Circle's leader, Schlick himself was not informed of it at all, let alone asked to contribute or to comment.

*The Scientific Worldview* meant to make things as clear as possible, since that is the purpose of any manifesto. And indeed, as a summary of the Vienna Circle, the text is still unsurpassed. In just a few pages it neatly described the group's historical background and its highest mission—a collective crusade against all metaphysical and theological doctrines. Only the results of experimentation and logical analysis were admitted—nothing else. The manifesto listed the names of the Circle's members, and the problems to be tackled: namely the foundations of mathematics, physics, geometry, biology, psychology, and the social sciences. An encyclopedia could not have done better.

At the end of an extremely hot week in July, the struggle was at last over, and Carnap wrote in his diary: "Finished typing the brochure. In the evening, a thunderstorm at long last."

In his ever-energetic manner, Neurath settled matters with the publisher. In September, shortly before leaving by train for Prague, Herbert Feigl picked up the first batch of copies, fresh from the printer.

## MANIFESTO IN BLUE

Summer was giving way to fall, and Feigl's mission was to attend the 1929 meeting of German physicists and mathematicians. Prague was now in Czechoslovakia, no longer in Austro-Hungary, but it still had a large German-speaking population as well as a German-speaking university. It remained an important center of the German-speaking intellectual world.

At this major physics congress, the Schlick Circle and a similarly oriented Berlin group wanted to attract attention by holding a satellite meeting on the topic "The Theory of Knowledge of the Exact Sciences." Philipp Frank wrote: "The German Physical Society was not exactly enthusiastic about the idea of mingling such a serious scientific meeting with something as frivolous as philosophy. But since I was head of the local organizing committee, they could hardly turn down my earnest wish."

Frank himself gave the main conference's opening plenary address. Before a packed house, he slammed metaphysical statements as being nothing but petrified leftovers from long-abandoned physical theories. This was strong language for most of the physics professors making up the audience. They were unfamiliar with the petulance of philosophers; indeed, some may have held German idealism to be sacrosanct.

Frank's wife had the impression that her husband's words "fell upon the audience like drops falling into a well too deep for anyone to hear them hitting the bottom. Everything seemed to vanish without a trace." That may or may not have been the case, but right after Frank's talk came a talk by his friend Richard von Mises, who had built up a large institute for applied mathematics in Berlin, and who, on the side, had also been developing his positivist philosophy. He sang his own variation on Philipp Frank's anti-metaphysics theme, and then, as the session broke, the manifesto's first copies were handed out at the door.

After that opening salvo at the main meeting, Hahn, Carnap, Neurath, Feigl, and Waismann all gave talks at the satellite meeting. Their texts were published in the first volume of *Erkenntnis* ("Knowledge"), a philosophical journal that formerly had been named *Annals of Philosophy,* before being taken over by the Vienna Circle and its Berlin counterpart. As icing on the cake, a Vienna Circle book series was launched as well—*Writings for a Scientific Worldview,* edited by Moritz Schlick and Philipp Frank.

Now armed with its own journal, its own book series, its own meetings, its own propaganda brochure, and perhaps most importantly of all, its own catchy name, the Vienna Circle catapulted itself from its initial private phase into the glare of public view.

Schlick, on returning to Europe from sunny California, headed straight to Italy's sunny Lake Garda for a vacation, and there, in the mail, he received a copy of the brand-new manifesto, dedicated to him, and elegantly bound in blue leather. Carnap had forewarned him of it in a letter: "And now, out of the blue, I am sending you your personal copy, bound in blue, but I hope it will not give you the blues. Please do not judge the contents with harshness, but rather, with your usual warmth and tolerant spirit. It was written jointly by Feigl, Neurath, and myself, and with more good will than skill."

Carnap suspected—quite correctly—that Schlick would be irked by certain self-promoting passages and dogmatic claims in the manifesto; indeed, to Schlick's taste, there was too much of Neurath behind the blue cover.

## MR. WITTGENSTEIN IS NOT AMUSED

Hans Hahn was apparently not all that happy with the brochure either, despite being one of its co-authors. According to Menger, Hahn had been asked for his comments only quite late in the game, when there was hardly any time left for substantial changes. He thus ended up by simply signing on the dotted line. It was "one of those compromises he occasionally found himself making, for the sake of dear peace." Tellingly, both Schlick and Hahn published their own independent philosophical agendas within a year.

The idea that philosophers might fully agree on a joint proclamation is basically a contradiction in terms. It says something about Carnap and Neurath, though. And Neurath might have been enough of a politician to foresee that the book's subtitle, *Der Wiener Kreis,* would be understood as a claim of collective authorship by the entire Vienna Circle. Indeed, the frontispiece of the manifesto discreetly avoided listing the names of the actual authors.

What had happened was that the intense time pressure had not permitted the coterie of writers to ask all the group members for their consent. Thus not only Schlick was surprised by the manifesto's content, but others were as well. For instance, Karl Menger, after reading the manifesto, made it clear that he did not wish to be identified as a *member* of the Vienna Circle, but only as someone "associated with the Circle." As a matter of fact, the brochure's authors had taken pains to distinguish between Circle *members* and mere "associates." After Menger's expression of reticence, other members of the Circle, including Kurt Gödel and Viktor Kraft, also asked to be downgraded. One assumes they felt they had been implicitly represented as accomplices in the team effort, and they did not appreciate being steamrollered in that manner.

To no one's surprise, the most vehement reaction came from Ludwig Wittgenstein. The group had debated long and hard as to whether or not he should be labeled as "associated with the Circle" (and of course calling him a "member" would have been out of the question). In the end, they had decided that Ludwig Wittgenstein should be listed, along with the great Albert Einstein and the great Bertrand Russell, as "one of the three leading representatives of the scientific worldview."

Well, Wittgenstein was not one to be cajoled by such blatant flattery. "This is a deeply troubling affair," he wrote to his friend Friedrich Waismann, "and it irks me to think that once again, a good cause has been turned into a pretext for patting oneself on the back.

"It is precisely because Schlick is no ordinary man that his friends should refrain from making him, and the Viennese school that he represents, appear ridiculous through their boasting, well-intended though it may be. When I say 'boasting,' I mean any kind of self-satisfied posturing. 'Rejection of metaphysics!' As if that were a new idea! Any praise that a school of thought heaps upon itself stinks, just as does all self-praise."

And then some well-meant advice for Waismann: "Always act honorably! Don't ever do a favor for a clique (Hahn, Carnap, etc.), if at some later date you (and others) will have to apologize—with a half-smile—for your act. Rest assured that *I* do not smile to think that possibly— in fact, probably—some asinine thing has been done to honor a man whom I highly respect, and that *you* were involved in it."

And yet, Wittgenstein was not inhuman, and his letter to Waismann ended up with a gracious growl:

"That you have married is a fine thing, and I congratulate you."

## THE ART OF PHILOSOPHY

Moritz Schlick, the leader of the Vienna Circle and the man whom Wittgenstein respected and did not include in "the clique," delineated his own philosophical position in *The Turning Point for Philosophy*. This essay appeared as the first contribution published in the new journal

*Erkenntnis*. It did not mention the manifesto, but it shared the manifesto's emphasis on analyzing language as a technique for poking holes in metaphysicians' claims.

Showy posturing was not part of Schlick's makeup, but he knew how to bring about a fine crescendo. To make his later points more effective, he began his essay on a subdued note, observing that "from time to time prizes have been established for essays describing the progress that philosophy has made" since this or that moment in history. "Such challenges inevitably betray a certain degree of doubt," he went on. Indeed, one gets the impression, said Schlick, of being asked, albeit in an indirect manner, whether philosophy has actually made any progress at all since the specified moment.

"This peculiar fate of philosophy has been described and bemoaned so often that it has become a time-worn cliché. Silent skepticism and resignation seem to be the only appropriate attitudes. Two thousand years of experience serve merely to demonstrate that all efforts to stop the bitter fighting among so many diverse factions and to save philosophy can no longer be taken seriously."

After this introduction came the audacious turn:

> I allow myself to refer to this philosophical anarchy, which has so often been pointed out by others, so as to leave no doubt that I am fully aware of the scope and the import of the convictions that I wish to express in what follows.
>
> Specifically, I am convinced that we now find ourselves at a major turning point for philosophy, and that we are objectively justified in asserting that the time is ripe for us now to put an end, for once and for all, to the useless conflict among rival schools of thought. I believe that we are already in possession of methods that, in principle, could eliminate all such conflicts. All that remains for us to do is to resolutely apply these methods.

Schlick's turnaround had been radical. In a lecture just a few years earlier, he had claimed that Nietzsche, the idol of his youth, was not among the greatest philosophers, despite his brilliance, since Nietzsche had failed to create a philosophical system. But now, such systems were objects of Schlick's scorn; they were precisely what had been overcome. Schlick expressed this as follows: "Philosophy is not a set of statements. It is not a science. But what is it, then? We see in philosophy not a system of knowledge, but a system of actions: philosophy is that activity through which the meaning of statements is revealed or determined. Through philosophy, statements are explained; through science they are verified. The latter is concerned with the truth of statements, while the former is concerned with what they actually mean."

Wittgenstein himself couldn't have put it any better. Though not a part of the Vienna Circle, he was nonetheless, in Schlick's eyes, its focus. "It is easy to see that doing philosophy does not consist in making statements, and that conferring meaning upon a statement cannot be done simply by making more statements. This process of explaining meaning cannot keep on going ad infinitum. It always will come to an end when one actually points something out—when one exhibits what is being talked about—thus, in physical acts, not just more words. Therefore, explaining meaning always involves deeds. It is these deeds or acts that constitute philosophical activity."

In the end stands the deed. In his lecture course, the professor of philosophy explained Wittgenstein's 4.112 ("Philosophy is not a body of doctrine but an activity") with these words:

The difference between the task of the scientist and the task of the philosopher is that the scientist seeks the truth (the correct answers) and the philosopher attempts to clarify the meaning (of the questions).

The method of science is observation and experiment, combined with calculation and inference; through this method one establishes the set of true propositions about the real world. The method of the philosopher, by contrast, is reflection; the philosopher looks upon the given statements, observations, and calculations, and explains what they mean. To do philosophy is not to give a list of true propositions. It is, rather, an art   an activity leading to clarity.

Philosophy as an art form—this vision of Schlick's seemed light-years away from Rudolf Carnap's comparatively pedestrian idea of aligning philosophy with the sciences and getting philosophy to imitate their style.

## HAHN REACHES FOR THE RAZOR

Hans Hahn, too, saw a turning point in the work of Wittgenstein.

In spite of his lifelong passion for philosophy, the mathematician had so far authored nothing in this field. But now, at age fifty, he felt ready. The Circle's journal *Erkenntnis* published the talk he had given at the Prague meeting: "The Significance of the Scientific Worldview, in Particular for Mathematics and Physics."

Hahn opened his presentation by defining his terms. The term *scientific worldview*, he explained, is proposed both as a creed and as a contrast. The term *creed* may seem strange, coming from the mouth of a confirmed freethinker, but yes: "It helps us to confess our faith in the methodology of the exact sciences, especially mathematics and physics, and our faith in careful logical inference (in contrast to bold flights of fancy, mystical intuitions, and emotional ways of relating to the world), and our faith in the patient observation of phenomena, isolated as much as possible, no matter how negligible and insignificant they may appear in themselves (as opposed to the poetic, imaginative attempt to grasp wholes and complexes that are as vast and as all-encompassing as possible)."

But the term *scientific worldview* was not just a set of interrelated faiths; it was actually a declaration of war! In Hahn's words:

This term serves to contrast our worldview with philosophy in its usual sense, as a doctrine about the world that claims to stand side by side with the scientific disciplines, or possibly even above them. In our opinion, anything that can be sensibly said at all is a proposition of science, and doing philosophy just means examining critically the propositions of science to check whether they are or are not pseudo-propositions (that is, whether they really have the clarity and significance ascribed to them by the practitioners of the science in question); and it means, further, exposing as pseudo-propositions those propositions that lay claim to another type, and a higher degree, of significance than the propositions of science.

Almost simultaneously with this article, Hahn's little pamphlet called "Superfluous Entities: Occam's Razor" was published in the same series as the Vienna Circle's manifesto. He opened his pamphlet by referring to "the bewildering variety of philosophical systems," and he then separated philosophies that are turned *toward* the world from ones that are turned *away* from it. Epicurus and Hume, he said, are turned toward it, while Kant and Plato are turned away from it.

Philosophies turned away from the world were not to Hahn's taste. According to him, however, they could still be found

in the systems of German idealism—and how could it be otherwise? The Germans are known, after all, as the nation of thinkers and poets. But a new day is slowly breaking, and the liberation is coming from the same land that gave birth to political liberation—namely, England: the English, after all, are known as the nation of shopkeepers. And it is surely no accident that one and the same nation gave the world both democracy, on the one hand, and the rebirth of a philosophy turned toward the world, on the other; nor is it an accident that the same land that saw the beheading of a king also witnessed the execution of metaphysics. Yet the weapons of a philosophy that is turned toward the world are not the executioner's sword and axe—it is not as bloodthirsty a beast as that—though its weapons are sharp enough. And today I want to talk about one of these weapons—namely, Occam's razor.

The English scholastic William of Ockham (ca. 1287–1347) had stated that one should never assume more entities than necessary. This principle became known as Occam's razor. Mach had adored it, for it epitomized the economy of thought that he had always sought.

Hahn, too, embraced the principle, and put it to work in his essay. Superfluous, he said, are all the "shadowy half-beings" encumbering our brains, such as universals, empty space, empty time, substance, Thing-in-Itself, the beyond, and of course gods and demons.

"Away with them all!" wrote Hans Hahn.

## WITTGENSTEIN, GOD, AND MATHEMATICS

Hahn wanted to relegate all metaphysical and theological gimcrackery to the dustbin of mystifications, hoping thereby to gain a less obstructed view of the *real* problems of philosophy. As a young lecturer, he had already written to his friend Paul Ehrenfest: "I have received more than a few compliments for my philosophical talents, and deep in my heart I know, and cannot dispute the fact, that I have a gift in this regard. And I can only tell you: I am persuaded that genuine problems lie behind these questions, and that claims to the contrary, no matter how often they are repeated, are nothing but silly verbiage, based in part on ignorance and in part on ineptitude."

What were the "genuine problems" that captivated Hahn? He formulated the most fundamental problem in this manner: "How is the empiricist position compatible with the applicability of logic and mathematics to reality?"

The idea here is that our limited experiences in the world can never provide us with any generalization whose truth is absolutely certain. To have total certainty, we would have to observe all instances to which the general statement might ever possibly apply. Take, for instance, the statement "Cats do not bark." It seems to be true, but it is perfectly conceivable that a cat might someday start to bark—even if no one has yet observed such a thing. But perhaps cats are actually barking all the time when no one is listening.

Mathematics, on the other hand, consists of statements that are always true. It is not conceivable that one of these days "Two times two equals five" will hold. And so, why is one kind of knowledge secure and the other one not? And since mathematics is based on logic, this leads to the next question: Where does the certainty of logic stem from?

According to Hahn, if logic were the science of the most general properties of the world, then empiricism would indeed face insuperable difficulties. But in fact, "Logic does not say anything about the world; rather, it has to do with the ways in which I talk about the world. The so-called 'propositions of logic' are merely indications that show how something that we have said could be said in various other equivalent ways."

Hahn further wrote: "It was Wittgenstein who uncovered the tautological character of logic, and who stressed that the so-called logical constants (such as 'and,' 'or,' etc.) do not correspond to anything in the world. Logic deals solely with the way that we *talk* about the world. "The certainty and the universal validity—or rather, the irrefutability—of a proposition of logic flows precisely from the fact that it says nothing whatsoever about objects of any kind."

Hahn appears to anticipate Wittgenstein's use of "language games" when he writes: "Whoever does not accept logical inference does not hold a different opinion from mine about the behavior of objects, but rather refuses to talk about these objects following the same rules as I do. It is not that I cannot convince such a person, but rather that I have to cut off our conversation, just as I would refuse to keep on playing Tarot with a partner who persisted in claiming that it is possible to take a Fool card with a Moon card."

As Hahn saw it, what holds for logic holds also for mathematics: it, too, consists of nothing but tautologies. To be sure, this view had been vehemently rejected by many mathematicians, such as Henri Poincaré, for whom the word *tautology* smacked of "triviality."

"And indeed," wrote Hans Hahn, "it hardly seems believable at first sight that the whole of mathematics, with its hard-earned theorems and its frequently surprising results, could be dissolved into nothing but tautologies. But this argument overlooks just one minor detail—namely, the fact that we humans are not omniscient. An omniscient being would, of course, know instantly everything that is implied when a set of propositions is asserted. Such a being would know immediately, on the basis of agreements about the use of numerals and the sign '×,' that what is meant by '24 × 31' and '744' is the same thing; for an omniscient being needs no logic and no mathematics."

And on another occasion: "An omniscient subject needs no logic, and contrary to Plato we can say: God *never* does mathematics."

## THE EMBODIMENT OF CHILLY CLARITY

Hahn's friend from their student days, Philipp Frank, wrote: "It could be said that Hahn was in a certain sense always at the center of our Circle. He always formulated its key ideas without worrying about irrelevant divergences of opinion. No one knew better than he did how to describe our Circle's basic tenets in a way that was simple and yet careful, logical and yet incisive."

The noun *Hahn* means "rooster," and his friends called him *Hähnchen,* or "broiler" (the students presumably did this too, but only behind his back). His talks and papers were of supreme clarity. To deliver his daily lectures, which he always prepared with meticulous care, he had developed a peculiar technique and carried it to its limits. His favorite student Karl Menger wrote: "He proceeded by taking almost imperceptible steps, following the principle that a mathematical proof consists in tautological transformations; yet at the end of each hour, he left the audience dazzled by the sheer number of ideas he had managed to cover."

One of his students fondly remembered the "chilly clarity" of Hahn's teaching style, and another, Karl Popper, wrote in his final essay, "Recollections of a Grateful Student": "The personal impression I gained was that of a strikingly disciplined man. Of all the mathematicians at the institute, he was the one who seemed to me the embodiment of mathematical discipline. Hahn's lectures were, for me at least, a revelation."

Shortly before assuming his duties in Vienna, Hahn finished the monumental 865-page first volume of his *Theory of Real Functions*. In this book, the foundations were developed in such painstaking detail that some of the most basic concepts of calculus, such as *derivative* and *integral,* had to be relegated to the second volume. According to Hahn, the publication of that sequel was "immediately pending" in 1921. But in fact, what appeared in 1932 was merely a totally overhauled version of the first volume, still lacking the concepts of derivative and integral. The second volume was published only fourteen years after the death of its author.

Hahn's investigations of infinite-dimensional spaces became more and more seminal. As he grew older, he came to embody ever more fully the stereotype of the world-renowned professor. Moviegoers know that this cliché would be incomplete without a set of harmless eccentricities and a pretty daughter.

Hahn's daughter Nora was by then a student at Max Reinhardt's actors' academy. After she graduated, Hahn wrote to his old friend Paul Ehrenfest in Leiden, Holland: "My daughter has become an actress and already has an engagement for next season (her first season was in Graz, and now she is in Brünn). She mostly plays serious roles (in the theatrical jargon, she is a "sentimental" actress) and she is devoted body and soul to her work." Later, Nora Minor—this was her stage name—did very well for herself in film comedies.

As to Hahn's eccentricity, it was his interest in parapsychology. Such an interest does not jibe with the stereotype one has of a coldly reasoning mathematician. As might be expected, Hahn was a prominent member of the freethinkers' association, and his political leanings were far to the left. But to the great chagrin of his brother-in-law Otto Neurath, Hahn attended spiritualistic séances.

To Hahn, the critical investigation of parapsychological phenomena seemed fully compatible with a scientific worldview. He based this opinion on two arguments. First, it is obvious that some people possess faculties that most others are lacking. For instance, only one hundredth of one percent of people are born with the trait known as perfect pitch. Given the existence of such rare innate traits, why should we discard out of hand the possibility that a medium could have spiritual faculties that most of us lack? Secondly, according to Hahn, it is precisely the incoherent babblings uttered by a medium during a trance that indicate that the babblings were emerging straight from the medium's subconscious, rather than being part of a well-prepared hoax.

Like Carnap, Hahn considered it a serious challenge for believers in the scientific worldview to figure out how to distinguish science, in a precise fashion, from pseudoscience, superstition, religion, and hocus-pocus. And Hahn and Carnap were not alone. An open-minded interest in spiritualist claims was shared by a good number of perfectly rational people, such as the legendary magician Harry Houdini (1874–1926) and the world-famous author Arthur Conan Doyle (1859–1930), as well as by impeccable scientists such as Guglielmo Marconi (1874–1937), William James, and Alexander Graham Bell (1847–1922).


## Much Ado About "the Nothing"

Despite being in such respectable company, Carnap, merely by owning a book on parapsychology, incurred Wittgenstein's wrath. In the latter's eyes, both Carnap and Hahn belonged to the "clique of busybodies" within the Vienna Circle. But Carnap did not need Wittgenstein; he found plenty of other friends in Vienna. His lecture courses were popular, attracting well over one hundred students. And professionally he was very active, writing not only a succinct *Introduction to Logistic* (a no-longer used term for mathematical logic) but also a fervid debunking of metaphysical claims, titled *Pseudo-Problems in Philosophy.*

In a memoir of his life, Carnap described his evolving reaction to metaphysics: "I came to hold the view that many theses of traditional metaphysics are not only useless, but even devoid of cognitive content. They are mere pseudo-sentences—that is to say, they seem to make assertions [ . . . ] while in fact they do not make any assertions and are therefore neither true nor false."

Indeed, metaphysical propositions are not empirically testable, nor are they intended to be tautologically true. Therefore, Carnap concluded, such propositions can only be meaningless. They just *look* like propositions. Their grammatical structure is flawless, but their logical validity is questionable, if not worse. The propositions of metaphysics use words such as *principle* or *God,* which are every bit as meaningless as the nonsense word *babic;* or else they use meaningful words arranged in a meaningless way, as for instance "*Caesar is a prime number.*"

Carnap wrote:

> If someone asserts "There is a God," or "The primary basis of the world is the unconscious," or "There is an entelechy which is the leading principle in living organisms," we do not react by saying: "What you just asserted is false," but instead we ask: "What do you mean by these statements?" And then it becomes apparent that there is a clear-cut distinction between two classes of statements. To one class belong statements of the sort that are made in empirical science; their meaning can be determined by logical analysis, or, more precisely, via a systematic reduction to extremely basic statements about empirical data; the other class of statements, to which the above-quoted sentences belong, reveal themselves as utterly lacking in meaning, if one takes them in the way that the metaphysician intends them.

The metaphysician falls for an illusion, misguidedly believing that there must be thoughts behind the grammatical fabric of the words. According to Carnap, this hope is similar to the magical thinking of a Sioux Indian who, by naming his son Mighty Bison, earnestly hopes to confer real strength on the boy. Just like the Sioux, the metaphysician is the victim of a superstition—namely, a belief in the magical power of words.

Even questions as innocent-seeming as "Is the external world real?" are not genuine problems, but mere pseudo-problems, for there is no way to answer them conclusively. "Everything that is beyond the factual must be considered meaningless."

Since science can answer, in principle, any question that can be meaningfully posed, there will remain no unanswerable questions. Or as the manifesto had confidently put it, "The scientific worldview knows no unsolvable riddle."

"In science," Carnap insisted, "there are no depths," presumably meaning deep chasms full of endless mysteries. In later years, however,

Carnap mellowed a bit, and even went so far as to admit: "Yet there still remains the emotional experience common to human nature, which is sometimes disconcerting."

Thanks to his all-out rejection of metaphysics, Carnap became the figurehead of the Vienna Circle; in so doing, however, he seriously hurt his chances for obtaining a professorship in Germany. Acerbic quips by the young lecturer, such as "Metaphysicians are musicians lacking all musical ability," did not endear him to those who held chairs in philosophy departments. Indeed, one common viewpoint in such venues was that clarity and depth preclude each other, and that pure philosophical thought, as long as it is pure enough, is able to transcend the haughty but pitiful scientific disciplines.

These were the days when German philosopher Martin Heidegger's star was just starting to ascend—Heidegger (1889–1976) being the thinker who once famously asserted, "Making itself intelligible is suicide for philosophy." His lecture *What Is Metaphysics?* was published in 1929—the same year as *The Scientific Worldview*. Heidegger's program was the diametric opposite of the Vienna Circle's. Science, including logic, was treated with condescension, at best.

During his cure in Davos in 1928, Carnap happened to run into Heidegger, who was giving a course in philosophy there. This chance encounter took place in the beautiful Alpine setting that novelist Thomas Mann had dubbed "*Der Zauberberg*," or "The Magic Mountain" (the title of his epic novel set in a tuberculosis sanatorium in Davos), and it marked a decisive moment in the history of philosophy. Metaphysics and language analysis turned their backs on each other and would henceforth follow radically different pathways.

From that moment on, whenever he wanted to hold meaningless ways of using language up to ridicule, Carnap would turn to Heidegger's book *What Is Metaphysics?* and pick a few juicy examples from its pages. From Davos, he wrote to Schlick that he had met "a huge metaphysical cloud" at whose core he had discovered—nothing. That is to say, *no thing*, whatsoever.

Heidegger was deeply involved with (some might say "obsessed with") the concept and the word *nothing*, which he liked to precede with the definite article *the*. Here is a passage taken from Heidegger's own metaphysical manifesto: "Where do we look for the nothing? Where do we find the nothing? We know the nothing. Fear reveals to us the nothing. All that we have feared, and for whatever reason, was actually—nothing. Indeed: the nothing itself, as such, was there. What then is the matter with this nothing? The nothing annihilates itself."

That last sentence, one possible way of rendering Heidegger's famous "*Das Nichts nichtet*" in English, is often translated as "The Nothing nothings," which is pleasantly ungrammatical. The German version is grammatical, if a bit of poetic license is granted. And its melody is flawless. However, Heidegger's sentences are not factual.

David Hilbert, too, selected a statement by Heidegger as a target for mockery. The metaphysician had claimed that "the nothing is the sheer negation of the totality of the being." The mathematician scoffed at this utterance, describing it as being "so instructive because, despite its brevity, it illustrates all the main ways of violating the principles that I set forth in my theory of proof."

Heidegger was fully aware that his questions and answers concerning "the nothing" were unscientific and possibly nonsensical. He was quite ready to admit that science and metaphysics were incompatible. All the worse for science, felt Heidegger; but fatal to metaphysics, felt Carnap.

Heidegger did not give a tinker's damn. Assuming the stance of an infallible pope, he declared: "The supposed sobriety and superiority of science become laughable if science fails to take the nothing seriously."

In his essay *The Overcoming of Metaphysics Through the Logical Analysis of Language*, Carnap fastidiously documented how meaningless nonsense, such as "The Nothing nothings," is produced through false analogy with sentences such as "The rain rains." The language loses traction and starts skidding out of control. It is, after all, but an imperfect instrument, through which, according to Hahn, "the primitive traits of our ancestral forebears smirk." The "it" in the sentence "it rains" implies an entity out there that does the raining. An expression such as "nothing is outside" suggests an entity that lurks outside—namely, the notorious Nothing. However, as was explained in *The Scientific Worldview*: "Ordinary language uses just one part of speech, the noun, to cover objects (such as 'apple'), qualities (such as 'hardness'), relations ('friendship'), and processes ('sleep'); in so doing, it leads one down the wrong pathway of seeing abstract functional notions as if they were concrete physical objects."

A logically correct symbolic language avoids such confusions. Carnap expected redemption through artificial languages—not through Esperanto, which he valued for other reasons and which Wittgenstein loathed, but rather through carefully devised formalized languages, such as had been designed by Peano, Russell, and Frege, or—many years later—through computer languages, whose birth Carnap lived long enough to witness.

Carnap's time in Vienna was, as he later wrote, "one of the most stimulating, enjoyable, and fruitful periods of my life." He had close connections with most of the members of the Circle, and he engaged in discussions with them, on almost a daily basis, in one coffeehouse or another.

Part of Carnap's sociability may have been due to the fact that he had left his family behind, in Germany. He was divorced. Although, while in Vienna, he still kept in close contact with his ex-wife, Elisabeth, and their children, he led a footloose and fancy-free bachelor's lifestyle, and his private life was definitely less sober than his philosophy.

Much like Russell, Carnap had developed skeptical views on marriage. Moreover, sexually open relationships were quite in vogue at the time. Blanche Schlick complained to her husband that Carnap was a bad influence on his life, and Moritz Schlick, who in no way was averse to the other sex (he did not call himself an Epicurean for nothing!), eventually had to ask Carnap to be more discreet in his letters to him: even within his own home, it seems, Schlick could not be sure of the privacy of his correspondence.

Carnap's "collective eroticism" (as one of his female friends called it) came to a halt one day when, at the end of a class period, his

FIGURE 6.2  Student Ina Stöger, officially and in private.

student Elisabeth Stöger, a slender girl with alluring eyes, gave him a book. Carnap played it cool for a while. Some weeks later, Fräulein Stöger asked her professor whether he had enjoyed the book; she wanted to know this, as she wrote to him, "for technical reasons." "What sort of technical reasons?" Professor Carnap wondered. Fräulein Stöger replied that his voice reminded her of an old girlfriend of hers. She added that she certainly did not belong to that set of young women who were on the lookout for better grades. Carnap still pretended to be obtuse, but after some weeks of this coquetry, the two became a couple.

Elisabeth Stöger preferred being called by her middle name, Ignazia, or Ina for short. Another female friend of hers, the photographer Trude Fleischmann, loved to take portraits of Ina's constantly changing face. After his first night with Ina, Carnap confided to his diary: "She has two faces: a serious one seen from straight on, 'Ignazia,' not relaxed, and with a somewhat artificial expression; and a kinder and lovelier face, especially when viewed from below, and to the left. In bright daylight at the breakfast table, I got somewhat startled by the wide-awake 'Ignazia face.'"

Soon Ina moved in with Carnap.

## THE FILTH OF PURE PHILOSOPHY

Change came as well to Otto Neurath's household: soon after Otto's return from Germany, his son Paul was brought back from the children's home. From the start, the ten-year-old boy got along well with his blind stepmother. The Neuraths resided in Margareten, a workers' district in Vienna, and Neurath wanted to turn his son into a true proletarian; in fact, the boy was already proficient in the working-class dialect. But raising a child can prove a frustrating task and lead to unintended outcomes: Paul ended up as a scholar.

The apartment of the Neuraths was spacious, smoky, and dark. Olga had a passion for smoking cigars. Running water was found only in the hallway. On the other hand, there was plenty of room for Otto's library—some twenty thousand books, at that point. And the stream of visitors never ceased. Discussions were held day and night on a huge couch surrounded by piles of books and full ashtrays.

Heinrich Neider (1907–1990), a Marxist student and junior member of the Vienna Circle, described his first visit there: "So I arrived at Neurath's house, at that time an old and extremely seedy place. The smell was awful. The door was opened by a blind woman: Frau Neurath. She led us to her husband, who was asleep; she had to shake him to wake him up. He was a huge person, as large as an elephant. I was introduced to him. His first question: 'What are you studying?' I replied: 'Philosophy, pure philosophy.' To which he said: 'How can you do such a filthy thing? Why not theology, while you're at it?'"

Thomas Mann's son Golo, the historian, met Neurath as a lecturer in a youth camp, and reported similar diatribes from the old firebrand:

Don't read any Kant, don't read any Schopenhauer—it's science you've got to tackle! You've got to break out of all those old eggshells, such as metaphysics, idealism, and all the rest! An intellectual floats like a blob of grease in a plate of soup, throwing up one smokescreen after another. That's what philosophers love most to do! Their turgid language—"manifestation," "emanation," "negation of a negation"—gives them away in a split second. If a proletarian were to read or hear the typical bloated talk of philosophers, he wouldn't understand a thing, and he would assume that he was just too stupid! But he isn't!

Neurath had given up all hopes for an academic career. But he had managed to convince the Social Democratic government of Vienna to establish a Museum for Social and Economic Affairs, with himself serving as director.

One day, the flow of visitors to the Neuraths' home brought along pretty Marie Reidemeister. Mietze, as her friends called her, was preparing to take her exams for a teacher's diploma in mathematics and physics in Braunschweig, Germany, but together with some friends, she had decided to make a brief hop to Vienna. Her brother Kurt, the knotmeister, had told her stories galore about the Schlick Circle and its indefatigable Otto Neurath, allegedly the funniest man in Vienna, and indeed Neurath did not let his visitors from Germany down. The hulking man with the bold nose and the twinkling eyes regaled his guests with a panoply of jokes till all hours of the night.

The next day, Otto and Mietze went to the *Heustadlwasser*, a small lake in the Prater, Vienna's sprawling recreational park. There they rented a rowboat and paddled about for a few hours. The moment he returned home, Otto shared the good news with Olga: "Guess what—I kissed Mietze today!" "Hmm . . . That didn't take you long!" replied his wife pleasantly. Marie briefly returned to Braunschweig, to take her examinations. Then she moved to Vienna.

## Neurath's Unity of Science

Not everyone fell for the charm of the blustering giant. The restrained, distinguished Schlick, for example, found it hard to put up with Neurath's intrusive ways, and never once invited him to his home.

"I can't have a man with such a loud voice in my place," said Schlick. "Here, one plays Mozart and afterward converses softly. What would a man with such a loud voice do here?"

A favorite topic of Neurath's was "the unity of science"—and he was unwilling to draw any dividing line between the natural sciences and the humanities. He saw them all as parts of one grand unified and noble structure.

"Our endeavor"—so proclaimed the manifesto—"is to link and harmonize the achievements of individual researchers in their respective fields of science. From this approach there naturally follows an emphasis on collective efforts." Neurath kept returning again and again to this theme, indeed harping on it.

Neider reported that even the overly polite Moritz Schlick once introduced a seminar talk by Neurath with the words: "Herr Neurath has declared himself ready to give a lecture today. He states that he wishes to speak on the unity of science. I cannot imagine that such a topic would be of interest to anyone in this room, but I nonetheless cordially extend an invitation to Herr Neurath to begin his talk."

Neurath gulped, but then did as told. After all, nobody could stop him from indulging in his passion. Valiantly he sailed forth. And this is how it came out:

It was only at the end of the nineteenth and the beginning of the twentieth century that scientific work in the various disciplines had progressed enough to aspire in earnest to a kind of unity, in which all concepts are formed according to one method, and one method alone. This method proceeds by using specific rules for systematically reducing every given assertion down to individual sensory experiences which everyone can check for himself.

Only through the joint work of generations of scientifically oriented people will this huge collective task eventually be accomplished.

The liberation from tradition-bound modes of thinking has been achieved by men such as Marx, who conceived social life as down-to-earth and subject to experience, or Mach, who reduced everything physical down into individual sense-data.

Marxism, in Neurath's view, was fighting the good fight, cheek by jowl with the scientific worldview. Everything that Neurath said came out sounding like a manifesto—and a communist one, to boot.

Neurath invited the members of the Vienna Circle to attend a weekly course on Marxism. His missionary zeal, however, met with only modest success. Schlick never went, for example. In Neurath's eyes, this was proof of the reactionary class-consciousness of the bourgeoisie. He looked at everything, as Karl Menger wrote, "through the often distorting lens of socialist philosophy. I have never seen a scholar so consistently obsessed by an idea and an ideal as Neurath."

A brief sample of what Neurath wrote about Bertrand Russell will give the idea: "The pathway of the socialist proletariat is prepared by all those who promote anticapitalist doctrine and keen-witted lucidity! Like few others, the Englishman Bertrand Russell has contributed to both. During the war, he experienced the full brunt of British justice, as he had earnestly opposed the militaristic flood. Antiwar activist, fighter for his convictions. And the same man today is renowned the world over as the leader of exact philosophy, the worldview that ferrets out the logical-mathematical structure of objects and their interactions, and everywhere helps experience come into its own."

So far, so good—at least if you can follow it. But Russell had recently returned from a visit to the Soviet Union, and he had precious few good things to report about the state of that Union. Neurath took this as a personal affront: "For Russell, Marxism means the uniformization of ideas, as well as contempt of reason, because it preaches class war, which is something Russell would rather do without. For him, Marxism is essentially the enemy."

And that led to this: "In his social views, Russell is nonscientific: he does not investigate connections; he does not describe facts by logical means, but merely expresses his personal wishes; he believes in the power of reason as a sociological fact, without ever submitting it to a deeper historical analysis. The same man who, as an exact philosopher, has displayed a critical and astute analytical mind beyond comparison, does not deem it necessary to engage in deeper studies when it comes to social topics. The anecdotes of a globe-trotter have turned into the presumptuous doctrine of a scientific petty bourgeois."

When, some ten years later, Russell was invited to deliver the inaugural address at a congress on the unity of science organized in Paris by none other than Otto Neurath, he readily accepted. That he had been called "a scientific petty bourgeois" by Neurath did not seem to faze Bertrand Russell, who, after the death of his brother, had become the third Earl of Russell and enjoyed a permanent seat in the House of Lords.

# *Tangents*

*Between-wars Vienna: The Circle embraces life through out-reach. Ernst Mach Society soon seen as leftist pocket in right-wing university of left-wing town in right-wing country. Tensions mount without cease. Jewish writer gunned down, who saw Exodus coming. School prayers abolished. Adult education hits all-time high. Neurath turns from small-scale gardeners to rows of little men, creates new museum as window on present. Pictorial statistics give joy to proletariat. Viennese architects work between squatters and Bauhaus. Mathematics and statistics inspire Vienna's writers.*

## LET NO ONE STAND APART!

The new public phase of the Vienna Circle began around 1929. However, the Thursday evening meetings remained an entirely private affair—an academic *Privatissimum,* as it was called. You had to be invited by Moritz Schlick.

But many members of the Circle wanted to do more than just engage in learned discussions and afterward wind down in a coffeehouse, chitchatting above the soft clicks of billiard balls in the background. As Carnap later wrote, "All of us in the Circle were strongly interested in social and political progress. Most of us, myself included, were socialists."

Most, but not all: for instance, Schlick and Menger were not. The manifesto of the Vienna Circle acknowledged this minority, albeit a bit condescendingly: "Of course, not every single adherent of the scientific worldview will be a fighter. Some, savoring solitude, will lead a withdrawn existence on the icy slopes of logic; some may even disdain mingling with the masses. However, their achievements, too, will take their place in the historical development."

The so-called left wing of the Circle was committed to advancing this "historical development"; it included Hahn, Carnap, and, of course, Otto Neurath above all. Since these were precisely the authors of *The Scientific Worldview,* we read: "The Vienna Circle is not satisfied with its collective work as a closed group. It is also trying to make contact with other contemporary active movements, as long as they are well-disposed toward the scientific worldview and turn away from metaphysics and theology."

To be sure, an academic discussion circle was not the optimal instrument for forging strategic alliances with "other contemporary active movements." Indeed, what had paved the way for the Vienna Circle's public phase was the setting up of a formal society—the Ernst Mach Society. The inaugural meeting of this organization took place in November 1928 in the Ceremonial Hall of Vienna's Old Town Hall, and Otto Neurath delivered the keynote talk: "Ernst Mach and the Exact Worldview."

It goes without saying that Otto Neurath was the guiding spirit behind the whole movement. But the first steps had been taken by others. The initial proponents of the Ernst Mach Society, as revealed by police records, were members of the Austrian Union of Freethinkers, none of whom belonged to the Vienna Circle. This, however, changed overnight when by chance Otto Neurath was called in. He seized the first opportunity for a complete takeover, and before they knew what had hit them, the erstwhile club of atheists had mutated into an offshoot of the Vienna Circle.

Moritz Schlick was elected president of the society, Hans Hahn vice president, Otto Neurath secretary, and Rudolf Carnap vice secretary.

"Contact with other contemporary active movements" was achieved by appointing two of Red Vienna's Town Counselors to the advisory board: the celebrated anatomist Julius Tandler, at the time absorbed in overhauling Vienna's health system, and the pediatrician Josef K. Friedjung, a member of Sigmund Freud's Psychoanalytical Association.

The activities of the Ernst Mach Society were well defined. As it states in the manifesto: "We are witnessing how the spirit of the scientific worldview is increasingly penetrating all forms of personal and public life, of education, of child-rearing, and of architecture, and how it helps shape economic and social life according to rational principles." This was the key point at which leverage had to be applied.

When the manifesto of the Vienna Circle appeared, it already was reporting on the activities of the Ernst Mach Society during its first year, and it concluded with the exhortation: "Let no one stand apart!"

The fact that the Ernst Mach Society had sprung into being from a group of staunch atheists could not help but provoke antithetical reactions in clerical circles. The conservative daily *Reichspost* observed its activities with disdain and suspicion: "A propaganda outpost for antireligious ideas." Hans Hahn, it reported, had dared to criticize religious institutions in a lecture on "superfluous entities," in which he had alluded to superstitions associated with the number seven, such as the seven sacraments and the seven cardinal sins. The daily paper called Hahn to order. A mathematician should stick to mathematics, rather than dabble in Catholic doctrine. Moreover, it was hardly reassuring that a certain Dr. Carnap was scheduled to speak next, with the bland title "On God and the Soul" but the telltale subtitle "Pseudo-Problems in Metaphysics and Theology." This did not augur well.

One of the first lectures given under the auspices of the Ernst Mach Society was Philipp Frank's "The Scientific Worldview Abroad: Impressions of a Traveler to the Soviet Union." Such a title made it easy to pigeonhole this radical band as a group of "fellow travelers." The fact that the liberal academic Moritz Schlick was its figurehead president did not pull the wool over the astute clerics' eyes. Moreover, Otto Bauer, head of the Austro-Marxists, had also lectured in the Ernst Mach Society; the moment that the society's secretariat moved into the Old Town Hall, which at the time was a stronghold of Red Vienna, all the worst suspicions of the right-wingers were confirmed.


## Pulp Fiction

From the armistice of 1918 on, Austria was divided, town versus countryside. Vienna was solidly in the hands of the Social Democrats, whereas the rural areas were dominated by Christian Socialists. With Marxists here and clerical conservatives there, the gaps seemed unbridgeable and only kept growing. After the war, there had been a short-lived coalition of the two camps, but it was shattered in 1920.

The leader of the Christian Socialists was the priest Ignaz Seipel (1876–1932), who was not just a prelate but also a professor of theology. By means of a strict fiscal policy, the cleric managed to get a grip on inflation and introduce a new, stable currency. By pledging that Austria would never unite with Germany, he allayed French fears and received economic help from the League of Nations. But then the number of unemployed more than doubled, and most of the middle class lost everything they owned. As a result, the nationalist circles— the so-called *Völkischen*—turned increasingly to the far right. In that corner, radical splinter parties (such as the German National Socialist Workers' Party and the National Socialist German Workers' Party— note the subtle nuances there!) were biding their time. The latter party already backed Adolf Hitler, and the former was destined soon to join his ranks.

All manner of zany ideologies sprang up near this fringe. The world was clearly run by a handful of Jews: the Protocols of the Elders of Zion proved this. History is an eternal struggle between superior races and subhuman species, otherwise known as *Untermenschen:* this is what the runes reveal. The present-day cosmos is the outcome of a fierce fight between fire and ice. Some conceded that the Earth is a globe but insisted that we live on its inside: the world is hollow. Each and every one of these creeds loathed "professorial science." And all of them longed for some sort of unique secret knowledge to set them apart; if such knowledge defied common sense, so what? Look at how that rapscallion Albert Einstein got away with it!

Each of the two major political movements—the Christian Socialists and the Social Democrats—had its own "self-defense force" organized along military lines: the Schutzbund on the left and the Heimwehr on the right. Both of these private armies were busily marking their territories, by parading through the streets and trying to intimidate their opponents, and the confrontations were turning increasingly nasty.

The ideological conflicts seemed unsurmountable. One side preached the salvation of the immortal soul, the other side class-consciousness. Every small disagreement could be used to further exacerbate the dispute. For example, cremation was strictly forbidden by the Catholic Church, with transgressors being excommunicated. In response, militant atheists promptly founded associations for the incineration of their bodies, with poetic names such as The Flame. Priests predictably refused to administer the last rites to members of what they termed "societies for the deflagration of corpses." Anyone who changed their mind on their deathbed left behind a body that was claimed by both sides.

The city government of Red Vienna erected a splendid crematorium, just facing the Central Cemetery. Exactly one day before the first cremation was to take place, a Christian Socialist minister revoked its operating license. The Social Democratic town mayor Karl Seitz, totally unfazed, went right ahead with the cremation. Instantly hauled into court, he was soon acquitted. The rightist conservative daily *Reichspost* screamed bloody murder, and riots broke out.

The crazily tension-filled atmosphere of postwar Vienna fueled the "penny novels" written by Hugo Bettauer (1872–1925). From the hand of this colorful and highly controversial penpusher came several thrillers every year, bearing titles such as *Unrestrained, The Fight for Vienna, The Rule of the Fist,* and *Vienna Unchained,* many of which

FIGURE 7.1  Bettauer's novel *Town Without Jews*.

carried progressive social messages. A number of his novels were turned into films, and thanks to one of them—*The Joyless Street*—a young Swedish actress, Greta Garbo, made her name. This promptly became the subject matter of Bettauer's next book.

Bettauer's novel *Town Without Jews* was particularly provocative. The plot: An iron-fisted Austrian chancellor—easily recognizable as Ignaz Seipel, the Catholic prelate—decides that all Jews have to leave Vienna. He admires the brilliant qualities of their race, he says, but just as a gardener can admire the shining colors of beetles yet must remove them from his flowerbeds, so he feels bound to protect his people by expelling the Jews from Austria. They are simply too smart. So the Jews all emigrate; within a short time, all the banks fail, and the town becomes extraordinarily poor, and also stupid, to boot—but not so stupid as not to change its mind, in the end. The Jews are permitted reentry, and within an eyeblink, money and talent have made a full comeback. Wouldn't you know?

Bettauer also edited a magazine called *He and She: A Weekly Magazine for Eros and a Better Life*. The Catholic press was predictably enraged: "sheer pornography," it opined. As for its editor, well, he was "a perverted beast coming straight out of the cesspool." Worse even than Arthur Schnitzler! The anti-Semites, of course, spotted the usual Jewish love of smut and filth.

One day in March 1925, the young ex-Nazi Otto Rothstock, a dental technician by trade, entered Bettauer's office, locked the door, and killed the writer with several shots. Then he waited calmly for his arrest. He offered no resistance.

The right-wing newspapers flipped into paroxysms. They all conceded that a murder was a murder, admittedly. But the victim, this Bettauer, had had it coming. With his shabby smut and his shocking novels, he had perniciously misled the youth of Austria. No one could possibly feel surprised that his life had been terminated in such a brutal way. After all, Bettauer had committed his flagrant crimes in the open. Hundreds of thousands of upright citizens had condemned them—but to what effect? None whatsoever! The upright citizens' righteous anger had turned into helpless resignation when it became clear that the police and the courts of justice were mere stooges of the communists and the Jews, and unwilling to act. And thus, little wonder that an idealistic young man with a pure soul, despairing at the ever-mounting flood of lewd pornography, had felt called on to act as a judge on his own. After all, the shameful scandal had been allowed to drag on for far too long.

Rothstock's defense counsel was a certain Walter Riehl, who happened to be the leader of the German National Socialist Workers' Party (not the one aligned with Hitler—at least not yet). Riehl could in fact boast, and truthfully so, that he had already raised the swastika banner in 1907. This upstart named Hitler had plagiarized *him*—yet for some weird reason, nobody seemed to care.

The jury for the Bettauer case listened to Riehl's arguments. The crime was undoubtedly an act of murder, but the perpetrator was certifiably insane, non compos mentis; and the person who had derailed his mind was none other than his victim, Bettauer himself. The murderer was thus committed to a psychiatric hospital. Eighteen months later, Rothstock was a free man again. He moved to Hannover, reentered the National Socialist German Workers' Party (this was Hitler's own party), and thereafter worked as a dentist until well in the 1970s.

As it happens, Bettauer's novel about the expulsion of Jews as well as his assassination seem like uncanny anticipations of the fate of the Vienna Circle.

In Vienna, a new expression gained currency: "A Rothstock will show up!", meaning some righteous spirit who, with a well-placed bullet, finally puts an end to scandalous deeds that have led to public outrage. The belief that violence was ultimately the only solution to political tension kept growing.

In the summer of 1927, that tension exploded in a frightful discharge. A jury had acquitted several ex-soldiers who, in the wake of a demonstration, had shot an unarmed man and a child. When the verdict was announced, savage rioting swept through the streets of Vienna. The city's main courthouse, known as the Palace of Justice, was set on fire. Neither the police nor the Social Democratic town politicians were able to control the crowd. The events of that fateful day claimed more than eighty lives.

By the next morning, all was quiet again. But it was an ominous silence. Austria was heading toward a civil war. At first, however, it remained a cold war.


## A Phalanx in a Bear's Lair

The University of Vienna turned into one of the most fiercely contested arenas of the political infighting. A large majority was nationalistic, meaning right-wing. Only a few professors, headed by Hans Hahn, openly declared their sympathies for Social Democracy. This was a time when *town* was further left than *gown*.

Many faculty members supported plans to prevent any further encroachment of Jewish professors and students. They demanded a *numerus clausus* ("closed number," similar to a quota), meaning an upper bound of, say, 10 percent. In particular, the Austrian chancellor Seipel favored such a measure, which he termed "self-defense anti-Semitism." In 1930, the rector Wenzeslaus Gleispach, a Nazi, introduced rules that classified students according to "language and origin" and that implied a de facto limit on Jews.

The Constitutional Court of Austria overruled this measure, but that did not impress nationalistic circles: since the constitution had been written by the Jew Hans Kelsen, it was clearly bunk. Fierce riots broke out at the university.

The so-called German Students' Union published blacklists of objectionable professors. These lists included Moritz Schlick, along with Hans Kelsen and Sigmund Freud. The bogeymen were the so-called *Ungeraden,* or "undesirables"—that is, anyone who was suspected of Marxist leanings or of having a Jewish background. Neither criterion applied to Moritz Schlick, but by association, he was seen as guilty.

The lectures of these "undesirables" were boycotted, and their appointments were blocked. Antisemitism became far more rabid than it had been during the *k.k.* dual monarchy. At that time already, the sociologist Max Weber had written that embarking on a scientific career was a "mad gamble"—*ein wilder Hasard.* As for the Jews, he added, one ought to tell them honestly: abandon all your hopes. The fact that many of them succeeded in spite of the virulent anti-Semitism that surrounded them was used to feed conspiracy theories.

At its yearly meeting in 1923 in Vienna, the German Students' Union demanded that all books by Jewish authors be marked with a Star of David. The director of the university library, himself a Jew, refused to do anything of the sort. As a result, he became the victim of a hate campaign.

Within the philosophical faculty of the University of Vienna, the "undesirables" were fought with cunning and resolve by an informal network of professors. This faction was led by the paleontologist Othenio Abel (1875–1946), who later boasted that he had "welded the anti-Semitic groups so tightly together that they could act as a phalanx."

FIGURE 7.2 The "Bears' Lair" in the University building.

The clandestine meetings of the nineteen professors constituting this phalanx were held in a biology room where taxidermic specimens and animal skeletons were stored, including the bones of a cave bear. It was therefore known as "the bear's lair."

Edgar Zilsel, one of the earliest and staunchest members of the Schlick Circle, was a victim of the bear's lair. He was an *Ungerader*—an "uneven," an "undesirable"—in every respect. Karl Menger wrote about him: "Zilsel was a militant leftist. I once heard him give a brilliant talk at a meeting in Warsaw. I have forgotten the details, but I remember that he spoke about some philosophical views that could not, in his view, be justified on rational grounds. He ended up almost out-Neurathing Neurath, by blasting the socioeconomic systems that he claimed motivated these views and were reinforced by them. I happened to be sitting next to the eminent logician Jan Lukasiewicz, who was overwhelmed by the talk and exclaimed repeatedly: 'What an intellect!'"

Edgar Zilsel had been studying philosophy, mathematics, and physics. In 1915 he submitted his PhD thesis on the philosophical aspects of the famous Law of Large Numbers in statistics. Already before World War I, Zilsel had been a mainstay of the Academic Union for Literature and Music. After the war, he became active in various Viennese centers for adult education, where he taught philosophy and physics full time. Later, at the Pedagogical Institute of Vienna, he instructed future schoolteachers.

Zilsel failed to become a lecturer at the University of Vienna. His *Habilitation* thesis on the development of the concept "genius" was deemed "too one-sidedly rationalistic"—just the kind of thesis one would expect to see coming from a Marxist Jew. Most of the professors decided not even to wait for the external reports: with a heavy heart, Schlick and Gomperz had to tell their protégé Zilsel to withdraw his application. When the external reports arrived, they were in fact favorable, but it was too late. The "well-known obstacles" had taken their toll.

In 1927, the Philosophical Society of the University of Vienna, which two decades earlier had been the favorite hangout for the *Urkreis* members Hahn, Frank, and Neurath, saw fit to transmute itself into a local section of the International Kant Society. Yes, Immanuel Kant, of all people! This preposterous philosophical annexation, or *Anschluss,* may well have been the final provocation that sparked the circle of anti-metaphysicians to found the Ernst Mach Society. What is certain is that from that time on, neither Hahn nor Neurath nor Frank ever lectured again in the Philosophical Society.

## New Men for a Better Future

With the economic crisis of 1929, the full brunt of which struck Austria one year later, political tensions were exacerbated. The atmosphere became hopelessly polarized. The state government was right-wing, the town government left-wing, and the university was turning increasingly to the far right. The Ernst Mach Society, as seen through hostile eyes,

FIGURE 7.3  Adult education: the Urania.

was nothing but a poorly camouflaged vehicle for vicious anticlerical propaganda emanating from Red Vienna.

The left-wingers of the Vienna Circle wanted nothing better. They eagerly moved up to the front lines of this battle, proudly taking a stand on such burning issues as public education and school reform. This was where the scientific worldview could most readily be turned into political action.

Adult education was a child of the enlightenment. In the eighteenth century, science had been a favorite topic in the intellectual salons; in the nineteenth, it required a larger space. In that regard, it was quite similar to music; soon, to the splendid concert halls of musical societies were added splendid public lecture halls for similarly high-minded societies devoted to popular science, such as the Urania.

Already before the Great War, the senior members of the Vienna Circle had been actively engaged in public lectures and courses. Austro-Marxists saw in education the key to raising the consciousness of the common people. Knowledge is Power. Education means Freedom. There was much talk of a New Man. (There could also have been talk of a New Woman, but linguistic equality was not on anyone's mind at the time.)

The Volksheim ("People's Home"), located in Ottakring, and the Viennese Society for the People's Education were showcase projects for Red Vienna. They were created not in order to provide professional skills but to raise public awareness of the working class and to help educate workers in the noblest sense.

Before World War I, it had often been university professors who freelanced as teachers in institutes of adult education. Now, with the economic situation continuing to deteriorate, not only did these institutes promise to improve the lot of the workers, but they also offered jobs for scientists who could not find a position at a university. There were almost four hundred of these "schools of socialism" in Vienna. Several thousand public lectures were delivered each year at the Urania alone.

The breeding of the "new men" (and women—Austria was among the first countries where women could vote) was to begin at school already. The minister of education, Otto Glöckel, drew up an all-encompassing plan for school reform. His first measure was to cancel the ritual of daily prayer. It almost became his last measure, too. A storm of outrage swept through the country. The Catholic newspaper *Reichspost* discerned "school Bolshevism" taking place. And it wasn't long before Glöckel had been replaced as minister of education. Undaunted, however, he carried on with his plans, now as president of the Vienna School Board.

Hans Hahn was a member of the same school board, and in newspaper articles, he valiantly fought for Glöckel's proposed school reform. Many of his demands had been made by Ernst Mach already, but to little avail.

## Hieroglyphs for the Museum of the Present

Whenever Otto Neurath gave a lecture in the Ernst Mach Society, the hall was filled to the last seat. He was a brilliant speaker. But Neurath had never earned the right to give lectures at the University of Vienna—another instance of the "well-known obstacles." This did not in the least diminish his rhetorical verve. He loved exclamation marks. Here is a brief sampler:

"The ultimate consequence of empiricism: science without philosophy!"

"Toward liberation from oppression by metaphysics and theology, hand in hand with liberation from social pressures!"

"Empiricism and the unity of science, alongside social behaviorism and social Epicureanism, are the hallmarks of the present era!"

And the concluding sentence of the manifesto was surely due to Neurath: "The scientific worldview embraces life, and life embraces it!"

Over and above his inexhaustible activity as a writer and platform speaker, Neurath was able to disseminate his views by another means: pictures. Soon after getting out of prison in Bavaria and arriving in Vienna, he had become the director of a museum again. The Museum for Social and Economic Affairs was entirely his own creation.

This museum was different from any other. It did not keep things from the past. Rather, it aimed to shape the future, by depicting social interrelations. For Neurath, this meant "serving the proletariat." He wrote:

"The proletariat, as a class, is eager for the truthful representation of social facts; whereas the bourgeoisie, by nature, is fearful of giving free rein to statistics and other sciences."

For Neurath, statistics was indisputably the key to understanding society, and therefore: "Statistics is pure joy for the international proletariat in its relentless struggle against the ruling classes!"

To convey a clear sense of "pure joy," statistics must shed its image as an occult science, Neurath said. Statistics had to be immediately accessible, without any prior knowledge. For this, a pictorial language was necessary. Indeed, he wrote:

"Contemporary people receive a great deal of their information and their general education through pictures, illustrations, slides, and films."

This was, of course, well before television, PowerPoint, and YouTube.

"Up till now, the method of pictorial representation has been underdeveloped. Our aim is to create pictures that can be understood without words, if possible."

FIGURE 7.4 One of Otto Neurath's exhibitions in the Town Hall.

"We have to create symbols that can be 'read' by all of us, just as we all can read letters, and just as experts can read musical notes. This requires the creation of a set of 'hieroglyphs,' which can be used internationally." This may seem obvious today, when icons in airport concourses, on automobile dashboards, and in user interfaces are ubiquitous, guiding us wordlessly through intricate labyrinths, but at the time it was a radical proposal.

"We must try whenever possible to use the same symbol for the same object in the whole museum, and in all exhibitions. If there is any place in the world for concrete representation of abstract designs, it is here."

In the early twentieth century, abstract art had emerged seemingly out of nowhere. Everything was turning increasingly abstract—even music and mathematics. "The more abstract an art becomes, the more it becomes art," intoned Robert Musil.

Abstract art frequently gives off a flavor of elitism and restriction to a secret cabal—even a touch of mysticism, in the hands of Vassily Kandinsky. But with Neurath, the goal was altogether different: his concrete fashion of representing abstract ideas by drawing "rows and rows of little men," as they were sometimes humorously called, led to the development of the so-called Vienna method of pictorial statistics. This method was designed to be able to convey to the masses the elusive complexities of social and economic situations.

The crux of the Vienna method was simple: "A larger quantity is represented by a larger number of symbols (rather than by a larger symbol)." In Egyptian art, important personages, such as pharaohs and generals, were represented by larger figures. But a statue twice as tall as another statue has four times the surface area, and eight times the volume and mass. So what is meant? Two, four, or eight?

If Neurath wanted to show, for example, that in ancient Egypt there were eight times as many workers as unemployed people, he would draw eight workers, each exactly as large as the unemployed person. Anything that could distract from this truth, such as facial traits, for instance, was left out of the picture.

The politicians of Red Vienna soon understood that they had been given a marvelous medium for propaganda and informational campaigns. Neurath easily persuaded them that his museum was "an adult education institute for social enlightenment," a "window on the present." It was housed in the monumental main hall of the Town Hall, and it also had a smaller offshoot—the *Zeitschau,* or "Time Show"—at an even more frequently visited spot in the city's center. A steady stream of new exhibits displayed facts about education, nutrition, sanitation, agriculture, and urban construction, or even about the intangibles of economics, in a precise and transparent way.

The young Marie ("Mietze") Reidemeister had returned to Vienna right after obtaining her diploma. From 1924 onward, she held a job as a designer, or "transformer," as it was called, at the museum. In her private life she served as Neurath's muse, as it were. This seemed to be all right with Neurath's wife, Olga.

The German graphic artist Gerd Arntz (1900–1988) soon became one of the mainstays of Viennese pictorial statistics. Neurath had met him in Düsseldorf in 1926, after the ban against Neurath's reentering Germany had been lifted. At first, Arntz was unwilling to move

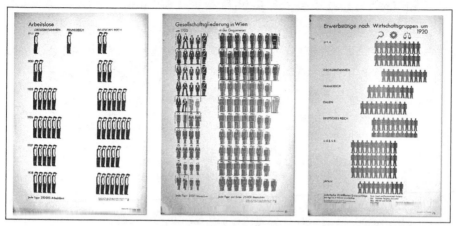

FIGURE 7.5 The Viennese method of pictorial statistics, also known as Isotype.

to Vienna. He wanted to collaborate in Otto's museum, but only by mail. Neurath insisted on having him always at hand, right there in his Viennese museum.

Arntz came to Vienna, albeit reluctantly, for a few months in 1928, and then in 1929 he moved there for good. His change of heart was great for the museum, as he created thousands of new pictograms and developed a specific, highly original style. Soon, the museum was drawing international attention. It became a showcase for Red Vienna.

Thus while the Vienna Circle was embroiled in heavy discussions of Ludwig Wittgenstein's picture theory of language ("A proposition states something only insofar as it is a picture"—4.03), Neurath's museum simply jumped into the use of pictures from the start; while his philosophical colleagues were engaging in arcane intellectual analyses and critiques of language, Neurath went one step further, trying to communicate entirely without language.

"Whatever can be shown by a picture must not be told in words."

With untiring zeal, Neurath kept hammering out his slogans:

"Words divide, pictures unite!"

"Picture pedagogics is marching on!"

"Remembering simplified pictures is better than forgetting exact numbers."

"Whoever leaves the most out is the best teacher."

And Neurath had certainly not forgotten his idealistic "plan of all plans": "A pictorial survey of the world economy is not just a scholarly representation of important facts; it is also the first step toward a planned world economy."

## GREAT ARCHITECTS FOR SMALL HOUSES

It had cost Otto Neurath no little effort to create his Museum for Social and Economic Affairs. His connections with the mainstream politicians of Red Vienna were friendly, but in some ways the "red giant," no longer having a flaming red mane but now somewhat balding, remained a marginal figure, too flamboyant for the run-of-the-mill Social Democrats, who, being far more conformist, fit easily into the power structure.

Neurath's political views were frequently at odds with the party line. After a brief flirtation with full socialization, or nationalization, the Austro-Marxist leaders had turned away from it. Thus the prospects were poor for someone who aspired to be a director of central economic planning. Having been dispatched back to Austria by the Bavarians, Otto Neurath was obliged to look around for other fields of action. He first became the secretary-general of the Austrian Union for Settlers and Small-Scale Gardeners, of all things! Nothing could sound more harmless and even petty bourgeois—but Neurath instantly sniffed the revolutionary potential that was latent in this movement.

Indeed, faced with a truly desperate shortage of housing and food after the war, many Viennese had settled in the outskirts of the town, constructing primitive dwellings on their own. These settlements were erected illegally, somewhat like the favelas in Brazil or the shantytowns in India, South Africa, and many other countries. But after some initial friction, the municipality of Vienna decided to assist the settlers' movement. The local politicians had also sensed, though more slowly than Neurath, its revolutionary potential.

Brilliant architects such as Adolf Loos, Josef Frank, and Margarete Lihotzky started supporting the settlers' enterprise. By that time, Loos had already garnered international fame with his buildings and writings, and within Vienna he enjoyed an almost cultlike status. Smooth surfaces, clean lines, tasteful design, and—no more stucco, please! Back in 1908, his book *Ornament and Crime* had touched a sensitive nerve in the modernist movement. Its title was often misquoted as *Ornament Is Crime,* and that slogan turned into a dogma. After the Great War, people had neither the mood nor the money for ornaments anyway.

Ludwig Wittgenstein greatly admired Loos. Loos returned the esteem, and he presented Wittgenstein with a copy of his book *Spoken into the Void,* with a nicely penned dedication to the philosopher and fellow architect. But after just a few personal encounters, Wittgenstein suddenly severed all contact, declaring that "Loos has become an insufferable philistine."

Philistine or not, in 1920 Adolf Loos became the honorary architect-in-chief of the Settlers' Union and head of the settlement office of the City of Vienna. "Great architects for small houses," he decreed.

A large-scale social experiment began. Some of the pilot projects involving self-help, public goods, and cooperative property were radical enough to satisfy even an Otto Neurath. For instance, while the settlers were constructing a housing estate on the Rosenhügel, they were not yet assigned their own plots. Only after the estate was finished were the plots assigned, by drawing lots.

Otto Neurath had known architect Josef Frank since the time of the *Urkreis,* through the latter's brother, the physicist Philipp Frank. As for the young Margarete Lihotzky, the first woman to become an architect in Austria, she fell for the incorrigible seducer Otto Neurath and became his lover for a short time. Soon after their fling had ended, she married her colleague Wilhelm Schütte.

Margarete Schütte-Lihotzky had a gift for functional design. Her 1930 Frankfurt Kitchen, modeled along the lines of the kitchenettes in American railroad dining cars, and designed to optimize the efficiency of kitchen work, became the prototype for all prefabricated kitchens. A constraint she imposed on her design was that there should be no place for a servant or helper in the kitchen. In matters of class consciousness, Margarete could hardly be bested.

Soon the wild settlers were thoroughly organized, indeed domesticated. Otto Neurath wove together a tight network of unions and societies. His old passion for self-sufficiency and barter was rekindled, and he decreed: "All small-scale gardeners ought to be settlers, and all settlers small-scale gardeners."

In 1921, to promote a clearer understanding of the economic basis for building a home, the municipality of Vienna had organized a huge exhibition in the park in front of the Town Hall. Otto Neurath used this as the starting point for a new Museum for Settlement and Urbanism. From there it was but a small step to creating his Museum for Social and Economic Affairs.

In the meantime, however, Red Vienna had opted for a huge housing project based on tall apartment buildings rather than low garden towns. This meant great buildings for small people. In frustration, Adolf Loos withdrew from the enterprise. He was not willing to build "palaces for people's apartments."

As for Neurath, he had discovered that he was a great fan of architecture, and was fascinated in particular by the German Bauhaus. Was tubular furniture not the very emblem of the New Objectivity? Neurath was invited to the opening of the New Bauhaus in Dessau. Shortly after that, Herbert Feigl and Rudolf Carnap also lectured there, and all were overwhelmed by the enthusiastic response of the avant-garde artists of the new age—great names such as Walter Gropius and Mies van der Rohe, Paul Klee and Vassily Kandinsky, László Moholy-Nagy and Josef Albers.

## ADVERTISERS OF MODERN TIMES

Josef Frank worked as architectural adviser for the Museum for Social and Economic Affairs. He was among the best-known of Viennese architects in the generation following that of Adolf Loos and Josef Hoffmann. Together with Neurath, he participated in the Austrian

Werkbund, an association of artisans and artists that had been created only a few days ahead of the Ernst Mach Society. Frank gave the inaugural lecture at the Werkbund, titled "Modern Worldview and Modern Architecture."

Among the architects of his time, Josef Frank belonged to the liberal modernist school. He criticized the formalized rigor of New Objectivity, an art movement that elevated "matter of fact" to a lifestyle and that was clearly in tune with the no-nonsense approach of the Vienna Circle. Frank also scorned the elitist aspirations of the strident "advertisers of Modern Times." He preferred a lively, individual design over the modular functionalism of Le Corbusier's "machines for living." One of the most important tasks for an architect, in Frank's eyes, was to reject the unified style that was then taking over architecture.

Presumably, Robert Musil shared Frank's view when he wrote: "I couldn't stand living in an apartment built according to norms. I would feel as if I had ordered myself from an interior designer."

Like Neurath and Loos, Frank opposed the municipal tenement complexes that were being favored by the city of Vienna. He once wrote: "What is happening here looks as if it had been randomly thrown on the road, and complacent stupidity grins out from every window."

He saw no future in tall apartment buildings. Instead, he proposed rows of detached single-family houses, each with its own garden. But since the City of Vienna stubbornly insisted on its building program, Frank nonetheless ended up participating in the construction of tenement complexes. An architect has to make a living, after all. But deep in his heart, he remained unconvinced: "A house in the open air was and is the ideal form of living, and thus the basis of domestic life in every culture."

Frank's goal was not to create an all-encompassing work of art, or *Gesamtkunstwerk,* but to bring comfort into people's lives. In a similar vein, Otto Neurath wanted to find out "how people will live in a maximally happy way in real homes in the near future. The optimal technical solution may have little or nothing to do with the solution yielding maximal happiness."

As vice-president of the Austrian Werkbund, Josef Frank organized the Viennese International Werkbund Settlement, which opened in 1932; aside from himself, its architects included Adolf Loos, Josef Hoffmann, and Margarete Schütte-Lihotzky, the only woman. The houses were built by the GESIBA, a municipally owned enterprise of Red Vienna. A few years later, GESIBA's dynamic director, one Hermann Neubacher, became the mayor of Vienna, under the ominous banner of National Socialism. Fate takes strange twists.

## The Discreet Charm of the Twentieth Century

One up-and-coming employee of Otto Neurath's Museum for Social and Economic Affairs was a purebred proletarian—as purebred as they come. Rudolf Brunngraber (1901–1960) was the illegitimate child of an alcoholic bricklayer and a menial laborer from Favoriten, Vienna's working-class district par excellence. Brunngraber had trained to become a teacher but couldn't find a job in any school, and so he wound up eking out a living as a day laborer, lumberjack, signboard painter, and fiddler in movie houses. Eventually, he became a commercial artist.

Brunngraber found not only a job in Neurath's museum, but here he also found the extraordinarily enthralling subject of his first novel—namely, statistics.

Brunngraber wrote a novel with the lyrical title *Karl and the Twentieth Century*. This imaginary Karl was the illegitimate child of an alcoholic bricklayer and a menial laborer from Favoriten. Karl had trained to become a teacher, but couldn't find a job in any school, and so he wound up eking out a living as a day laborer, lumberjack, signboard painter, and fiddler in movie houses. Time and again, poor Karl ended up on the street, flat broke.

Here is a gripping excerpt from the novel:

There are fourteen unemployment offices in town, each one doling out welfare payments to 800 unemployed people every day. For the country as a whole, this means over 200 million shillings a year. But

Karl, after standing in line for nearly three hours, gets just 16 shillings a week, and each month an extra 3 shillings' allowance for his rent. On April 5, he collects his first paycheck. And from April 3 on, he attends a course for occupational retraining at the industrial district commission. The course takes up the whole morning, and at the end of ten weeks, he'll be qualified to be a store-window decorator.

Unlike the author who created him, this touching hero finds no Otto Neurath to offer him a job, and so, at the novel's end, he flings himself under a train.

The obituary for the victim is a miniature gem of New Objectivity. Brunngraber fastidiously describes how to assign a value to Karl's corpse "exactly according to the quantities of raw materials contained in the body. Thus the fatty tissue of a human suffices for the production of seven bars of soap. Using the iron within a human body, one can make an average-sized nail. The sugar is enough for half a dozen carnival cookies. With the chalk one can whiten a chicken coop. The magnesium yields one dose of milk of magnesia. Using the sulfur, one can rid a dog of its fleas. And the potassium is enough to make a cap for a child's cap gun."

Brunngraber's novel appeared in 1932 and was printed full-length in the *Workers' Daily.* The soberly engrossing odyssey of a permanently unemployed man became a best seller.

The Austrian Association of Social Democratic Writers elected Brunngraber as its chairman. This honor did him no good, however, for shortly afterward the new Austro-Fascist regime disbanded the association and banned him from publishing his works. But strangely enough, in neighboring Nazi Germany his books were well received. His novel *Opium War* became one of the major publishing success stories of the Third Reich. The fact that the British were the villains in that story was probably helpful. Dr. Goebbels even sent a plane to Vienna to airlift the brilliant author to an artist's reception in Berlin.

After World War II, the politically flexible Brunngraber returned to Austria's Socialist Party. The Second Austrian Republic had decided to learn from the mistakes of the First. Strengthening national solidarity was among the top priorities of the new coalition government. It was felt that this job should not be left entirely in the hands of Austria's triumphant downhill skiers, one of the few sources of patriotic pride. The ministry thus entrusted Rudolf Brunngraber with a lucrative task: he was to write the script for the film *April 1, 2000,* together with Ernst Marboe, a member of the other party, the Christian Socialists. Indeed, one of the lessons learned from the mistakes of the First Republic was not to make war on each other, but to work hand in hand, in a happy team; and this meant two persons per job, one from each party.

The two authors of the film script did their duty with bravado. Austria's most popular actor, Josef Meinrad, who was rightly viewed as the embodiment of the ideal Austrian (being honest, modest, and utterly charming to boot), played the part of the chancellor. The film showed how he managed to free Austria after fifty-five years of occupation by Allied troops. It was a deft example of Austria performing cosmetic surgery on its past. In the film, the chancellor appealed to the better nature of the occupying forces by recounting Austria's history from the start, Mozart and all. His tale admittedly left out a few less pleasant moments: Adolf Hitler, the Nazis, the two world wars—these episodes were simply glossed over. It was a fine example of that old Viennese proverb, "Whereof one cannot speak, thereof one must be silent." In the film, the occupying powers seemed to have mysteriously landed from another star. Rudolf Brunngraber always had a fine nose for the Zeitgeist.

In another novel written by Brunngraber in 1949, a few years after Neurath had passed away, the author's former boss reappears as the huge, bald, potbellied director of a museum, "the very picture of a condottiere, as if he had stepped out of a painting by Castagno," smooth-shaven, with a jagged nose jutting out like a cube from his face, and shoulders like mountains. Brunngraber added: "Under closer scrutiny, his elephant eyes seemed equally cunning and friendly, and his mouth was as coquettish as a girl's."

This novel was titled *The Path Through the Labyrinth,* and in the end the museum director is arrested, summarily tried, and shot by his opponents.

## ANOTHER NOVEL FOR THE CENTURY

Aside from Rudolf Brunngraber, three other Austrian novelists were connected to the Vienna Circle. The personal ties were mostly superficial, as these writers and thinkers all tended to frequent the same coffeehouses, but occasionally there was a link that ran somewhat deeper. These three authors—Robert Musil, Hermann Broch, and Leo Perutz—took considerable pride in drawing artistic inspiration from the Circle's theme songs of exact thinking and the scientific worldview. As a matter of fact, all three had studied mathematics, and each in his own way made sure that no one ever lost track of that fact.

Robert Musil had turned down an academic career for the sake of literature. His early success, however, proved hard to match. When the war came in 1914, Musil had hailed it as a deliverance; but in fact, it brought him only "five years of slavery." As a captain in the reserves, he wasted away the bulk of World War I copyediting the Tyrolean military newspaper. After the war, he tried to make a living as a freelance author, writing novellas, essays, and a couple of plays, but he garnered no more than mediocre success. Now he was planning his big breakthrough: to crown his literary work with a great Austrian novel, planned to be a full mirror of his epoch.

Initially, the hero of *The Man Without Qualities* was to be called Anders, and he was going to be a lecturer in philosophy. But soon, Musil had second thoughts about this profession for his hero: "Philosophers are terrorists who, having no armies at their disposal, instead subjugate the world by locking it into a system."

This was why Musil decided, in an early draft, "to turn Anders into a mathematician." These words from his notes were framed by bold lines, and supplemented with: "He derives his mathematical style from Nietzsche: his thinking is cool, sharp as a knife, mathematical." Later, Musil changed the name of his hero: Anders became Ulrich. "The one thing that can be said with certainty about Ulrich is that he loves mathematics on behalf of all those who cannot stand it."

In Musil's comedy *Vincent and the Mistress of Important Men,* the main character is likewise a mathematician, but this one turns out to be a con artist who swindles his rivals out of their money in order to finance a gambling system "which does not contradict the laws of probability."

The hero of *The Man Without Qualities,* far from being a con artist, was going to be "the intellectual of the future," someone who would fearlessly face "a mysticism as clear as day." In Musil's eyes, mathematics was a formidable weapon. Musil's characteristic aggressiveness comes out clearly in chapter titles that he devised but eventually rejected, such as "Logician and Boxer" or "Mathematics, the Science with the Evil Eye." Some aspects of this "profusion of audacity" and the "devilish dangerousness of the mathematician's intellect" had already given food for thought to young Törless, the hero of Musil's first novel: "If people were truly conscientious, mathematics would not exist."

What Musil admired in mathematicians was their "sense of possibility." This trait, as he saw it, was the ability to dream up all sorts of hypothetical entities, and to take imaginary things just as seriously as real things.

In spite of working relentlessly and with tremendous self-discipline for more than twenty years, Musil never managed to finish his novel of the century. His own "sense of possibility" kept on getting in the way. New variations constantly sprang up in his mind, only to be rejected in favor of others. Even his novel's title was chameleonic: it was, by turns, *The Spy, The Redeemer, The Twin Sister*—and new perspectives constantly kept on turning up.

The novel's plot—at least this much is certain—centers on the so-called Parallel Campaign, which takes place in 1913. The goal of this campaign is to prepare the festivities scheduled for 1918, the jubilees of the coronations of two monarchs: Austria's Franz Josef (who then would have been on the throne for seventy years) and Germany's Wilhelm II (a mere thirty years!). A special committee is on the lookout for good ideas as to how to celebrate the great event. Ulrich, who is

FIGURE 7.6 Musil's friend Richard von Mises is ready to take off.

professionally pledged to a scientific worldview, suggests the creation of a General Secretariat for Precision and the Soul. Precision was indeed Musil's mantra; as for the soul, he defined it as "everything that crawls away and hides whenever there is talk of algebra."

In 1930, the first volume of *The Man Without Qualities* was at last published. The critics wrote paeans in praise of the work. Musil's financial situation, however, did not improve in the least, and the high expectations now attendant on the next volume slowed down the author's pace even further. He grumbled that even as he worked on it, his novel threatened to become a historical one.

Because he felt that by living in Vienna he was too close to the world of his novel, Musil moved back to Berlin in 1931. There, however, most of his social life was spent in the salon of Richard von Mises, the Viennese mathematician and philosopher who had built up in Berlin, after 1920, the first and foremost center for applied mathematics. This was the same person who had insisted on *not* being counted as a foreign member of the Vienna Circle, and it had cost him no little effort to convince Otto Neurath to indulge him in this wish.

Von Mises became living proof of the fact that the predilection of Viennese novelists for mathematics and its practitioners was returned in kind. In his school days, Richard von Mises had been friends with the writer Hugo von Hofmannsthal; moreover, he was a patron of the celebrated poet Rainer Maria Rilke, as well as of Peter Altenberg, a witty and bohemian writer who always had his mail delivered to Café Central. Von Mises established a foundation whose sole purpose was to support Robert Musil. Unhappily, though, this came to an abrupt end with Adolf Hitler's rise to power. The Nazis' literary taste was different.

And so, in 1933, Robert Musil chose to move back to Vienna. There too, a society was founded to support him; but once again, it lasted only a short while. After the *Anschluss,* the writer emigrated to Switzerland, because his home country left him, as he wrote, "with not enough air to breathe."

Musil died during World War II, in a sunny garden near Geneva, almost destitute and virtually forgotten. He had just completed his most lyrical chapter, "Breath of a Summer Day." To read it is to understand what is meant by "mysticism as clear as day."


## A YOUNG ENTREPRENEUR POACHES ON MUSIL'S TURF

The Viennese writer Hermann Broch (1886–1951) also chose a mathematician as the protagonist of one of his novels. In *The Unknown Quantity,* this young scientist dreams of discovering a logic that has no axioms. "With the enlightenment of someone who knows much more than he reveals," the hero voices his conviction: "Logic and mathematics, however, are identical." "Yes, yes, it's called 'logistic,'" replies his counterpart, unimpressed, "just another newfangled idea."

Hermann Broch had inherited a large textile firm. Like Musil and Wittgenstein, he had studied engineering before turning to mathematics and philosophy. As an heir-to-be, he had been obliged by his father to obtain a diploma from the Technical College for Textile Industry. Broch hated this "spinning school," as he mockingly called it. On the side, he enrolled in classes at the University of Vienna.

From early on, Broch considered himself to be a born mathematician, and he was not in the least discouraged by his poor marks in mathematics at school. The first poem in his *Collected Works* is titled "Mathematical Mystery," and it sounds as if it had been written by a romantic schoolboy infatuated with both mathematics and rhymes. Its opening four lines will suffice to get across its flavor:

| | |
|---|---|
| *Auf einsame Begriff gestellt,* | *Built solely out of pure thought,* |
| *Ragt ein Gebäude steil hinauf,* | *A structure of imposing height* |
| *Und fügt sich an den Sternenhauf,* | *Soars up to join the starry light,* |
| *Von ferner Göttlichkeit durchhellt.* | *And sparkles in God's distant vault.* |

The vague metaphysical aspirations of the young thinker were dashed when he became aware of the crisp positivism of Ludwig Boltzmann. Torn between mathematical, philosophical, and literary leanings, Broch decided to forgo his studies for a while and devote himself fully to the textile industry, as owner and manager of a large plant in nearby Lower Austria.

Nevertheless, Hermann Broch, armed with his pipe and his clever repartee, and always surrounded by beautiful women, remained a well-known fixture of the Viennese coffeehouse scene. Indeed, that is precisely how he entered literary life, in 1920—not as an author but as a character in a play. He starred, in fact, as the Young Textile Entrepreneur in Robert Musil's comedy *Vincent and the Mistress of Important Men*. In real life, however, the Young Entrepreneur was not so young (he was pushing forty), and aside from his wealth and his dubious reputation as a man-about-town, he didn't have a great deal to brag about.

But Broch had not forgotten his youthful ambitions. He therefore sold his firm and reenrolled at the University of Vienna, plunging himself back into mathematics and philosophy. He attended lectures given by Menger and Hahn, side by side with another elegantly clad heir from the textile industry, but one who was twenty years his junior. This was a shy fellow named Kurt Gödel.

Some of the sentences jotted down in Broch's notebooks, such as "Mathematics is the paragon of a tautological discipline based entirely on itself," sound like an echo of the discussions in the seminars of Schlick and Carnap. However, what Hahn considered a profound revelation appears here as almost a throwaway remark. Indeed, Broch's familiarity with the ideas of the Vienna Circle did not prevent him from holding quite different views.

Nevertheless, he conceded to the positivists an "awareness of their affliction," since he, no less than they, realized that philosophy, even if it was treated mathematically, would never be able to deal adequately with "the enormous area of the mystical-ethical." This sudden realization triggered a therapeutic catharsis in Broch's mind, for it convinced him that art was superior to mathematics and philosophy. From that moment on, his endless vacillations were a thing of the past.

As a result, Broch was now able to concentrate fully on literature; and whereas up to his forty-fifth year he had not written anything of substance, he now became a tireless and immensely productive writer—indeed, far too productive, said the jealous Musil, who branded Broch as a literary hack.

As soon as he had read the synopsis of Broch's first novel, a trilogy called *The Sleepwalkers,* Musil noted sourly: "It seems to me that there are points in common between Broch's intentions and mine, and in certain details the similarities seem quite far-reaching." The future literature Nobel Prize winner Elias Canetti, at that time an unknown chemistry student in Vienna, confirmed that Musil viewed Broch's trilogy "as a copy of his own enterprise, which had occupied him for decades, and the fact that Broch, shortly after starting his book, had already completed it filled him with the deepest distrust."

Unperturbed, Broch kept poaching on Musil's territory. The main figure in his next novel was a mathematician who had recently received his doctorate, just like Musil's "man without qualities." And while Musil struggled on with the second volume of his masterpiece, becoming ever more bogged down, Broch finished *The Unknown Quantity* in one straight shot. He took only a few months—from July till November 1933—to complete his "attempt to write something popular." And while he was at it, he also polished off a film script for Paramount Pictures, to go along with it; to his chagrin, however, Hollywood was not interested.

In the ensuing years, during which the political madness of the crowds was relentlessly growing, Broch penned essays, dramas, radio plays, fragments of novels, and even resolutions for the League of Nations. When the *Anschluss* came in 1938, Broch was imprisoned for three weeks in a jail at Bad Aussee, in the idyllic center of what, in a flash, had ceased to be Austria. Eventually, via Scotland, he was able to emigrate to the United States.

On the steamer crossing the Atlantic, Broch made out a faintly familiar face: it was that of the young mathematician Gustav Bergmann (1906–1987), who had been a junior member of the Vienna Circle between 1927 and 1931. The two emigrating shipmates became friends for life. Apart from his long conversations with the author, who now looked more distinguished than ever, with his pipe and double-breasted overcoat, Bergmann had something else to do during the journey: Otto Neurath had asked him to write a paper on his memories of the Vienna Circle. By then, in 1938, the Circle was already part of the past.

## MATHEMATICAL PROPS

Hermann Broch was not the first writer with whom Bergmann had come into contact. A few years earlier, together with his friend Hans Weisz, Bergmann had written a letter to Leo Perutz (1882–1957), who was one of the most successful Viennese authors during the wars.

The letter followed up on a short story by Perutz, called "The Day Without an Evening." Just as Musil and Broch had done, Perutz, too, used a mathematician as his main figure. This personage, named Botrel, was a mathematical genius, no less; Perutz didn't do things by halves. Young Botrel, so the story went, had lived quietly from one day to the next, a student showing no signs of ambition or diligence. And then one day, for no reason at all, he became embroiled in a quarrel that spiraled out of control, culminating in a duel to be conducted with pistols. Three days before the duel was to take place, Botrel began to write obsessively. It was as if a dam had broken and water was gushing out fast and furiously. Needless to say, Botrel perished in the duel, but to the great surprise of posterity, from the notes that he had feverishly jotted down until minutes before the fatal shot, there emerged new mathematical theories of unprecedented depth—so deep that, to the present day, a committee of the Academy of Science is working on the publication of Botrel's oeuvre.

The story was sheer invention, of course—although if one looked only a little deeper, it was not even that, since a remarkably similar episode had occurred in France a hundred years earlier, which wound up in the romantic death of twenty-year old Évariste Galois, who was an actual mathematical genius, and who did indeed write down a plethora of brilliant new mathematical profundities on the last night of his life, and Galois's jottings took the mathematical world a few decades to fully understand.

The letter that the two students, Bergmann and Weisz, wrote to Leo Perutz must truly have pleased the author:

> Two young mathematicians who have read your essay *The Day Without an Evening* with great interest would like to ask, with your kind permission, whether the episode rendered by you with such tastefulness has any factual background; for indeed, although the events sound as if they were taken from a novel, the details concerning Cayley curves, cubic circles, etc. are so different from the simplistic absurdities used by other authors as mere "mathematical props" that they support our conjecture that Mr. Botrel must really have existed. With the most humble request for further information (such as the place where Botrel's notes are stored, etc.), we hereby sign, with great respect . . .

It is not known whether Perutz ever replied to this letter. But it is easy enough to explain where he had picked up his "mathematical props." Indeed, just like Musil and Broch, Perutz had studied mathematics. He earned his daily bread as an actuarial mathematician, and in fact a minor mathematical formula is named after him. And Perutz made sure that this distinction was known to all the regulars in all the coffeehouses that he frequented.

In contrast to Robert Musil and Hermann Broch, Leo Perutz had no philosophical ambitions; nevertheless, he came to occupy a niche, albeit a small one, in philosophy. In one of Perutz's fantasy novels, *The Master of the Apocalypse,* an unusual color called "drummet red" is mentioned. The merest glimpse of this color causes its viewer such unimaginable horror that it instantly leads to delirium followed shortly by suicide. Rudolf Carnap, in his article "Pseudo-Problems in Philosophy," uses the sentence "The color 'drummet red' exists, which causes horror" as an example of a statement that, though true, is not verifiable. Schlick was fond of this example: "I read your 'Pseudo-Problems' the very day it arrived, and with great enjoyment. Best of all is 'drummet red'—splendid!"

Hermann Broch quickly recognized the talent of Leo Perutz, and in 1920 he wrote an enthusiastic review of the latter's novel *The Marquis of Bolibar.* Robert Musil, too, knew Perutz quite well but could not put up with his brash temper. It was not so much that Musil disliked Perutz the writer—though he had no great fondness for him—but mostly his distaste focused on the mathematician who somehow had managed to get a formula named after him. Curses!

Eventually, their enmity became public: indeed, the *Prague Daily* had published an anecdote to which the hypersensitive Musil reacted, four days later, with an acid reply: "Some time ago, this paper published an anecdote with the following approximate content:

"A well-known scribbler named Robert Musil once approached the great writer and mathematician Leo Perutz with the request: 'Would you mind writing something about mathematics, or perhaps about a related topic such as ethics, for my newspaper?' Whereupon the writer and mathematician Perutz, without batting an eyelash, replied: 'You know what? I will write you something about the moral basis of the isosceles triangle!' This exchange took place at the height of the big flap about Einstein."

And Musil continued: "I will gladly refrain from challenging the claim that I am, as a writer, the exact opposite of the great Leo Perutz. But I happen to know a few things about mathematics. Yes, I have even written, on occasion, about certain connections between mathematical and ethical thinking; not in the traditional manner, it is true, but I am pleased to be able to state that another manner does exist."

Indeed, that topic was of central importance to Musil. He once wrote: "We may already today have reached the point where our morality has split into two fragments. I might say: into mathematics and mysticism."

Musil's mysticism impressed Perutz as little as his wrath. Perutz simply kept writing one best seller after the other. His novels were reprinted, translated, and adapted into films. Alfred Hitchcock and Ian Fleming were among his admirers. But when in 1933 Hitler's Germany banned his books, he lost most of his readers. And five years later, he lost his homeland as well.

In 1938, Leo Perutz emigrated to Palestine. "A cool good-bye to Europe," he somberly noted in his pocket calendar as his steamer was sailing out of the lagoon of Venice.

# 8

# *The Parallel Circle*

*Vienna, 1929–1932: Young mathematicians develop their own version of Vienna Circle: Mathematical Colloquium. Karl Menger writes play, defines dimension, seals letters. Kurt Gödel throws monkey wrench into Hilbert's program: axiomatic systems necessarily incomplete, and consistency not provable. John von Neumann deeply impressed. Menger hails the new logic.*

"Karl? He Will Become a Professor!"

Karli Andermann was a bright little lad. "Karl? He will become a professor!" said his first-grade teacher to the proud mother. At age ten, Karli easily passed the entrance examination for the *Gymnasium*—the secondary school.

A new chapter in life began, with new schoolmates. It seemed like the right time to change the youngster's name: from now on, he would be Karl Menger. Thus the child, born out of wedlock, was legitimized in 1912 *per rescitum principis* ("by an act of the emperor"). Court Counselor Professor Carl Menger now took on ten-year-old Karl as his own son.

The boy soon understood that along with his new name came a splendid reputation. But he was not satisfied to bask in his father's glory; he wanted to make a name for himself—or rather, an initial: K. Menger, not C. Menger.

199

Karl's father, Carl Menger (1840–1921), belonged to the legendary generation of Boltzmann and Mach, and could hold his own in that illustrious company. He had founded the Austrian school of national economics, and his theory of marginal utility marked a turning point in economic thinking. Already as a thirty-year-old lawyer and journalist, Carl Menger had noticed that no classical theory of economics could explain how the market determines prices, so he decided to do something about it.

Without any guidance, he wrote his book *Principles of National Economics*. When it appeared in 1871, the self-taught economist became one of the leaders in his field, almost overnight.

The ideas he came up with must have been in the air, for at nearly the same time, Léon Walras in France and William Jevons in Great Britain also hit on the same approach. For all of them, the central tenet was that the value of any particular good is determined not by production costs, working hours, or raw materials, but by the *demand* for one additional unit of the good. In the words of Friedrich von Wieser, a disciple of Carl Menger: "Man values goods not for themselves, but for himself." This was a fresh new approach that allowed economists to derive models from first principles, independently of local social structures or historical contingencies.

Within a year of the appearance of his book, Carl Menger had earned his *Habilitation*—that is, the right to lecture at the university. In 1876, he was designated as the scientific tutor for Archduke Rudolph, the Crown Prince. Menger instructed the future heir to the Austro-Hungarian throne in political economics and took him on educational journeys. Of all the countries they visited, it was Britain that most impressed both teacher and pupil.

Upon completing his stint as tutor to royalty, Carl Menger was appointed to a chair at the University of Vienna. Almost daily, he and his brother Anton, a professor of civil law and a famous social theorist, met in one of the lively coffeehouses on the Ringstrasse to read the papers and to discuss current events. Sometimes they were joined by their brother Max, a well-known deputy in the Reichsrat, Austria's parliament.

FIGURE 8.1  Father figures: Carl Menger and Arthur Schnitzler.

Disciples of Carl Menger, and their disciples in turn, could be found in all the universities in the monarchy. As for Carl himself, he was working on a theory of money while spending his own money on oodles of books, just as Otto Neurath's father did. Furthermore, Menger was planning to write, should he ever find the time, a philosophical treatise. His preliminary notes, dating way back to 1867, read like a harbinger of the Vienna Circle: "There is no metaphysics . . . Kant rejects metaphysics and replaces it by a Critique of Pure Reason. I say there is no pure reason. There is no riddle in the world to be solved. There is only incorrect consideration of the world."

A young journalist named Hermine Andermann began working on a catalog of Carl Menger's vast private library. She was not daunted by the hugeness of the task; in fact, it turned into a labor of love. One thing led to another, and soon enough, her employer, a sixty-year-old bachelor, became the father of a boy. Not long thereafter, he applied for early retirement from the university: like the slightly younger Boltzmann, he suffered from severe neurasthenia.

After finishing elementary school, the younger Menger entered the newly founded secondary school in Döbling, a residential district in the western outskirts of Vienna. The school was privately financed, and had a modern curriculum. Here English was the first foreign language, rather than the usual French. Soon, however, French and English lessons were interrupted. The Great War had started. The school was forced to move, because its buildings were needed as a military hospital. But the teachers—all elderly men who did not have to enlist—faithfully maintained the strictest educational standards.

The expectations facing Karl Menger at school were as high as, if not higher than, those at home. In a class just a couple of years ahead of the lad were two students who later went on to receive Nobel Prizes: Richard Kuhn (Chemistry 1938) and Wolfgang Pauli (Physics 1945).

## "Megalomaniac and Eccentric Traits"

While still at school, Karl Menger decided to write a play. This seemed like a promising avenue to attain early fame. With an ironic tip of the hat to Dante, he titled his work *The Godless Comedy*. The central character was the apocryphal Pope (or Popess) Joan, who had passed herself off as a man until she gave birth to a child. The hopeful young dramaturge had the lucky chance of being able to show his manuscript to Austria's most famous playwright, as his classmate and good friend Heini (short for Heinrich) was the son of Arthur Schnitzler. Karl was not the only one in his class to have a famous father.

Would history repeat itself? A few decades earlier, Hugo von Hofmannsthal, then just a schoolboy, had shown some of his poems to Arthur Schnitzler, who gave them high praise, and with that green light from a known master, the writers of Young Vienna had found their new wunderkind. But fate was not so kind to Karl Menger; his play sank without a trace.

Despite this setback, Karl's excursion into the dramatic arts gave him a small role in Arthur Schnitzler's diaries. Several entries therein allude briefly to the young would-be author. Here are a few excerpts:

"27.10.1919. At the table, Menger, Heini's colleague, who recently sent me a plan for a play on Pope Joan; serious, gifted boy (though as to literary, I have doubts)."

"19.1.1920. Afternoon, a colleague of Heini's, Menger, to whom I said a few words about his failed first dramatic attempt about Pope Joan ('miracle')."

"11.11.1920. Reading in the afternoon. Menger's play. Talented fellow, but not literary."

"13.11.1920. Karl Menger (Heini's classmate), about whose play (Pope Joan) I had to say quite a few not so favorable things, was here. He has no literary ambitions, just wants to write this one play meant to oppose religion, Catholicism, superstition. His true calling: physicist . . . In reply to my question about long-term plans: 'I would most like to kill myself.' Certainly a very talented but possibly not quite normal young man."

In the meantime, this "possibly not quite normal young man" had enrolled in mathematics, physics, and philosophy at the university and was lucky enough to get tickets to all three lectures given by Albert Einstein in January 1921. He was already quite familiar with the theory of relativity. In fact, over the summertime, the "certainly very talented" youth had discussed it in an exchange of letters with his former schoolmate Wolfgang Pauli, his senior by a small margin. Incidentally, Pauli owed his middle name, Ernst, to his godfather—none other than Ernst Mach. What a small world Vienna was!

At that point, Pauli had received his doctorate from the University of Munich, summa cum laude, and in record time. Talk of a wunderkind: though merely an eighteen-year-old, he had already written a landmark paper on the theory of relativity for the prestigious *Encyclopedia of Mathematics*. He was not in the least intimidated when he engaged in intense technical debates with far older luminaries such as Bohr, Einstein, and Max Born. It is reputed that one time, when Einstein had just given a talk on a daring new idea, the teenage prodigy blurted out, in front of the entire lecture hall, "*Was Herr Einstein gesagt hat ist nicht so blöd.*" ("What Mr. Einstein just said is not so foolish.") This remark by Pauli might be taken as the definition of chutzpah.

Karl Menger, like Pauli, studied physics, but he had not quite yet shelved his literary designs. Arthur Schnitzler noted in his diary:

"2.11.1921. Karl Menger . . . read out loud a new scene from his play (between Pope Joan and the Heretic). Talented person, may be a genius—but with megalomaniac and eccentric traits."

A few years later the tune sounded a little different:

"17.1.1928. At the table, young Menger, just back from Holland, waiting for a professorship here. Only 25, he seems already to enjoy Europe-wide fame, and I always feel his genius in a field which, however, is inaccessible to me."

## The Sealed Letters, Part I

In the seven years it took him to mutate from schoolboy to professor, Karl Menger endured many ups and downs.

His father had died shortly after the Einstein lectures. By then, Carl Menger was over eighty years old and was universally renowned as the founder of the Austrian school of economics. The disciples of this great school included, among many others, Eugen von Böhm-Bawerk, Friedrich von Wieser, Ludwig von Mises, and Friedrich August von Hayek. (The new republic had abolished the old aristocratic prefix *von*, but for some odd reason, none of these economists seemed to have noticed it.)

Among the papers left behind by Carl Menger, his son discovered documents shedding light on a cause célèbre that had deeply shaken the old monarchy. In 1878, an anonymous pamphlet called "An Exhortation to Austria's Aristocratic Youth" had made the rounds. It compared the hedonistic and shallow lifestyle of Austria's youth with its counterpart in the United Kingdom, and asserted that "the kingdom's leading position is due, among other things, to the political and economic activities of the nobility."

This "exhortation" would have raised few eyebrows had rumors not spread to the effect that it issued from no less than Archduke Rudolph. Eventually, a liberal journal reprinted some of its most scandalous parts, claiming that the authors were Carl Menger and his pupil, the Crown Prince. Censors lashed out left and right and doused the flames as quickly as they had flared up, but the affair put great distance between the emperor and his heir apparent. The coldness was to last for a decade, until the gory suicide of Rudolph in the hunting lodge of Mayerling. This sordid episode provided another field day for press censorship.

In the documents left behind by his father, Karl Menger discovered definitive proof of the joint authorship of the pamphlet by the archduke and his tutor. Karl wrote a newspaper article on his find, and he deposited the documents in a sealed envelope at the Academy of Science in Vienna.

The father had also bequeathed to Karl a huge number of notes concerning the planned new edition of his *Principles of National Economics*, on which he had been working ever since his retirement, and whose preface had been written a full thirty years earlier. It now fell to the son to complete the new edition.

"Horrible trouble and toil," young Karl noted in his diary, "mitigated only by knowing that I have completed a great work for science, perhaps even for humanity." In addition, he wrote an article for Austria's leading newspaper, the *Neue Freie Presse,* to explain how the devastated economy of the dismembered Austrian state could be resuscitated. The newspaper editor added: "We are publishing an article by the son of the master of national economics, without agreeing with all its conclusions."

The highest-ranking ministry official, Wilhelm Rosenberg, whose fiscal policy had been sternly rebuked by young Menger as a step toward hyperinflation, invited the nineteen-year-old to lunch. Over the meal, Rosenberg explained that Austria's postwar economy was not so easily cured of its ailments as Menger appeared to think. Even the famed economist Josef Schumpeter, during his brief spell as minister of finance, had run miserably aground.

The Sealed Letters, Part II

Playwright, physicist, economist, and even journalist—young Karl Menger had tried his hand at many tasks before he hit on his true métier: mathematician. The key event happened during the third semester of his studies. The newly appointed professor Hans Hahn was giving the first class of his seminar, an introduction to curves. Everybody, said Hahn, has an intuitive conception of a curve, yet a mathematically impeccable definition had so far proved elusive, and some eminent mathematicians doubted whether it was possible at all.

For a long time it had been thought that a curve is what one obtains by drawing an uninterrupted, continuous line with an idealized infinitely thin pen. But Giuseppe Peano and David Hilbert had shown, to the astonishment of their peers, that a continuous line made in that way could follow such a tortuous pathway that it could pass through every single point in the interior of a square, or of a cube (a "space-filling curve," as it was called), and yet nobody would claim that a solid square or a solid cube was a curve.

Hahn used paradoxical-seeming concoctions of this type to warn about the pitfalls of intuition: "For it is not true, as Kant maintained, that intuition is a pure means of acquiring knowledge a priori; rather, it is merely a force of habit, rooted in mental inertia."

According to common sense, a curve is something one-dimensional, a surface is two-dimensional, and so on. But how is *dimension* defined? By the number of coordinates needed to determine the position of a point? This raises several problems. Most sets of points cannot be defined by equations; they are more complicated than is commonly thought. And even the simplest sets, such as a square, for instance, do not conform to such a definition. Indeed, the examples of "space-filling curves" show that a single number, rather than two, is enough to determine the position of a point in a square. (Think of the infinite decimal expansion of, say, pi; the digits in odd positions could be taken as determining the $x$-coordinate of a point, while the digits in even positions determine the point's $y$-coordinate. Or vice

FIGURE 8.2 The first steps in the construction of a space-filling curve.

versa—given the *x* and *y* coordinates of any point in a square, you can interweave them to make just a single real number.) This realization was counterintuitive and gave rise to deep questions about what *dimension* really means.

Almost instantly, young Menger discovered a new approach to the riddle of the nature of "dimension." The basic idea was surprisingly simple. To split a three-dimensional object, such as a wooden block, one uses a saw; the cut is two-dimensional. To split a two-dimensional object, such as a sheet of paper, one uses scissors; the cut is one-dimensional. To split a one-dimensional object, such as a wire, one bends it back and forth with a pair of pliers; the cut is zero-dimensional. Menger's suggestion was to make the definition run the other way around, by first introducing sets of dimension zero (those where the cut is empty), then of dimension one, then two, then three, each time using the lower dimensions, which had already been defined, to define higher ones. (There is no reason, of course, to stop at three.)

To be sure, working this crude idea out in a mathematically precise way took a great deal of effort, but Menger soon convinced Hans Hahn that his approach was going to be successful. Feverishly, he continued with his investigations.

Feverishly—and soon overcome by real fever. The university classrooms were unheated. No money, no coal. The shaky health of the young man broke down. The diagnosis: pleurisy.

A few months later, Karl Menger emerged from a long convalescent leave and at last threw himself back into his work with the same old passion. Karl Popper, who in 1921 had attended mathematics classes at the University of Vienna, recalled, more than seventy years later: "At the institute, there was also Karl Menger, of my age, but obviously a genius, filled with new and splendid ideas."

Soon, however, came more rough blows. Hahn discovered a mistake in one of Menger's key proofs, and Menger fell ill again. This time, the diagnosis was tuberculosis, an illness that had killed many during the postwar years. Physicians had even taken to calling it *Morbus Viennensis*. This was a scary time.

Students of mathematics learn at an early age about Norway's Niels Hendrik Abel and France's Évariste Galois, two mathematical geniuses from the nineteenth century who died tragically young, before their work had found deserved recognition: Abel died at twenty-seven from tuberculosis, and Galois while only twenty, a few days after the calamitous duel described earlier. And now Karl Menger had to wonder if a similar "romantic" fate awaited him. Before departing anew for the sanatorium, he deposited another sealed envelope at the academy, this one sketching out the mathematical ideas that he had not yet had the time to work out in full.

## RATHER TROUBLING

Karl Menger spent almost a year and a half in the sanatorium of Aflenz in the region of Styria. Later, he compared his rustic health resort with the *Magic Mountain* so alluringly described by Thomas Mann. This was certainly looking at things through rose-colored glasses! But in fact, he had used the time well. By the time he returned to Vienna, he had

FIGURE 8.3 Young Karl Menger as a man for all seasons.

completely patched up his proof and could confidently submit his work as a PhD thesis.

During his absence, a trusted friend had regularly sent him lecture notes from all the mathematics courses he had missed during those three semesters. This friend, Otto Schreier (1901–1929), had attended the same high school as Karl, one class ahead of him, and was as talented as anyone in that school.

In 1923 Schreier traveled to Marburg to attend the annual meeting of German mathematicians. Aspiring young mathematicians used to deride these conferences as "slave markets," because, naturally enough, the older professors always eagerly inspected the new young crop. Those who stood out could hope to be hired as scientific tutors or possibly even as assistant professors.

Otto Schreier did very well for himself at that year's slave market: he was offered a job in Hamburg. But for his friend Karl Menger, who did not attend, he had some unfortunate news, which he sent on a postcard: "I will of course tell you everything in detail as soon as we are together again. But one thing I must mention to you right away, although I am afraid that you will find it rather troubling. A young Russian, Mr. P. Urysohn from Moscow, gave a lecture on dimension theory. As far as I can tell, his results are essentially the same as yours, and he seems to have found them at roughly the same time (or perhaps a bit earlier)."

"Perhaps a bit earlier"! This was a devastating blow for Menger. While he had been recuperating on his "Magic Mountain," Pavel Urysohn had submitted a synopsis of his results to the *Comptes rendus de l'Académie des sciences de Paris.* This type of official announcement of new scientific results was far more likely to impress the community than a sealed envelope deposited by an unknown student at the academy in Vienna.

In desperate haste, Menger completed his manuscript. Hahn arranged for speedy publication in the *Monatshefte,* the Austrian mathematical journal. And in a footnote at the end of the paper, the ominous Urysohn was referred to. Menger sent reprints to leading mathematicians all over the world.

As fate had it, that same summer, somewhere off the coast of Brittany, Pavel Urysohn drowned in the treacherous breakers of the Atlantic. The friend who was with him, Pavel Alexandroff, another young Russian mathematician, could do no more for him than recover his corpse from the surf. An article giving all the details of Urysohn's dimension theory appeared posthumously.

It so happened that Otto Schreier also met with an early death soon after that, from blood poisoning, at age twenty-eight. He ranks, like Galois, among the major figures in the history of modern algebra. By contrast, Karl Menger, who had feared an early demise, lived to be eighty-four, and he was hounded, all the way until the end of his days, by the question of who really deserved the credit for the deep advances in dimension theory. Ah, a human ego is such a frail thing!

## The Excluded Middle

In 1924, soon after getting his PhD, Karl Menger received a Rockefeller grant and used it to visit the Dutch mathematician L. E. J. Brouwer in Amsterdam. Brouwer was the foremost expert in Menger's area of mathematics. In addition, he was the head of the new school of logic called *intuitionism*. This group had a view on the foundations of mathematics that was in radical contrast with Hilbert's formalistic approach.

According to Brouwer, mathematics is entirely a creation of the human mind. It consists solely of mental objects—points, numbers, sets, and so forth—all of which have to be constructed. Such constructions require explicit recipes. It is not possible to deduce the existence of a mathematical object merely by showing that the assumption that the object does *not* exist leads to a contradiction. That sort of approach to mathematical existence, which was standardly used by almost all mathematicians, and which had been explicitly endorsed by David Hilbert, constituted, in Brouwer's view, a blatant misuse of the Law of the Excluded Middle.

This classical law states that there is no middle ground between true and false. If a given statement is false, then its negation is true, and vice versa. But Brouwer had the audacity to question the hallowed principle. Let us look at an example.

Either a real number can be written as a quotient of two integers (a fraction such as 2/7, say) or it cannot. In the former case, it is a rational number; in the latter case, it is irrational. The number $\sqrt{2}$, for instance, is irrational, as we have known ever since Pythagoras and his crew, some two thousand years ago. Consider the claim that there exist two irrational numbers, $x$ and $y$, such that $x^y$ is rational. Here is a lovely little proof: Take the number $\sqrt{2}^{\sqrt{2}}$, which is an irrational to an irrational power. Is this number rational? If so, we are done. And what if it is *not* rational? Well, then it is irrational, so let's now consider the following number: $(\sqrt{2}^{\sqrt{2}})^{\sqrt{2}}$, which has the form $x^y$, where (by hypothesis) both $x$ and $y$ are irrational. A rule governing exponents—namely, $(a^b)^c = a^{bc}$, learned in high school algebra—tells us that this quantity equals $\sqrt{2}^2$, which is of course 2, which is obviously rational. So we are done again! Either one way or the other way, we have found our desired irrational numbers $x$ and $y$.

Let us spell the reasoning out one more time. There are two cases to consider. Either $\sqrt{2}^{\sqrt{2}}$ is already rational, in which case we are done—or it is irrational, in which case $(\sqrt{2}^{\sqrt{2}})^{\sqrt{2}}$ is rational, and so we are again done. We don't know *which* of the two pathways gives us our desired pair of irrational numbers *x* and *y*, but to a normal person, that lack of knowledge doesn't matter. The proof is airtight; if our *original* choice of values for *x* and *y* doesn't work, then our *alternative* choice for *x* and *y* works. The only hitch is that we haven't specified *which one* of the two alternatives gives us our desired *x* and *y*—but why should that matter? One or the other of them will turn the trick.

For Hilbert and nearly all other mathematicians, the proof as stated is flawless, because we've set up a fork, and we've shown that one or the other branch of the fork must lead us to the desired result. It's just that we don't know which of the two branches will work, but that seems irrelevant, since we have proven that one or the other of them *has* to work. But that kind of forking proof was not good enough for Brouwer and his disciples! To them, this was a classic case where one could not draw any conclusion.

Today, we know that $\sqrt{2}^{\sqrt{2}}$ is irrational, but that is not the point. The point is that Brouwer raised doubts about the use of the excluded middle, and he found followers. Brouwer claimed that it was particularly worrisome to apply the law, self-evident though it seemed, to infinite sets.

Roughly speaking: If some apples are stashed in two bags, then either both bags contain apples, or one bag does not; there is no "middle" located in some elusive zone between those possibilities. This is immediately obvious. But mathematicians use a similar argument to deal with *infinitely* many apples: Either both bags contain infinitely many apples, or one of the bags contains only finitely many (possibly none). If we can exclude, for whatever reason, one of the alternatives, then the other must hold; there is no middle way.

This seemingly self-evident truth was precisely what Brouwer cast doubt on. It is easy to check whether a bag is empty or not—but how does one check whether its content is finite or infinite? We can count, to be sure; for instance, by removing one apple after another from the bag in question. If this process comes to an end after a while, meaning that the bag is now empty, then we can safely infer that its content was originally finite. But as long as we are still in the process of counting, we cannot tell whether the number of apples is infinite or finite.

Brouwer claimed that we should beware of infinity, unless there is a well-defined, constructive procedure on which to rely. As long as neither the statement A nor the statement non-A has been proven, we cannot say that one of the two must be true.

Brouwer's constructive constraints made logical reasoning in mathematics even harder than it was already. Hilbert immediately raised his voice in protest: to preclude a mathematician from using the Law of the Excluded Middle was like "forbidding a boxer to use his fists." Nevertheless, many first-class mathematicians joined Brouwer's camp.

A camp? Mathematics prides itself on the fact that it knows no opposing camps, creeds, or congregations. Such scandalous schisms were better left to theologians or philosophers. However, there it was: a philosophical disagreement had crashed the party uninvited.

Young Karl Menger tried his hardest to explain to the Schlick Circle what intuitionism was all about. Brouwer's arguments, however, struck them as obscure—and certainly the diametric opposite of what they called "intuitive." Thus Menger, who was very open-minded, went on a pilgrimage to learn more about intuitionism at the feet of the master.

L. E. J. Brouwer was an imposing figure, with sharp, angular features. While still a student, he had written a paper called "Life, Art, and Mysticism," which was a radical attack on traditional logic. He now lived in an artists' colony in a suburb of Amsterdam. In fact, his home was just a small, austere hut in a garden, and it contained nothing but a desk, a bed, and a piano.

Apart from Menger, there were other young mathematicians in the mystic's entourage, including the athletic Pavel Alexandroff, who had rescued Urysohn's dead body from the sea. Brouwer had taken on the task of publishing the posthumous writings of the drowned Russian, with an introduction of his own.

Karl Menger admired Brouwer deeply. He became the latter's assistant at the University of Amsterdam and lectured there. When Menger's mother died unexpectedly, Brouwer cared touchingly for his protégé.

But after some time, the vexatious question of priority raised its ugly head and led to deep discord. Karl Menger opined that his own role had not been given enough consideration in Brouwer's account of dimension theory. Brouwer, however, was not ready to emend his views. He was, like some knights of yore, famous for his fights. For his part, Menger was not willing to avoid a collision course either. Thus his position in Amsterdam became highly unstable and distressful.

In the midst of this mess, out of the blue appeared a solution coming from, of all places, Vienna. In 1927, Kurt Reidemeister, the master of knot theory, accepted an offer from Königsberg, thus vacating his position as associate professor of geometry in Vienna. The empty slot was offered right away to the twenty-five-year-old Karl Menger. This was an exceptional career move. Even his former schoolmate Wolfgang Pauli had not reached professorship much faster.

However, the quarrel with Brouwer kept plaguing Menger. The letter that he had written, sealed, and deposited in the academy in order to bolster his claims was formally opened in front of witnesses, and its contents were officially displayed—and yet nobody seemed to give a hoot. Menger's next step was to publish a stream of papers on the argument, and he even pressed Hans Hahn to send long letters to Brouwer in support of his claims.

In that same period, Menger and Hahn invited the Dutch mathematician to give talks in Vienna. After all, the scientific worldview required the strict separation of personal and scientific matters. Brouwer accepted the invitation with pleasure, as he savored every possible opportunity to attack Hilbert's formalism. Thus came about his famous Vienna lectures, which rekindled Wittgenstein's smoldering philosophical passions.

Moments before Brouwer's first lecture began, Hans Hahn introduced himself, in the bustle of the crammed lecture hall, to the author of the *Tractatus*. Wittgenstein "thanked him with an abstract smile and with his eyes focused at infinity," wrote Menger. He had witnessed the scene with dismay. The young geometer, nettled by Wittgenstein's obvious lack of interest, felt vicarious resentment for the snubbing. He resolved never to impose himself on anyone! As he later wrote: "I have always tried to avoid making the acquaintance of someone who did not appear to be interested in making mine."

Menger felt convinced that he knew the reason lurking behind the rebuff. He ascribed Wittgenstein's icy aloofness to a resentment directed against all Viennese mathematicians. Well, there wasn't much that could be done about such an irrational prejudice. And so things were left as they were.

This explains how, although mathematics was soon to assume center stage in Wittgenstein's thinking, there was never any personal contact between Wittgenstein and the Viennese mathematicians: not with Hahn, nor with Menger, nor with the quiet, frail student with horn-rimmed glasses named Kurt Gödel.

Like Wittgenstein, Gödel, too, was sitting in the lecture hall on that memorable occasion and was taking in Brouwer's dark words, and for him too they marked a turning point in his life.

## LITTLE MISTER WHY

Kurt Gödel was born in 1906 in Brno. The town, named Brünn at that time, is located in southern Moravia, just a couple of hours by train north of Vienna. Some guidebooks of the day described it as "the Manchester of the Austrian monarchy." Gödel's father, a native of Vienna, had moved to Brünn and had managed, by dint of his technical skills and his entrepreneurial mind, to rise from the status of a penniless school dropout to that of co-owner of a large textile firm.

Together with his elder brother Rudolf, Kurt grew up in carefully sheltered surroundings, in a villa on the southern slope of the Spielberg, Brno's dominating landmark. The yard of the Gödels' house opened directly onto the town park, crisscrossed by lovely paths. And nearby was the monastery where Gregor Mendel had grown his peas.

Kurti's clever and cultured mama was delighted by the eager inquisitiveness of her youngest: in the family, he was hailed as "Little Mister Why." A notebook that recorded his first encounter with mathematics, clumsy though it was, was carefully preserved.

The boy became a brilliant pupil with strong scientific interests. His grades were always the best possible ("very good"), with one single exception, when he had to be content with a mere "good"; this was in mathematics, oddly enough. But at that time, Kurt was only eleven. From then on, he never lapsed again.

In 1924, Kurt enrolled in theoretical physics at the University of Vienna. The theory of relativity still held the world in thrall, and the university had some noted experts in this field. Indeed, physicist Hans Thirring (1888–1976) and mathematician Josef Lense (1890–1985) had just shown that a sphere that is rotating (like the Earth) gives rise to a different gravitational field from that of a sphere that is not rotating. This wonderful discovery was an early vindication of what Einstein termed Mach's principle; the effect was so microscopic, however, that it took eighty years for the prediction to be verified by experiment.

Kurt Gödel quickly mastered the arcane labyrinth of general relativity. However, like Karl Menger four years earlier, he turned from physics to mathematics. He was fascinated by the introductory lectures given by Philipp Furtwängler, one of the most eminent number theorists of his time. Because of a severe illness, Furtwängler had to use a wheelchair, and he had to be carried into the lecture hall. Even so, his lectures were awe-inspiring in their elegant perfection, attracting far more students than there were seats.

Another lecture course that cast its spell on Gödel during his first year at the university was called *Overview of the Main Problems of Philosophy*, by Heinrich Gomperz. The students told each other, in awe, that this bushy-bearded philosopher had played a key role in bringing Ernst Mach to Vienna. But those glory days now seemed like ancient history.

Before long, the talent of Kurt Gödel attracted the attention of Professor Hans Hahn, and as early as 1926, the gifted young man was invited to the meetings of the Schlick Circle. There, Gödel met two of Schlick's students, Herbert Feigl and Marcel Natkin. The three of them became close friends.

Karl Menger describes Kurt Gödel: "He was a slim, unusually quiet young man. I never heard him speak in these meetings or participate in the discussions, but he evinced his interest by slight motions of his head indicating agreement, skepticism, or disagreement."

The motions of Gödel's head indicating disagreement must have been barely noticeable. But decades later, it became clear that his opinions, already at that time, were diametrically opposed to those prevailing in the Vienna Circle. However, the young man was loath to engage in heated debates; instead, he simply listened in friendly silence while forming his own views. What interested him most was Hahn's seminar on Russell and Whitehead's famed *Principia Mathematica*. He purchased for himself a copy of those three heavy tomes. It cost him a small fortune, but then again, Kurt was hardly a pauper. And not too much later, the investment paid off royally.

Each year, Kurt and his brother Rudi moved to a new flat. They made sure always to be near the university, and always in elegant four-floor apartment buildings dating from the turn of the century. Rudolf Gödel was studying in the illustrious faculty of medicine in Vienna where, every few years, another of its professors would collect a Nobel Prize, and Sigmund Freud would grumble that he had been passed over once again. Ah, such a frail thing is a human ego!

## COMPLETENESS

Every four years, there took place an International Congress of Mathematicians. In 1928, it was held in Bologna, and Hahn and Menger gave talks. David Hilbert headed up a large German delegation. This was remarkable, because Germany had been excluded in 1920 and 1924—a condition that the mathematicians of the victorious entente powers had insisted on. But at last the ban had been lifted.

Being Dutch, Brouwer had not been directly affected, but he was outraged by the unfairness. He felt that the Germans should take revenge—perhaps by closing their mathematical journals to French submissions, or at least by boycotting the Bologna Congress. On both points, though, Hilbert took a different stance, and this settled the issue for Germany's mathematicians. From then on, Brouwer fumed against

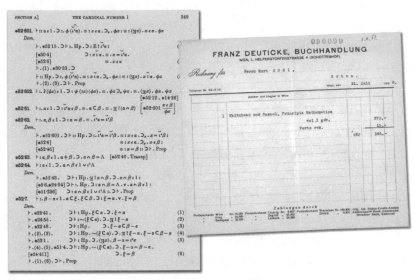

FIGURE 8.4 Gödel buys himself a page-turner:
Volume I of *Principia Mathematica.*

Hilbert even more than against the French. Not that there was much love lost: Brouwer had always sharply disagreed with Hilbert's formalistic approach to mathematics.

At the congress, Hilbert lectured on his program for proving the consistency of mathematics. A great deal had happened since the Paris congress of 1900. At one point, Hilbert had even believed he had found his Holy Grail, only to discover that he had made a crucial mistake. But he kept on seeking a proof. One year before the Bologna Congress, he published his *Principles of Mathematical Logic,* co-authored with his disciple Wilhelm Ackermann. The book was, like all works by Hilbert, a masterpiece of concise elegance, and Kurt Gödel studied it with great interest.

In their book, Hilbert and Ackermann had described a series of open problems, and young Gödel proceeded to solve two of them. They both concerned first-order logic. (Roughly speaking, this means a formal reasoning system that can prove propositions concerning "all entities," but not propositions concerning "all properties.") Gödel was able to show that the axioms and rules of inference of first-order logic were sufficient

D^r Kurt Gödel

FIGURE 8.5 Kurt Gödel, PhD.

for deriving all generally valid propositions, and that conversely, none of these axioms was a consequence of the others. In other words, every single axiom was needed, and no other axiom was needed.

Gödel submitted his proofs to Professor Hans Hahn. Hahn's verdict didn't come right away. The semester holidays arrived, and Gödel returned to his parents' home in Brno.

Then Hahn delivered his judgment. The proofs were correct. The paper could be submitted as a PhD thesis, and Gödel was asked to publish his results in the *Monatshefte*.

When Hahn approved the thesis, he wrote in his report that the work of Gödel "fulfills all the requirements for a PhD thesis." However, this was a vast understatement. To solve any problem that had been posed by David Hilbert was a supreme achievement for any mathematician, and the attendant glory was on the same level as being knighted.

A few months before Gödel earned his doctorate, in February 1929, his father suddenly died. The new widow decided to move to Vienna to be near her two sons. The elder one, Rudolf, had recently completed his studies and was now a professional radiologist. The three of them lived in a spacious flat on Josefstädter Strasse, not far from Max Reinhardt's celebrated theater.

The fresh new mathematics PhD Kurt Gödel was unfortunately unable to find a position at the university, but fortunately, he did not need one. As a well-off private scholar, he could do as he liked. He thus got into the habit of working till late at night, sleeping till late in the morning, and then strolling over to the mathematics institute on Strudlhofgasse. There, he could almost always be found in the library, correcting exam papers on behalf of Menger and Hahn, or helping students prepare their seminar talks.

## INCOMPLETENESS

Brouwer's lectures had given Gödel much food for thought. The Law of the Excluded Middle suggests that every open problem always has a clear-cut answer, one way or another. To Brouwer's mind, this was equivalent to making the unquestioned assumption that every mathematical problem has a solution. Hilbert certainly thought this was the case. But was it not possible that the formal systems so beloved by Hilbert might turn out to be too weak to encompass the whole of mathematics? Carnap, who met his quiet young colleague Gödel every few days—usually in a coffeehouse, usually to talk logic—recorded in his diary:

"(23.12.1929) Gödel. On the inexhaustibility of mathematics. He has been inspired by Brouwer's Viennese lectures. Mathematics not completely formalizable. It seems he is right."

At that point, it was still "It seems," but half a year later Gödel had come up with miraculous new insights, giving him a proof that dispelled all doubts. Again, the moment is captured in Carnap's diary:

"Tue. August 26, 1930, 6–8.30. Café Reichsrat. Gödel's discovery: incompleteness of the system of *Principia Mathematica. Problems with the proof of consistency.*"

Late in the summer of 1930, a few members of the Vienna Circle had met in the Café Reichsrat, just behind Parliament, to plan their joint trip to Königsberg on the Baltic. The annual meeting of the German Union of Mathematicians—that ritual "slave market"—was slated to be held there in September. After the success of the Prague meeting the previous year, it had been decided to hold another satellite meeting. Once again, the Vienna Circle had someone on the spot to organize the event. In Prague, this had been Philipp Frank. Now, Kurt Reidemeister was eager to step in and help out as the on-site agent.

The main subject of the satellite meeting, this time, was to be the foundations of mathematics. The three great rival schools were to confront each other: the *logicists,* whose goal was to reduce mathematics to logic; the *formalists,* who were looking for an ironclad proof that mathematics was contradiction-free; and the *intuitionists,* who redefined mathematics by insisting on everything being explicitly constructible, and restricting the use of the Law of the Excluded Middle. Each of the three factions had a famous chieftain: Russell, Hilbert, and Brouwer, respectively. Ironically, though, it turned out that none of those three was able to attend the satellite meeting. Even Hilbert, who was in Königsberg at that time, was unable to participate, being fully occupied with the German mathematics congress.

Each of the three contending approaches, therefore, was represented by a trusty deputy: Rudolf Carnap spoke for logicism, Arend Heyting for intuitionism, and John von Neumann, Hilbert's favorite disciple, stood in for the formalist viewpoint. In addition, Friedrich Waismann had decided to present Ludwig Wittgenstein's views.

Waismann's project was ill-starred, however. On his way to Königsberg, he fell sick. The last part of the journey had been by steamboat, and a violent storm had raged. To make matters worse, Wittgenstein had insisted that Waismann start out his talk by announcing that Ludwig Wittgenstein had declined any responsibility whatsoever for any view Waismann might wish to attribute to him. This was hardly an appealing way to begin spreading the gospel.

Strangely enough, in their lectures, none of the members of the Vienna Circle breathed a word about Kurt Gödel's monumental new result. Even Gödel did not talk about his incompleteness theorem, focusing instead on the completeness theorem that he had proven in his PhD thesis one year earlier. Only in the final discussion did he mention, almost in passing, a corollary of his incompleteness result. This was right before lunch, at the end of the meeting.

All at once, the Hungarian John von Neumann could no longer devote any thought to food. Quite the contrary, he had found food for thought, having immediately grasped the enormous significance of Gödel's offhand remark. Now he intently quizzed Gödel on his proof. Dr. von Neumann (János, Johann, or Johnny—he answered to all three), a vivacious bon vivant and born entertainer, was already being hailed at that time as a mathematical superstar. Though only twenty-six, he had made seminal contributions to set theory, analysis, and the foundations of quantum theory.

Johnny understood everything at first try. There was nobody in the world who could think any faster than he. And what he had just realized, in a blinding flash, was that Gödel's discovery had totally exploded his prior worldview. All at once, von Neumann understood that *there exist true mathematical statements that cannot be derived by formal means from a set of axioms.*

A few weeks later, von Neumann wrote to Gödel to point out that his proof of incompleteness implied that Hilbert's program was not feasible. That is to say, if mathematics is consistent, then the statement "mathematics is consistent" is precisely one of those weird Gödelian propositions that are true but cannot be proved! On its surface, this sounds utterly paradoxical, but it turned out to be correct.

Gödel, however, had already reached the same conclusion, and by return mail he sent the page proofs of his paper to John von Neumann. The epoch-making article "On Formally Undecidable Propositions of *Principia Mathematica* and Related Systems, I" was published a few weeks later in the *Monatshefte,* the mathematics journal edited by Hans Hahn. The reason for the roman numeral *I* at the title's end is that Gödel had originally intended to write a second part, in which his proofs would be more fully spelled out, but thanks to the warm reception of "I" by von Neumann and others, it quickly became apparent that "II" would not be necessary. Part I was sufficiently clear to convince the leading lights of the mathematical world. For the instruction and benefit of philosophers, Gödel also wrote a short synopsis, which was published in *Erkenntnis,* the house journal of the Vienna Circle.

John von Neumann was tremendously impressed. Someone had actually thought faster than he had! Often, the Hungarian genius came up with a proof while he slept. Sometimes, when he woke up, he found that his dream proof was fallacious. But he was fond of saying that by the third dream at the latest, his proof was invariably correct. Twice already, he had dreamt that he had proven that mathematics was consistent. How fortunate it thus was, he chuckled, that he had not dreamt it a third time! Indeed, had von Neumann found a formal proof of consistency, that would have meant, thanks to Gödel's paradox-grazing result, that mathematics is *inconsistent* after all.

## GÖDEL NUMBERING

Gödel's proof occupies dozens of pages, yet it is based on a stunningly simple basic idea. In any formal system, mathematical statements are simply strings of symbols. Gödel found a systematic way of turning any such string into a unique integer (always a huge integer, as it happens, but that detail didn't affect anything). The given string of symbols determined its integer uniquely (that is, the string could be mechanically "encoded" as an integer), and vice versa—given a large integer, if it represented a string of symbols, then that string was uniquely determined (the large integer could be mechanically "decoded," if you will). These large integers were later called the *Gödel numbers* of their associated strings, and the encoding/decoding recipe was called *Gödel numbering.*

The next idea came from the fact that proofs in *Principia Mathematica* (or any other formal system) were built up regularly, in a way that could be mirrored in the world of numbers. Thus for any theorem there was a *theorem-number* that could be defined in terms of addition, multiplication, and other mathematical concepts. Therefore, provability of a string in a formal system corresponded to a purely mathematical property of a very large number, and this property could be talked about using the notation of *Principia Mathematica.* In other words, just as one can assert of a number $N$ that it is a square or a cube or a prime number, and prove all sorts of theorems about such notions (e.g., "There are infinitely many primes"), so one can assert of $N$ that it is a theorem-number, and there are all sorts of theorems about this more complex notion of "theorem-numberhood" (e.g., "There are infinitely many theorem-numbers"). And in this way, *Principia Mathematica* acquired the ability to talk (in code) about the provability or the nonprovability of strings in *Principia Mathematica* itself. Talk about a snake biting its own tail!

Gödel's coup de grâce was the construction of a special mathematical proposition G that asserted that a certain string having Gödel number $g$ is unprovable—meaning that it cannot be formally derived from the system of axioms of *Principia Mathematica*. And astonishingly, Gödel managed to arrange things so that the integer $g$ was precisely the Gödel number of the statement G ("somewhat fortuitously," as he slyly noted).

Proposition G thus asserts its own nontheoremhood, or unprovability, inside *Principia Mathematica*. You might paraphrase G as saying "I am unprovable inside *Principia Mathematica*." Now is G true, or is it false? Is it provable, or is it unprovable? Well, if G is actually proven, this leads to a contradiction; and conversely, if its negation, not-G, is proven, this leads to another contradiction. This sounds like a disaster, leading inevitably to a self-contradictory system—but wait! Perhaps neither G nor not-G is provable. In that case, the system (*Principia Mathematica*) is saved from being self-contradictory, but only at the price of being forever unable to decide which of G and not-G it "believes."

In short, if *Principia Mathematica* is consistent, meaning that it will never prove two statements that contradict each other, then neither G nor not-G can ever be proven by using its axioms and rules. And since G claims about itself that it cannot be formally proven, what it claims is true. Gödel's demonstration of G's truth, however, relied on the *meaning* of G. It was *not* a proof in the sense of *formal* provability; that notion is limited to proofs that are defined by following the formal rules of *Principia Mathematica*. Gödel had managed to show that his strange proposition G was true by thinking *outside* the rules of the system.

The writer Hans Magnus Enzensberger (born in 1929) compared Gödel's proof, in one of his poems, with the droll tale of the infamous Baron Münchhausen, who had claimed to be able to lift himself out of a swamp by pulling on his own pigtail. "But Münchhausen was a liar, whereas Gödel was right."

The sentence "I am unprovable" does not look like a normal mathematical statement at all. After all, mathematicians usually deal with numbers, figures, or functions—not with abstract philosophical-sounding notions such as *provability*. But via Gödel's recipe for encoding and decoding, the formal provability of a proposition was seen to correspond precisely to an arithmetical property of the proposition's Gödel number.

It was soon shown that in addition to Gödel's statement G, which, via Gödel numbering, truthfully asserts of itself "I am unprovable," there are infinitely many other true propositions that are unprovable but sound more familiar to mathematicians—for instance, there are unprovable true statements whose form is similar to the Goldbach conjecture, which Gödel mentioned at the Königsberg meeting. The example is instructive.

Back in 1742, an amateur mathematician named Christian Goldbach surmised that every even number greater than 2 is the sum of two prime numbers. You can check this claim out, in part. Thus: $6 = 3 + 3$, $12 = 5 + 7$, and so forth. By means of the fastest computers on Earth, Goldbach's conjecture has been verified for all even numbers up to 300,000,000,000,000,000 (seventeen zeros). This virtuosic feat does not prove the Goldbach conjecture, of course, or even come close to doing so: after all, there still remain infinitely many even numbers that have not been checked!

A computer program can examine one even number after another and check whether it is the sum of two prime numbers. If the Goldbach conjecture is wrong, then the computer will reveal this fact, sooner or later—we just have to wait until it reaches an even number that happens *not* to be the sum of two primes. But if the conjecture is right, then the computer will keep on grinding away, and we will sit there waiting till kingdom come, not knowing whether it will at some point hit an exception. Clearly, Goldbach's conjecture cannot be proven in this way. The computer program can *falsify* the conjecture (if it is indeed false), since that would take only a finite amount of time, but it cannot *verify* it (if it is in fact true), since that would take an infinite amount of time.

In a similar vein, a computer could inspect any formal mathematical sequence of formulas—such a sequence being, after all, nothing but a string of symbols—and check whether it is a formal proof. Such checking is a purely mechanical act, requiring no understanding of what any of the symbols stands for. The computer could inspect all possible such texts, one by one, arranged by length in an infinite list. This list would never come to an end. But if either the proposition A or its negation, not-A, were provable, then the computer would eventually hit on the corresponding proof: this proof, after all, has to occur somewhere in the list. This seems at first to suggest that all mathematical statements can be mechanically decided. Just start the computer churning and see which one it hits first—a proof of A or a proof of not-A. Piece of cake!

There is a catch, though—which is the fact that, thank to Gödel's incompleteness theorem, there exist true propositions for which there is *no formal proof.* If A is such a formula, then the computer will thus go on endlessly with its checking process, and it will never discover whether A is true or false, because neither A nor its negation has a proof!

This means that Gödel's incompleteness theorem foreshadows deep ideas about the absolute limitations of computer programs. Such limitations were first unearthed by the English mathematician Alan Turing (1912–1954) in 1936. What Gödel, Turing, and other brilliant logicians worked out over a period of years was that the process of mathematical thinking can never be entirely captured by purely formal axiomatic reasoning. In this sense, mathematics is an inexhaustible fountain.

A few decades after Gödel's and Turing's pioneering articles, a game of logical reasoning named Sudoku captivated millions. Its enormous popularity shows that logic can be highly entertaining. Moreover, Sudoku puzzles can be thought of as being analogous to formal mathematical theories à la Hilbert, as we shall now see.

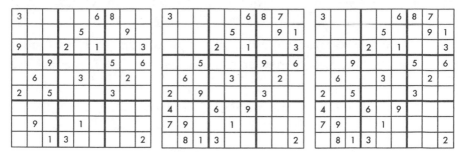

FIGURE 8.6 Some Sudokus: (a) incomplete, (b) inconsistent, and (c) "normal."

The goal in a Sudoku puzzle is to place the digits from 1 to 9 in eighty-one small boxes forming a 9 × 9 square, in such a way that no row, no column, and none of the 3 × 3 subsquares contains the same digit twice.

In a few of the boxes, the digits are placed, for free. In our analogy to formal systems, then, these preset digits correspond to the "axioms." We expect of a decent Sudoku puzzle both that it can be solved and that it can be solved in one way only. That is, for each of the eighty-one boxes, there is supposed to be exactly one unique digit that can go in it. If there is some box that cannot be filled in any way, then the "axioms" lead to a contradiction. Or if there is a box that can be filled in two or more different ways, then the "axioms" are not complete: the problem of discovering which digits go in which boxes is unanswerable. One would need more "axioms" in order to specify the digits uniquely.

An ideal Sudoku puzzle thus should be both consistent and complete. However, whoever expects the same from a formalization of mathematics will be disappointed. That is what Gödel showed in 1931.

Later, Ludwig Wittgenstein summed matters up as follows: "Gödel's theorem forces us to view mathematics from a new perspective." (Most scholars agree, however, that neither Wittgenstein nor Russell ever really understood Gödel's ideas.) With 20/20 hindsight, the idea that all the true statements (and no false statements) of mathematics could be churned out, one by one, by a computer program seems as strange as the alchemists' hope of finding the philosopher's stone. Mathematicians cannot be replaced by rigid automata. The same is presumably true for some other jobs too, but no one has proved it. (And as an aside: in 2013, the mathematician Harald Helfgott proved that every *odd* number is the sum of *three* primes. But we are still not yet at the stage of proving Goldbach's conjecture.)

Occasionally Gödel is misinterpreted. This, of course, is almost inevitable, and it is a fate that Gödel shares with Darwin and Einstein and other great pioneers. Some people claim that Gödel proved that mathematics is inconsistent. This is absolutely not the case. What he showed is that consistency cannot be *formally proven*. Analogously, you might find it impossible to produce a loophole-free legal argument to prove that you have never in your life committed a murder—never ever. In spite of that, we standardly presume people to be nonmurderers. The same holds for mathematics. Nobody seriously believes that mathematics is contradictory. To paraphrase the words of a French mathematician who neatly summed up the creed of his craft: God made mathematics consistent, and the Devil made sure that this cannot be proven.

## The Parallel Campaign

Karl Menger missed out on the dramatic events of the Königsberg meeting, because at the time he was touring the United States as a visiting professor. But the moment he learned about Gödel's discovery, he realized its enormous importance and he spread the news wherever he could.

Although he was barely four years older than Gödel, Menger had become a sort of mentor to Gödel, and an almost fatherly friend. When he returned to Vienna, he and Gödel slowly drifted away from the Vienna Circle. To their taste, it had taken on too much of the flavor of Wittgenstein and of Neurath; there was too much of a cult about the former, and too much politics in the latter.

Karl Menger did not wish to be counted among the leftists in the Circle. He was enthusiastic about the abstract art in the pictorial statistics streaming out of Otto Neurath's museum, but he could never warm up to Neurath's dreams of "full socialization," or nationalization. Such pie-in-the-sky fantasies were too far removed from the liberal worldview of his father and the ideas developed by the Austrian school of national economics. The latter's more down-to-earth approach was based on the needs and decisions of individuals, rather than on impossibly idealized collectives, classes, and masses.

As for Wittgenstein, Menger was certainly not prepared to share Schlick's almost religious veneration of the refractory genius. Menger deemed the first three quarters of the *Tractatus* to be impenetrably turgid. And anyway, how could one find so much to say about the unsayable?

At the conclusion of an evening in the Vienna Circle during which Schlick, Hahn, Neurath, and Waismann had held forth at great length about language, Menger wistfully said to Gödel, on their way home: "Today you and I have once again out-Wittgensteined these Wittgensteinians: we kept silent." To which Gödel replied: "The more I think about language, the more it amazes me that people ever understand each other."

The two young skeptics still showed up occasionally at the Thursday sessions of the Circle, but more and more rarely. Their interests shifted increasingly toward the Vienna Mathematical Colloquium, which had been founded by Menger. It was designed along the lines of the Vienna Circle—a parallel campaign, to use Robert Musil's term. And indeed, like the Circle, the colloquium had initially been set up by students attracted by the ideas of an inspiring young professor—in this case Menger, not Schlick.

The Vienna Mathematical Colloquium started up its own journal, the *Ergebnisse eines Mathematischen Kolloquiums* (Results of a Mathematical Colloquium). It was published once a year and was mostly based on the manuscripts of the talks. The most common topics of the articles it contained were dimension theory, mathematical logic, and mathematical economics.

Among the main participants in the colloquium were Abraham Wald, Franz Alt, Georg Nöbeling, and Olga Taussky. Nöbeling (1907–2008), who was German, was Menger's favorite student, and understandably so, since he had greatly extended Menger's early results on dimension theory.

The Romanian mathematician Abraham Wald (1902–1950) came from a family of ultraorthodox rabbis and had been home-schooled. By the time he took up mathematics, he was considerably older than the usual undergraduates. But he progressed all the faster. After only three semesters, he obtained his doctorate. Karl Menger, his professor, was of about the same age. He described his student as "small and thin, obviously poor, looking neither young nor old, a strange contrast to the lively beginners."

Franz Alt (1910–2011), the son of a Viennese lawyer, studied mathematics under Karl Menger. After completing their doctorates, both Alt and Wald were jobless, and to make ends meet, both had to give private lessons. Karl Menger helped them as best he could. The economic crisis was weighing down heavily on the town.

Olga Taussky (1906–1995) was the only female member of the group. After getting her PhD, she found a job in Göttingen, working as an assistant in preparing the publication of Hilbert's *Collected Works*. There she met Emmy Noether (1882–1935), the foremost female mathematician of her era. Noether was in fact the first woman in Germany ever to obtain the right to lecture at university. To be given this honor, however, she had to wait for the advent of the Weimar Republic. In prior years, her applications for a *Habilitation* had always been rejected as unsuitable—this in spite of the vigorous support of David Hilbert, who had famously quipped: "A faculty is not a public bath."

Now Olga Taussky was able to attend the lectures of Emmy Noether, who had developed a revolutionary new way of looking at modern algebra. Even before the Weimar Republic met its end, however, Taussky was advised, during a brief visit to Vienna, that she should abstain from returning to Göttingen. "Political reasons" had reared their ugly head. Olga belonged not only to the wrong sex but also to the wrong race.

Marooned in Vienna, Olga Taussky found no job. But Hahn and Menger decided to do something about it. They organized a series of public lectures on science. The entrance fees were hefty—roughly in the same price range as opera tickets. With the proceeds, the two mathe-

FIGURE 8.7 Karl Menger and Kurt Gödel visit Olga Taussky.

maticians were able to finance a stipend for their friend and colleague Olga Taussky.

The cycle of lectures was titled *Crises and New Foundations in the Exact Sciences,* and it proved remarkably successful. Hans Hahn gave a talk on the often misleading role of intuition in mathematics, Karl Menger spoke on the concept of dimension, Werner Heisenberg described quantum mechanics, and the physicist Hans Thirring speculated about the possibility of humans venturing into outer space. Would it ever happen? By all means, said Thirring, but certainly not in this century. But in the end, he lived long enough to witness humans walking on the moon.

## No Trifling Matter

In this brilliant constellation of unemployed young mathematicians, Kurt Gödel was the unquestioned star. The manifold profound implications of his incompleteness theorem were becoming increasingly clear. The German Ernst Zermelo, who once had been such a headache to Ludwig Boltzmann, thought for a while that he had discovered a mistake in Gödel's proof, but the latter was easily able to clear up the misunderstanding.

"My dear Göderl," wrote Gödel's friend Marcel Natkin, using an endearing diminutive, "unjustifiably, I am terribly proud . . . So, you have shown that all Hilbert-style formal systems contain undecidable problems. That is no trifling achievement!" At the time, Natkin was an up-and-coming photographer in Paris.

Soon Gödel stressed that his incompleteness theorem applied not merely to certain *specific* axiomatic reasoning systems, such as *Principia Mathematica,* but to *any* consistent axiomatic system in which the notions of *natural number, addition,* and *multiplication* are defined. In other words, there is no finite set of axioms from which the whole of number theory can be deduced.

In 1932, Karl Menger gave a public lecture on "The New Logic" at the University of Vienna, attracting a large crowd. In this manner, he became the first person to introduce Kurt Gödel's results to a general audience. This lecture of Menger's was one of the highlights of the cycle *Crises and New Foundations in the Exact Sciences.* To Kurt Gödel himself, this event was so meaningful that he kept his ticket (standing room only!) for the rest of his life.

In the spring of 1932, Kurt Gödel landed another coup. Since intuitionism restricts the use of the Law of the Excluded Middle, the set of theorems to which it gives rise has to be a subset of the set of all theorems of classical mathematics. It was precisely this limitation that had caused Hilbert to protest so vehemently. And yet Gödel now had shown, in a certain sense, the exact opposite. Intuitionism is only *apparently* narrower. If one reinterprets the formal symbols in a suitable manner, it turns out that the theorems of classical number theory become a subset of those of intuitionistic number theory. What, then, is the nature of mathematical truth and provability? With results like this, it was growing ever subtler.

The deeply ideological conflict about what is and what is not allowed in mathematics started to lose its pressure, like a tire punctured by a nail. Which interpretation holds is merely a matter of convention. Gödel's astonishing result coincided exactly with Menger's views. For a long time already, Menger had opined that dogmatic prescriptions about what is permissible and what is forbidden have no place in mathematics. All that matters is that one should specify precisely which ground rules one is going to use.

When Kurt Gödel reported on his new results to the Vienna Mathematical Colloquium, Oswald Veblen (1880–1960) was among the few in his audience. Veblen, the son of the famous Thorsten Veblen, who had made conspicuous consumption an issue, was one of the most eminent mathematicians in the United States. At that moment, however, he was touring through Europe on behalf of the newly founded Institute for Advanced Study in Princeton. His mission: to look out for suitable scientific rookies. Menger had convinced him to attend Gödel's talk. Veblen was deeply impressed and immediately put Gödel on his short list.

In 1932, Gödel applied for his *Habilitation* at the university. Some years earlier, Vienna's mathematicians had decided that it was necessary to let four years elapse between one's PhD and one's *Habilitation*. Luckily for Gödel, this interval was nearly over. He proposed "The Construction of Formally Undecidable Propositions" as the subject of the lecture that was required for the examination.

Hans Hahn summed up the lecture in his report: "A scientific breakthrough of the first order, generating the highest level of interest in expert circles. It can be said with certainty that it will hold its place in the history of mathematics. Thus it is shown that Hilbert's program for proving the consistency of mathematics cannot be carried through."

In the spring of 1933, Kurt Gödel became a *Privatdozent* (an independent lecturer at the university, but essentially without salary). Due to the terrible state of the economy, there was no hope of an appointment. Not that Gödel needed one: he had enough money to live on.

Moreover, he had been invited to spend the next year at the Institute for Advanced Study in Princeton. Veblen had not forgotten about him.

The research institute in Princeton was privately funded by Louis and Caroline Bamberger (a department store magnate and his philanthropist sister), who had decided to make their generous gift just at the right moment—shortly before the great crash of 1929. Among the first permanently appointed members were Albert Einstein and John von Neumann. Both scientists had had to flee from Berlin, as in 1933 the Nazis had come to power.

# *The Circle Squeezed*

*Vienna, 1930–1933: Latecomer Popper turns philosopher, advocates falsification, locks horns with Circle. Schlick locks Popper out, but accepts book by Popper while rejecting one by Otto Neurath. Neurath threatens to sue. Waismann's Wittgenstein book faces endless delays. Carnap moves to Prague, sees no morals in logic. Wittgenstein rages, feels plagiarized by Carnap.*

## POPPER'S PRIVATE REVOLUTION

For a while in his youth, Karl Popper had been apprenticed to an aged Viennese cabinetmaker with a vast store of knowledge of the sort that comes in handy for crossword puzzles. The old man used to say, with modest pride and a Viennese twang: "Go ahead, m'boy—ask me whatever you like: I know it all!"

Popper later wrote that he had learned more about the theory of knowledge from his dear, quasi-omniscient master than from any other of his teachers. "None did so much to turn me into a disciple of Socrates."

Socrates is reputed to have declared: "I know that I know nothing." Popper liked to add: "and frequently not even that." There is no secure knowledge. And yet, for someone allegedly holding such a modest view, Popper was remarkably opinionated and self-assured.

Karl Popper, who was born in 1902, was not supposed to become a cabinetmaker. After all, his father was among the most highly respected lawyers in Vienna, and the family lived in the center of the city, only one block away from Saint Stephen's cathedral, in a venerable old town house that had been renovated during the Baroque era. At that time, it had been owned by Samuel Oppenheimer (1630–1703), the banker who financed the emperor's campaigns against the Turks— an unbroken string of victories. To grow up in such a historical place was quite special.

When Popper turned sixteen, however, the great monarchy on the Danube was falling to pieces, and there were no more victories in sight. The young emperor Karl renounced all participation in affairs of state and moved out of the Hofburg, the old palace of the Habsburg Empire. On the nearby Herrengasse, a National Assembly constituted itself. The Republic was proclaimed on November 12, 1918, a cold and rainy day on which chilly gusts swept harshly down the Ringstrasse.

Karl Popper witnessed the riots and demonstrations of those gray days as if from a front-row seat. He heard the chanting of slogans just around the corner and the whistling of bullets just over his head. He decided to quit school, partly to stage his own private revolution, as he later wrote, and partly because his mathematics class was moving at a painful snail's pace. (Geometry was his favorite subject.) He enrolled in the university as a part-time student.

More than seventy years later, Popper wrote: "It was in the middle of the winter 1918–1919, most likely in January or February, when I first set foot, hesitantly and almost trembling, on the sacred ground of the Mathematical Institute of the University of Vienna in the Boltzmanngasse. I had every reason to be apprehensive."

Not having finished secondary school, young Popper was registered only on a provisional basis. The other students had all passed the *Matura* (the exam allowing one to graduate from secondary school), were properly enrolled, and had already progressed well beyond his level. For a change, the lectures moved too quickly for Popper, and he gave up in frustration.

He joined a communist youth group, but that was only a short-lived fling. After witnessing a demonstration that ended with the police shooting several young workers, Popper renounced his Marxist creed. He did not wish to contribute in any way to an intensification of the class struggle, he later explained.

In 1922, Popper belatedly passed his *Matura* exam. He no longer had any problems with the university now. He sampled courses all over the intellectual map: in history, literature, psychology, medicine, physics, and philosophy. But mathematics remained his favorite subject: "Only the department of mathematics offered really fascinating lectures," he wrote. "I learned the most from Hans Hahn. His lectures attained a degree of perfection which I never encountered again. Each lecture was a work of art: dramatic in logical structure; not a word too much; of perfect clarity; and delivered in beautiful and civilized language. Everything was alive, though due to its very perfection a bit aloof." Popper added: "Mathematics was a huge and difficult subject, and had I ever thought of becoming a professional mathematician, I might soon have been discouraged. But I had no such ambition."

And so, what precisely *was* the young man's ambition? As he was not yet sure, he dabbled in various occupations. As was alluded to earlier, for a while he tried his hand at being an apprentice cabinetmaker (like Ernst Mach, but at a much later age), and then he turned to social work with abused children in the clinic of the psychotherapist Alfred Adler (1870–1937), who had strayed from Freud's fold. Adler studied the defense mechanism called *compensation*, whereby people's choices in life are unconscious reactions against real or imagined deficiencies, such as a small man making up for his physical stature by aggressively striving for social status. (Popper, it might be mentioned, was on the small side in terms of stature.)

After a while, the young apprentice decided to become a teacher. He enrolled at the newly founded Pedagogical Institute in Vienna. This was a lucky break, for there he met a fellow student named Josefine Anna Henninger, who went by the name of Hennie. The two discovered a deep mutual affinity and within a short time were married. They lived with Hennie's parents in Lainz, a suburb of Vienna. Money was scarce. Popper's father had lost all his savings in the runaway inflation after the war, as had most of Vienna's middle class.

During his training as a teacher, Popper assiduously kept attending lecture courses at the University—in physics, mathematics, psychology, and philosophy. In 1928, he earned his doctorate with a thesis called "On the Problem of Method in the Psychology of Thinking." This work was supervised by Karl Bühler, who had joined the faculty at the same time as Moritz Schlick and whose Viennese School of Psychology had earned worldwide renown.

His dissertation, as Popper put it, was "a kind of hasty last-minute affair. I have never again even glanced at it."

Whoever does glance at it, however, is immediately struck by the critical acumen of the young author, and by the ferocity of his fighting spirit. No holds barred! Literally from the first line, Popper rejected the views of Moritz Schlick on knowledge and science. Such an unbridled attack was a high-risk strategy, of course, particularly since Schlick happened to be the second reader of the thesis.

Moritz Schlick, however, did not take up the challenge. He may have been too busy to lock horns with an unknown student. He merely assented to the high grade proposed by Bühler, going along with the latter's lukewarm words of praise, to the effect that "Popper's work is clearly of a secondary and literary nature. On the other hand, it expresses the well-read author's great spontaneity and facility in combining and comparing."

Thus the thesis was accepted. However, in the *Rigorosum* that followed (a grueling oral exam on the history of philosophy, required for the doctorate to be granted), Popper performed so poorly that he was sure that he had failed. But once again, Schlick and Bühler were generous. Popper wrote: "I could hardly believe my ears when I was told that I had passed with the highest grade: "*Einstimmig mit Auszeichnung*" [unanimously passing, with distinction]. I was relieved and happy, of course, but it took quite some time before I could get over the feeling that I had deserved to fail."

FIGURE 9.1 Teacher Popper on an excursion.

The next time he had to write a thesis, this one for a teacher's diploma, Popper returned to his old love: "Axioms, Definitions, and Postulates in Geometry." No problems arose in that familiar territory. Everything went swimmingly.

In 1929, Popper became a teacher in an elementary school, just as Ludwig Wittgenstein had done a few years earlier. The new Dr. Karl Popper instructed youngsters ranging from age ten to age fourteen in mathematics and physics. His interests slowly drifted away from the study of subjective processes of thought, and closer toward the logic of science. Everything seemed to be pushing him in the direction of the Vienna Circle.

## POPPER PUTS TWO AND TWO TOGETHER

Many years earlier, the master carpenter who "knew it all" had stimulated his young apprentice's spirit of fearless inquiry. And now, as a newly baked philosopher, Karl Popper turned his critical mind to what he saw as the most exciting facet of human knowledge—the turf on which knowledge is scrutinized and contested, on which it grows and develops: namely, science. What are the defining traits of science? This was a favorite topic of the Circle, and rightly so. Anyone who crusades for a scientific worldview should be able to separate science from nonscience.

Popper wrote: "It was only after my Ph.D. examination that I put two and two together, and my earlier ideas fell in place. My view implied that scientific theories, if they are not falsified, forever remain hypotheses or conjectures. This consideration led to a theory in which scientific progress turned out not to consist in the accumulation of observations, but in the overthrow of less good theories and their replacement by better ones—in particular, by theories of greater content."

Popper became a fervent opponent of the view that natural science is based on acts of induction. He firmly rejected the claim that induction—that is, the passage from specific local observations to the assertion of all-encompassing generalities—could ever provide certainty. We can repeat observations or carry out experiments until we are blue in the face, yet without ever being able, despite all that work, to assert, with total security, any general law.

The fact that induction did not have the same logical status as deduction had not escaped others, of course. Moritz Schlick, for instance, had claimed: "Induction is merely guesswork guided by method—a psychological, biological process whose study has nothing to do with logic."

Quite evidently, for these thinkers, induction could not possibly serve as the defining hallmark of scientific methodology. Where else, then, might one find that central defining essence of science? Trying to distinguish pseudoscience from science became a favorite topic with Popper, although he was not willing to devote his precious time to refuting standard pseudoscientific types of silliness, such as alchemy or astrology. Nor did he deign to consider the claims of parapsychology, nor the notion that the Earth was hollow, with humans living on its inside, nor the doctrine that the universe is the arena of a cosmic struggle between fire and ice, nor the thesis that the pure Aryan race is forever threatened by subhuman lechers. In the frantic 1920s, some of these lurid creeds attracted a following, of course, and sometimes even a considerable band of believers. But Popper did not turn to such far-fetched types of lunacy for his examples of pseudoscientific thinking.

Instead, as was his usual style, he picked his opponents from the heavyweight class. Marxism and psychoanalysis were the two hottest topics of debate in Vienna at that time. Thus it was against those two that Karl Popper launched his attacks. He was not going to accept either one as part of science; indeed, he was going to tackle them both head-on. For a while, even Darwin's theory of evolution aroused Popper's suspicions. Popper claimed that anyone who had acquired even a modicum of debating skills could easily shield these doctrines against all criticism, no matter how sharp. And this trait of invulnerability or undefeasibility, he pointed out, was precisely what disqualified such theories from being genuine branches of science.

A theory, no matter how grand, is always just a hypothesis, and cannot ever have a higher status than that, although a theory can change in status by being refuted outright, thanks to empirical observations. In Popper's view, then, it was the testability—or rather, the *falsifiability*—of a claim that constituted a far more reliable demarcation line between science and pseudoscience than did the use, or the lack of use, of that dubious process called *induction*.

In contrast to this view, most of the Vienna Circle held that a claim's empirical testability, or lack thereof, tells you whether the claim belongs to science or merely to metaphysics. In other words, to most Circle members, the importance of verifiability was that it draws a dividing line between meaningful and meaningless sentences.

Popper scornfully dismissed this view. In the first place, he was not greatly interested in endless debates about which sentences are meaningful and which are not. Meaningful for whom? And secondly, for him, verification meant not merely passing some test but rather the final confirmation of a theory, the definitive sealing of its truth, which for him was a nonsensical pipe dream.

As a matter of fact, no member of the Vienna Circle would ever have construed the word in that manner. But Popper was headstrong and was not prepared to listen to arguments about possible misunderstandings. "Never let yourself be goaded into taking seriously problems about words and their meanings." With this personal maxim, Karl Popper hoped to save philosophy from the clutches of those who were in love with the *critique of language*. This was a trend that Popper despised, seeing it as an aberration deflecting good thinkers away from the pursuit of real issues.

## Popper Circles the Circle

The first mentor of the young philosopher was Heinrich Gomperz, who as a young man had paved the path that had brought Mach back to Vienna. "Gomperz invited me from time to time to his house," as Popper wrote, "and let me talk." However, "we did not agree on psychoanalysis. At the time he believed in it, and he even wrote for *Imago*." This was the name of Sigmund Freud's journal.

Gomperz introduced the talkative young Popper to Viktor Kraft, a librarian at the University of Vienna, who for many years had been a member of the Vienna Circle. Popper also made friends with Friedrich Waismann, Schlick's librarian, and he gave his first philosophical presentation in the flat of Edgar Zilsel. He was plagued by intense stage fright, he reports, but this did not keep him from ruthlessly assailing the views of the Vienna Circle. And he did quite well in the follow-up discussion. As a result, other groups that formed a sort of halo around the Vienna Circle started to invite him to their meetings.

His acquaintance with Herbert Feigl marked a turning point for Karl Popper. As he later wrote, "the meeting became decisive for my whole life." Indeed, after one intense all-night discussion session, a thoroughly exhausted Feigl suggested to Popper that he might publish his ideas as a book. The thought had never occurred to Popper before that.

Apart from Feigl, though, everyone else was totally opposed to the idea of a book—in particular, Popper's wife, Hennie, who would have vastly preferred to use any spare time with her husband to go skiing and mountain climbing. The Vienna Woods and the nearby mountains of

FIGURE 9.2 Karl and Hennie Popper, here without typewriter.

the Rax and the Schneeberg were ideal for such excursions. "But once I started the book, she taught herself to type, and she has typed many times everything I have written since."

Indeed, the Poppers would always carry a typewriter along on their hiking trips. Whenever they were relaxing in the guest garden of some rural inn, Hennie would haul out the device and start typing away.

The title of Popper's book-in-the-making was *The Two Basic Problems of the Theory of Knowledge.* These two problems were easy to state: *induction* and *demarcation.* And Popper's two basic answers to them were equally stark: there is no such thing as induction, and falsifiability is what constitutes the demarcation line between scientific and nonscientific theories.

As Popper wrote: "From the beginning, I conceived the book as a critical discussion and correction of the views of the Vienna Circle." Increasingly, Popper became the "official opposition" of the Vienna

Circle, as Neurath acknowledged. But Popper remained an outsider: "I was never invited, and I never fished for an invitation."

On another occasion, he wrote: "I never was a member of the Vienna Circle, but it is an error if one assumes that my nonmembership in the Circle was a consequence of my opposition to its ideas. That is not true. I would have loved to become a member of the Vienna Circle. But the fact is, Schlick never invited me to participate in his seminar. Being invited was the only avenue through which one could become a member of the Vienna Circle."

That such an invitation never materialized appears to have been a serious blunder on the part of Moritz Schlick. Indeed, Popper's great talents had long been blindingly obvious. But Schlick feared that Popper's aggressiveness and obstinacy would destroy the atmosphere of goodwill that was so central to the spirit of the Vienna Circle. Schlick had witnessed Popper operating at full steam in December 1932, during a meeting of the Gomperz Circle.

It may be that Popper was suffering from stage fright on that occasion, but if that was the case, it did not in any way take the edge off his aggressive spirit. He immediately launched into a virulent attack on Wittgenstein's ideas. He criticized the latter's view that a proposition can be meaningful only if it describes a possible fact. Wittgenstein's claim that everything else was "unsayable" was branded by Popper as a suppression of free speech, which he likened to the dogmatism of the Catholic Church.

Schlick's patience was quickly used up, and he angrily stormed out of the meeting. He was willing to listen to any sort of criticism leveled at himself but not to sit by passively while savage attacks were directed against Wittgenstein.

Others, too, were irritated by the constant brashness of the young interloper. Kurt Gödel, who was by no means a devotee of Wittgenstein, wrote to Karl Menger: "Recently I met one Herr Popper (philosopher) who has written an endlessly long work which, so he claims, solves all philosophical problems. He tried eagerly to attract my interest. Do you think he is any good?"

But Popper was not one to be worried by the doubts of others. After he had finished writing his "endlessly long" book on the basic problems in philosophy, he managed to get Feigl, Carnap, Schlick, Frank, Hahn, Neurath, and Gomperz to read at least parts of it. And this manuscript, mind you, was a scathing attack on the Vienna Circle! The criticism was troubling—particularly in those spots where Popper was right.

Even Herbert Feigl, who early on had urged Popper to put his ideas into a book, found afterward that "the whole thing leaves an unpleasant aftertaste. Here is Popper, barely a step ahead of us, yet presuming to preach to us from far above. His undeniable verbal virtuosity, his untamable energy, and his untiring hunger for debates (which has cost me many a night's sleep) have the power to bludgeon any upright opponent to death. But what stands out most strongly in my memory as disagreeable is his constant arrogance—his lust to cash in on every one of his 'triumphs' and carry them home, to be sealed and confirmed again and again."

The ever level-headed Carnap proved, as usual, to be willing to listen to any criticism. He wrote to Schlick: "Gomperz says that on the points where Popper agrees with us in the Circle, his presentation is more accessible; and that on those points where he criticizes our views, he is in closer correspondence with the techniques used in the sciences. It seems to me that Gomperz is not completely wrong here. I personally think that we can truly learn something from Popper's commentaries."

"Popper is in a terrible hurry," groaned Moritz Schlick. But he was willing to live and let live; indeed, he accepted Popper's book for the series of *Writings on the Scientific Worldview*. And thus, after a drastic reduction in size, which was required by the publisher, the volume appeared in 1934, under the title *The Logic of Scientific Discovery*.

Nowhere else could Popper's first book have been published. Moritz Schlick was obviously fully able to appreciate the merits of Popper's thoughts—he just could not stand the author as a person. It might be noted that the full text of Popper's *Two Basic Problems* finally appeared in print only forty-five years later.

"It is an exceptionally intelligent work," wrote Schlick about *The Logic of Scientific Discovery*, "but I cannot read it with unalloyed pleasure, despite the fact that I think that the author is almost everywhere in the right, if interpreted sympathetically. However, his presentation appears to me to be misleading. Indeed, in his unconscious urge to try and make his own contributions as original as possible, he takes very minor examples of our group's positions (sometimes just terminological points), distorts them *ad libitum*, and then he paints these views, concocted by himself more than by us, as fatal blunders on our part on major issues of principle (and he sincerely believes that this is what they are). This warped way of doing things serves the whole perspective very poorly. With time, though, his self-esteem will decrease, no doubt."

Schlick's optimistic prediction was never verified.

## A Shock for Schlick

If Moritz Schlick found that he could not read Popper's book "with unalloyed pleasure," he soon fared even worse when faced with a manuscript on the scientific basis of history and economics, submitted by Otto Neurath. After reading it, Schlick felt he had no choice but to oppose its publication in the book series that was entrusted to him. He confided his feelings to his co-editor Philipp Frank: "I started reading it with great hopes, but I was in for a great shock! I found the presentation so tasteless and so unsuited to our aims that I am convinced that no one could ever take the text seriously, except completely blind supporters."

It is no trifling matter to reject the manuscript of a close colleague, especially one suffering from having been unfairly excluded from an academic career. And Otto Neurath, moreover, was not the sort to take "no" for an answer. He attacked Schlick's verdict, insisted on a written referee's report, and appealed for the support of Philipp Frank, his trusted comrade and friend from their student days together.

Schlick thus had to justify his decision. "After a most conscientious examination and most careful reconsideration," he wrote to Frank, he still could not shoulder the responsibility for accepting Neurath's book in his series; and this was "not because of the opinions it expresses—in fact, I mostly agree with the author's ideas—but because of its style, which lends a completely unscientific and unserious character to the book. . . . The writing is designed entirely for purposes of propaganda and persuasion. This is blatantly obvious from the fact that roughly one half of the sentences in it—I do not exaggerate—end with exclamation marks; and although these symbols can easily be replaced by full stops (and by now Neurath probably has done just that), the exclamatory character of these statements remains unchanged."

"Almost every argument," Schlick complained, "follows the same scheme: 'Such and so must be the case, because any believer in the unity of science rooted in materialism has to embrace this viewpoint'; or else 'Such and so is the case because any contrary opinion would be metaphysics and theology.' And when almost every page triumphantly exclaims that one can do very well for oneself without God or angels, then this becomes extremely boring for the sympathetic reader, and very dogmatic for the antithetical reader, and ridiculous for both."

The irritation of Schlick was heightened by the fact that he had been obliged to postpone his summer vacation because of this onerous duty. Neurath's obstinacy was the last straw. He had promised to submit to Schlick's book series a justification of the scientific worldview; but instead, he had delivered nothing but a rabble-rousing tract. Schlick sadly wrote: "Can anyone believe that an opponent will be converted by this? If Neurath actually entertains such hopes, then it can only be due to a certain childishness and naïveté on his part. If an author feels the need to crusade against metaphysics on every single page, a reader cannot help but wonder whether the author, deep down inside, ever fully rejected metaphysics, after all."

In a long letter added to the report, Schlick unburdened himself to Frank:

Unfortunately, when I apprised Neurath of my decision in a coffee-house—and you should imagine all of this happening in the friendliest-seeming way—my arguments were met with no understanding at all.

On the contrary, he immediately told me that my view could be explained as the residue of bourgeois prejudices still afflicting me; and although he admitted that my intentions were pure, he called me an aristocrat, and haughty to boot. Should the publisher refuse publication on the basis of my report, he would sue, he said, and maybe he would publish the book together with a note stating that I had rejected it. And all this was said, I remind you, in the gentlest tone of voice.

On the next day—that is, yesterday—he talked to me for hours (!) on the phone, again in a friendly-sounding manner, but the whole time leveling harsh accusations against me. In his view, I have transgressed against the etiquette that is customary in the republic of scholars, and any neutral party would agree that I am morally in the wrong. He says I have no right to act as a censor. He also used this occasion to tell me a few more basic truths. In particular, he accused me of being asocial; while outwardly coming across as sympathetic, I am actually aloof and totally lacking in inner warmth.

And then, in great seriousness, Schlick asked Frank if he would request Albert Einstein for his judgment on the matter. On second thought, however, even this option appeared problematic to Schlick: "Neurath would merely claim that Einstein, too, had been spoiled by his contacts with members of the Prussian Academy of Science, or something else of that general sort."

Fortunately, Philipp Frank, who had been friends with Otto Neurath ever since their student days, came up with a way to appease Neurath and to resolve the conflict without either party losing face. Soon, a greatly relieved Schlick was able to report to Carnap that Neurath "was planning to submit a completely new manuscript, which would provide an introduction to theoretical sociology, a topic so familiar to him that he could write it up in the blink of an eye."

This time, Neurath used periods instead of exclamation marks, and his new book soon took its place in the series. (The old one, a Marxist pamphlet, remains unpublished.) The erstwhile tempest in a teapot left lasting traces, however. Neurath was now more convinced than ever that Schlick's character was pervaded by bourgeois inhibitions. And Schlick, for his part, concluded: "I do not believe that tactfulness and good taste are traits that ought to be manifested solely by members of the 'bourgeoisie.'"

## THE SERIES WITHOUT A BEGINNING

Neurath's book *Empirical Sociology* was published as Volume 5, and Popper's *The Logic of Scientific Discovery* as Volume 9, of the series called *Writings on the Scientific Worldview*.

There was, however, no Volume 1—and this oddity was not due to absentmindedness. The missing volume was Friedrich Waismann's *Logic, Language, Philosophy*. The book had already been heralded in the Vienna Circle's 1929 manifesto as an "easily accessible presentation of the main ideas" of Ludwig Wittgenstein's *Tractatus Logico-Philosophicus*. And some years before, Schlick had even written an introduction to it; however, that essay was still lying inert in the drawer of his desk, collecting dust.

Friedrich Waismann had been one of the original students who instigated the Schlick Circle. For a student, he was actually quite old. Born in Vienna of a Russian father, he first enrolled in mathematics and physics, but at age twenty-six he turned wholeheartedly to philosophy, enraptured by the newly appointed Professor Schlick.

Moritz Schlick in return provided Waismann with a position as a librarian and "scientific auxiliary"; at first, this position was totally unpaid, and later it was just extremely poorly paid. Fortunately, by teaching at Viennese institutions for adult education, Waismann was able to supplement this modest income, but not by much. His lectures were renowned for their great clarity and liveliness. Moreover, he excelled in running Schlick's practical courses. For this reason, he wound up being unofficially entrusted with regular lecture courses, such as an *Introduction to Mathematical Philosophy*, a topic that suited him well. But he was not a *Dozent* (lecturer); in fact, he did not even have a PhD. Sadly, Friedrich Waismann was unable to muster the inner strength to write a dissertation or face an examination. A strange lack of willpower bedeviled him.

As soon as Wittgenstein had consented to meet with members of the Vienna Circle, Friedrich Waismann fell totally under his spell. Schlick therefore encouraged his assistant to write a sort of *Tractatus for Beginners*. Waismann was ideally suited to this challenge, having actively followed all the debates in the Circle from the start.

Among Viennese students it was well known that one could learn a great deal from Waismann about Wittgenstein's ideas on logic. "I greatly like his careful reasoning and the focused way he leads the discussions," wrote a visitor from Berlin, the young Carl Hempel (1905–1997).

In January 1928, Schlick reported to Carnap: "Waismann has written up Wittgenstein's basic ideas in a very nice essay." A few weeks later, however, Schlick had to revise this news: "Unfortunately, Waismann's text is not finished yet. It is a pity that he seems almost unable to overcome the inhibitions that stand in his way whenever he writes something down; otherwise, his lucid mind could really be most productive."

Wittgenstein, too, held Waismann's "lucid mind" in high regard. Evidently, this student was someone worth talking to; he was someone who asked smart questions and who knew to stay silent while the thinker was wrestling for an answer. Wittgenstein reported to Schlick: "He [Waismann] waited with the greatest patience while I squeezed explanations from my brain, under pressure, drop by drop."

After his return to philosophy, Wittgenstein taught at the University of Cambridge. Having received a fellowship, he was now a professional philosopher, a fact that on occasion caused him great anguish. He did everything to dissociate himself from that guild. For instance, he would not teach in a lecture hall, but only in his private dwellings in Trinity College. Soon, a devoted group of followers, or perhaps disciples, congregated around him. He convinced more than a few of them to give up their studies in philosophy.

Throughout the 1930s, Wittgenstein returned quite regularly to Austria between university terms, either to the family estate in the Hochreith, deep in the Vienna Woods, an hour by car from Vienna, or to one of the town houses belonging to his three sisters. When in Vienna itself, he granted permission to Moritz Schlick and Friedrich

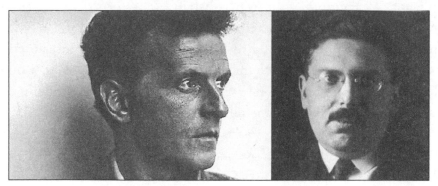

FIGURE 9.3 A tense collaboration:
Ludwig Wittgenstein and Friedrich Waismann.

Waismann to visit him in his quarters. They reported on these audiences in the meetings of the Vienna Circle, which Wittgenstein still kept studiously avoiding.

A young visitor from overseas, the philosopher Ernest Nagel (1901–1985) wrote: "Except to small, exclusive groups in Cambridge and Vienna, his present views are not accessible. . . . In certain circles, the existence of Wittgenstein is debated with as much ingenuity as the historicity of Christ has been disputed in others. . . . For various reasons, Wittgenstein refuses to publish."

Actually, Wittgenstein was keen on publishing his new thoughts, but in order to do this, he first had to arrange his rapidly growing flood of philosophical notes into some order. Over the years, thousands of pages of remarks and aphorisms had accumulated in his small room in Trinity College, and also in the mansions of his relatives. But since Wittgenstein kept revising his own positions and rearranging his notes, he was just as unable to bring his work to any conclusion as Robert Musil was with his "novel of the century," or Hans Hahn with his textbook on analysis—or, for that matter, Friedrich Waismann with his Wittgenstein book project.

At first, everything had seemed to augur well for Waismann's enterprise. The idea was simple: just as Moritz Schlick had once been the "prophet" of Einstein, so Friedrich Waismann could play that role for Wittgenstein. But Wittgenstein, as it turned out, was far more demanding than Einstein. Each time he took a look at Waismann's manuscript, he would insist on drastic changes; occasionally, he would insist on these changes being undone, or else he would demand that a completely new approach be taken. It gradually became apparent that Wittgenstein felt that most of the positions he had advocated in the *Tractatus* had been "overcome." That is, they were to be rejected! It thus seemed absolutely pointless, to the new Wittgenstein, to present his old ideas in a more accessible form, as Waismann had envisioned.

So then Waismann, always eager to do his best, proposed a different project: to write a book expounding Wittgenstein's *present* views. For this he was uniquely suited. Indeed, Wittgenstein had taken to talking with him for hours, and Waismann would regularly report the gist of these talks to the Circle. It was he who first reported on and tried to explain Wittgenstein's slogan, "The meaning of a proposition is the method of its verification." This Wittgensteinian proposition soon became a leading thesis of the Vienna Circle. But in time there arose a nagging question: How does one verify this thesis? It seems impossible. So does the thesis itself have any meaning?

Clearly, Waismann's notes could furnish the basis for a book on Wittgenstein's new philosophy, with all its open problems. Everyone was in agreement on this point. But the problem was, Wittgenstein seemed unwilling to accept any of Waismann's accounts, even if the latter could provide proof positive that he had taken the words down verbatim as they emerged from Wittgenstein's mouth. Again and again, Wittgenstein surprised everybody with one reversal after another. This was particularly strange, given that he had written: "The solutions to philosophical questions must never surprise. Nothing can be discovered in philosophy." It is true that he had added: "However, I myself have not understood this well enough, and have erred against it." But Wittgenstein was not the only one to deviate from the eternal canons dictated in his own *Tractatus*.

## CARNAP IN A FEVER

Rudolf Carnap was approaching his forties. He had an outstanding international reputation but still no stable job. In German-speaking countries, the institutes of philosophy were dominated by idealism; they were in no way drawn to the author of *Pseudo-Problems in Philosophy* and *The Elimination of Metaphysics*.

But the stalwart Philipp Frank managed to put things right. He figured out a way to set up, in the Science Faculty in Prague, a chair in philosophy that was tailor-made for Carnap. Frank's argument for creating the new chair was that modern quantum mechanics promised new answers to age-old philosophical riddles concerning determinism, probability, vitalism, and free will. In these areas, a philosopher could well be useful to the physicists. In a certain sense, the appointment of Carnap could be seen as an elegant quid pro quo: half a lifetime earlier, a philosophy chair had been created in Vienna to accommodate a scientist from Prague (namely, Ernst Mach), so now it seemed only fitting that a philosopher from Vienna should come to mingle with scientists in Prague.

Carnap assumed his position in the autumn of 1931. It was with a heavy heart that he left Vienna, and the Vienna Circle was equally sad to see him go. "I cannot bear to think of our Thursday evenings without you," Schlick sighed. And Carnap wrote, looking back: "My life in Prague, without the Circle, was more solitary than it had been in Vienna."

On a more upbeat note, Ina Stöger had moved with Carnap to Prague. For a long time they had debated whether it was correct for him to introduce her with the words "And this is my wife." After all, they were not married. But finally they decided to bow to convention and tie the conjugal knot. Carnap reported to Schlick: "On February 8 [1933], we were married by civil law. Thus my situation is in proper and conventional order again. Since marriage did not affect our real life in any way, the ceremony had little meaning for us; however, it was amusing that we and the Franks, who acted as our witnesses, had to listen to an official speech

FIGURE 9.4 Ina and Rudolf Carnap: at long last, a proper couple.

in Czech without understanding a single word of it, and even more amusing when we had to repeat a few Czech phrases, word for word." Czech, apparently, was not quite so easy to master as Esperanto.

Carnap frequently visited Vienna and stayed in close touch with his philosopher friends there. In his diary, the days in Prague are often left blank, while those in Vienna overflow with descriptions of meetings and lively discussions.

The most important outcome of Carnap's years in Prague was the completion of his second major work, *The Logical Syntax of Language*. This set of ideas had been conceived, and it then evolved, under the influence of Kurt Gödel, and it was directed against Wittgenstein's claim that it is impossible to talk about the logical structure of propositions. These things just "show themselves," Wittgenstein had said. Carnap was not convinced.

The decisive breakthrough occurred while Carnap was still in Vienna, a few months before his move to Prague. Carnap remembered: "After I had been thinking about these problems for several years, the whole theory of language structure and its possible applications to philosophy came to me like a vision during a sleepless night in January 1931, when I was ill. On the following day, still in bed with a fever, I hastily wrote down my ideas on 44 sheets under the title 'Attempt at a Metalogic.' These shorthand notes were the first version of my book *The Logical Syntax of Language* (1934)."

By "the logical syntax of language," Carnap meant language's formal rules, irrespective of the meaning of its symbols or the sense of its expressions. The distinction between syntax and semantics—that is, between the patterns of symbols and what they denote in the world— became an essential feature of Carnap's metalogic, which also drew inspiration from David Hilbert's way of viewing formal expressions as strings of symbols, regardless of their meaning. Taking this perspective allows one not only to carry out logical derivations in a particular system but also to move between different systems. Whereas Wittgenstein and Russell had investigated "the" logic and "the" language, as if these were uniquely defined, Carnap's more general notion of *logical syntax* allowed any number of different logical systems to be compared. None would have a privileged status with respect to the others, much as in relativity theory, there was no privileged reference frame.

This view led to Carnap's *tolerance principle,* a principle that rings familiar to mathematicians: whichever system one chooses is a matter of convention only. It is like choosing which axioms of geometry to use—a question of pragmatism and taste, rather than one of absolute truth.

"Our business is not to set up prohibitions, but to arrive at conclusions," Carnap wrote. "In logic, there are no moral standards. Anyone is at liberty to build their own logic, i.e., their very own form of language, as they wish. All that is required is that, if someone wishes to discuss a specific logic, they must state their methods clearly."

In his personal relations too, Carnap was an epitome of tolerance. His visitor Ernest Nagel was delighted: "He is one of those few with whom one need not agree to be understood."

However, the tolerant Carnap clashed fiercely with the zero-tolerance Wittgenstein.

SCHLICK DELIVERS THE MAIL

For the Vienna Circle, all knowledge must be grounded in experience. Thus we read in the manifesto: "There is knowledge only from experience, which rests on what is immediately given." But what is meant by "immediately given"? On this matter, the members of the Circle did not see eye to eye. Even the authors of the manifesto did not.

In Carnap's view, it was *sense-data* that were immediately given, playing the role for him that *sensations* had played for Ernst Mach. In Otto Neurath's opinion, however, this came perilously close to being a solipsistic viewpoint, based on private experiences. Neurath had no patience for this view, nor for other idealistic creeds. For him, it was the *facts* of the external world that were "immediately given," not the sense-data. In more concrete terms, a stone is a stone, even if you don't kick it.

Carnap was not averse to discussing such a viewpoint. Although he had included "the reality of the external world" among the pseudo-problems in philosophy, he was prepared to acknowledge that his approach and Neurath's might, in the end, turn out to be two sides of the same coin.

Another pseudo-problem, in Carnap's eyes, was whether the consciousness of another person is similar to one's own. In principle, we cannot ever personally feel the distress of Mr. K, for instance. We can only deduce from his behavior that Mr. K is distressed. One's belief that deep down inside, Mr. K feels things in just the same way as one does oneself is not verifiable, and hence must be irrelevant to science. In Carnap's words, it is "cognitively unimportant."

Neurath was pleased with Carnap's opinion, since it dovetailed nicely with his famous old hobbyhorse—unified science. The humanities do not require a separate place of their own. Everything that happens is a part of nature, and hence a part of the physical world; this includes all that is spoken and thought. Morals, societies, emotions, customs—everything is a result of physical law. Carnap reported that Neurath even once tried to translate all the concepts of Freudian psychoanalysis into a "physicalist language" based on purely behaviorist terms. This heroic endeavor led to nothing, but both philosophers clung tightly to physicalism.

FIGURE 9.5 Carnap meets abyss.

However, when Carnap wrote an article called "Physicalist Language as the Universal Language of Science," the scene suddenly darkened.

The moment that Wittgenstein stumbled across the essay, he was beside himself with rage. This man Carnap, whom for years he had refused to set eyes on, had had the gall not to give any credit to him; instead, he had cited only himself, and this "with outstanding conscientiousness." Angrily, Wittgenstein complained to Moritz Schlick, who relayed the complaint to Carnap, who in turn defended himself, arguing that in his previous writings, he had often given credit to Wittgenstein, but in the present case he simply saw no need to do so.

Schlick forwarded Carnap's reply to Wittgenstein, but it did nothing to appease the latter. Wittgenstein was particularly irked by Carnap's acerbic remark that Wittgenstein himself was not known for his care in crediting other authors. Was this not the pot calling the kettle black?

"One must be a great deal more profound than Carnap to write such a thing," thundered Wittgenstein. Moreover, "physicalism," he noted, was an awful term. Carnap in turn opined that only a psychoanalyst would be able to explain Wittgenstein's rage.

This whole heated exchange was routed back and forth via the intermediary of Moritz Schlick, who was trying to enjoy his vacation in Carinthia, in southern Austria. He had failed to heed the advice of his doctor, who had suggested to him not to have his mail forwarded. And so now, here he was, inextricably stuck in the thankless role of a go-between. What a rotten vacation!

Schlick wrote to Carnap:

> Enclosed with this letter, you will find a double letter from Wittgenstein, which he sent me with the request to fill in your address on the envelope and then send it along to you, unopened. However, I know exactly what it contains, and it is most painful for me to play the role of messenger. I had refused to forward to you Wittgenstein's letter to me, so now he's sending you a copy of it himself. You of course know my high esteem for both of you, and so you may well imagine how much I suffer from this unfortunate affair.

Schlick added: "In several ways at once, I am facing a riddle! How lucky that you are such a quiet and reasonable person. At this point, I am truly at my wits' end."

## No Stone Remains Standing on Another

Friedrich Waismann, too, suffered greatly from this "unfortunate affair." Indeed, Wittgenstein (of all people!) reproached him for not having completed his book long ago. Had such a book existed, it would have unequivocally proven Wittgenstein's priority! But instead, Waismann had spent far too much time and energy transmitting Wittgenstein's views to the Vienna Circle. Soon, wrote Wittgenstein, things would come to a point where his ideas would be considered as plagiarizing Carnap's. "This, of course, would be highly undesirable to me," Wittgenstein added bitterly.

Waismann was deeply wounded. For the next couple of semesters he nervously avoided making even the slightest allusion to Wittgenstein in the Circle—not even saying his name! Schlick, too, started watching his own words. Obviously, this kind of self-imposed taboo severely crippled the heretofore free flow of thoughts in the Circle. A hidden minefield had emerged into the open.

But then, quite unexpectedly, things took a turn for the better, and the world seemed bright again. In September 1933, Schlick and Wittgenstein spent a happy few days together in a small village on the Adriatic coast. Wittgenstein had discovered a new tool for philosophizing, the method he called "language games." These were thought experiments that revealed how deeply rooted all our concepts and propositions are in human activities.

Here are a few sample language games, each of which at first seems to be purely verbal, but when looked at more carefully, is seen to crucially involve a nonverbal activity as well: "To command and to obey an order; to produce an object according to a verbal description. To utter a request, to thank, to swear, to greet, to pray."

It would be artificial to interpret these activities in terms of the picture theory. If I say "Thank you," I may well be describing feelings of gratitude and obligation, but in many ways this is beside the point. "Thank you" is used as an instrument to make our social life flow more smoothly, or to use Wittgenstein's expression, as part of our "form of life."

Wittgenstein rejected what he saw as Carnap's erroneous claim to the effect that "logical analysis brings hidden things to light (in the fashion of chemical or physical analysis). If one wishes, for instance, to understand the word 'object,' one has to look at how it is actually used."

The meaning of a word, said Wittgenstein, is defined by the rules of its usage, just as is the meaning of a rook or a bishop in chess. The same holds for propositions: "A proposition should be thought of as being like a tool, and its meaning should be seen as residing in the way it is used." This was very different from what had become a central mantra of the Vienna Circle: "The meaning of a proposition consists in the method of its verification."

An ideal language for science, a language of the type desired by Carnap, interested Wittgenstein even less than Esperanto. He wanted to analyze *everyday* language, with all its hidden traps. The aim of his philosophy was therapeutic: "to show the fly an escape route from the fly-bottle." Again, language gives the clue: "A philosopher treats a question like an illness." And Wittgenstein noted that his own style of analyzing language was related to the psychoanalysis of the legendary old Professor Freud: both techniques "make the subconscious conscious and thereby harmless."

Once again, Schlick and Wittgenstein decided to restructure Friedrich Waismann's book—as usual without asking for the latter's consent. When he heard their new constraints, Waismann frustratedly described his plight to a sympathetic Karl Menger: "Now I am charged with constructing a series of examples leading from the simplest notions to the very toughest in all of philosophy: in this way, the solutions to philosophical questions will supposedly drop into one's lap like ripe fruits."

The idea was of course appealing, but to execute it was a daunting challenge. Waismann desperately tried to avoid tackling this monumental task but to no avail. One night, in the wee hours, he was awoken and instructed to come at once to Schlick's flat. Waismann obediently arose and made his way over in the darkness. Wittgenstein was waiting for him there. Waismann was told that he should at least give the new idea a try. Was this asking too much? Just a try! Waismann gave in at the end, naturally. However, he warned them, it would not be his fault if the book should never appear, or too late.

His premonitions proved to be exactly right. Wittgenstein's ideas kept on developing but ever more stormily. Time and again, the book was "all set to go to press," and then had to be withdrawn at the last moment and subjected to a radical new round of alterations. Whose book was it, anyway?

In a letter to Schlick, Waismann bemoaned "the difficulty of a joint work, since he [Wittgenstein] keeps following the latest inspiration of the moment, tearing down whatever he has erected before, so that one almost gets the feeling that it never makes the slightest difference how one arranges the thoughts, since in the end no stone remains standing on any other one."

At some point, Waismann and Wittgenstein came to an agreement that each of them should write their own book. But soon enough, this plan was dropped too, and once more Waismann had to face the formidable task all alone. But then, when he at last seemed to have succeeded, Wittgenstein decided to overhaul the whole thing one final time during the summer months.

Sarcastically, Neurath commented to his young comrade-in-arms Heinrich Neider: "Are the words of the Anointed One the same as those of the LORD? How is poor Waismann's work coming along? When will the LORD make his appearance, descending upon this mortal wretch during the holidays? Or will it instead come to pass that the LORD's revelations are published, so that He might deem the other's work to be no longer needed?"

Carnap, for his part, was kept up to date by Schlick: "Waismann's book is indeed close to completion; it needs only a few small amendments. And then Wittgenstein, whom I expect here in a week, wishes to add just a few annotations to the book."

However, a few months later: "I have to correct my last note about Waismann's book: All that Wittgenstein plans to do during the current month is sketch out a detailed outline; then it will fall to Waismann to fully flesh out the skeleton. I do not envy him this task."

Again and again, the plot took unexpected turns. Schlick wrote to Carnap: "The most recent development concerning Waismann's book is that it will be written not by him, but by Wittgenstein himself! I do not yet know how Waismann will react to this new twist, since I have only spoken with him by phone. I beg you not to talk about it for the moment, since this may well not be the last word about the matter, and too many rumors have already been spread about this unfortunate book."

Thus was discarded the sacrificial lamb Friedrich Waismann—for a while. But in the end, Schlick's intuition turned out to be right: this was still far from being "the last word about the matter." Of all people, it was Wittgenstein himself who summed the situation up the most precisely and concisely: "It is hell to work with me."

# *Moral Matters*

*Vienna, 1933: Austria and Germany abolish democracy. Dictators engage in showdown. Schlick sees perdition coming. Menger develops valueless ethics. Economist Morgenstern dismisses perfect foresight. Unemployed mathematician discovers economic equilibrium. Schlick claims ethics is science, not philosophy; finds meaning in life: it is youth. Also suggests: "Give happiness a chance!" Ex-student stalks Schlick, warning of homicide plus suicide.*

## EMERGENCY RULES

A former schoolmate of Ludwig Wittgenstein's had become the chancellor of Germany, and he had no intentions of stopping with that. At the top of Hitler's to-do list stood the annexation of his Austrian homeland. In January 1933, immediately after his accession to power in Berlin, he started in earnest with the *Heimführung*—the "bringing home" of Austria into the German Reich.

During the preceding year, the Austrian branch of the National Socialist Party had grown by leaps and bounds. The Austrian Republic was near the tipping point, with the Christian Socialists clinging to power by the narrowest of margins. Indeed, together with the right-wing splinter parties of the Landblock and the Heimatblock, their parliamentary majority hung on but a single vote.

The two leading politicians on the right, Ignaz Seipel and Johann Schober, had died within a month of each other. The new Austrian chancellor, Engelbert Dollfuss (1892–1934), was a staunch opponent of Nazism but not exactly a true-blue democrat. In the constitution Dollfuss had discovered a clause that allowed him to govern with emergency decrees, thus avoiding parliamentary debates that could easily turn tumultuous on occasion, with inkpots flying left and right. The basis for the chancellor's decrees was the nearly forgotten Enabling Act for War Economy—and never mind that the Great War was long since over.

Years earlier, Otto Neurath had recognized the great political power residing in the mechanisms of a wartime economy, and now those same mechanisms were being craftily exploited by Dollfuss, but toward utterly different goals.

In Germany, too, the constitution was being undermined by *Notverordnungen,* as the enabling acts were called. A most convenient pretext for such actions was provided by the Reichstag fire in February 1933. The arson was blamed on the communists, though other conspiracy theories flourished as well. In any case, the elections held soon thereafter gave Hitler almost 44 percent of the vote—not as much as expected, but quite enough. The National Socialists summarily outlawed the opposition.

In Austria too, at almost the same time, although in a different manner, parliamentary rule was abolished. An absurd sequence of procedural events provided the opportunity. During a tight roll-call vote—it happened to be about whether to prosecute striking railway workers—the speaker of the house, Karl Renner, stepped down from chairing the session, so as to be able to cast his vote. But then, to counter Renner's move, the second chair, who belonged to the opposing side, also stepped down. And then, in the heat of the moment, the third chair followed suit. And all at once, before anyone could think straight, there was no chair left—no one who could formally close the session, adjourn it, or call for another meeting. What now? The bewildered delegates eventually left the chamber and went home, totally confused.

For Chancellor Engelbert Dollfuss, this fluke happening was "a hint from God," and he took the hint and ran with it. He unilaterally declared that the Austrian parliament had abolished itself, and the next day, police forces prevented the delegates from entering the building.

From then on it was all smooth sailing. The cabinet swiftly imposed censorship of the press and prohibited political rallies. The left-wing paramilitary group known as the Schutzbund was dissolved. Step by step, the Social Democrats, nicknamed the Sozis, had to yield ground. Dollfuss was pleased as punch: "Nothing unnerves the Sozis more than this kind of calculatedly slow tactic."

But the Nazis were far more dangerous than the Sozis: they were clamoring for general elections in Austria. And this was what Dollfuss was dead-set against. One didn't have to be a democrat to fear the rise of the Nazis.

Dollfuss ordered that the leader of the Austrian National Socialists be deported to Bavaria. In retaliation, Hitler imposed the so-called *Tausend-Mark-Sperre:* any German heading for Austria had to fork over one thousand marks for permission to cross the border. As a result of such highway robbery, German visitors in Austria suddenly became scarce as hens' teeth—a fatal blow for tourism.

It was not long before Hitler's stormtroopers went on the rampage in Austria, and a wave of terrorism swept all across the country. Day in and day out, phone booths and bridges were blown up, rail lines and power lines were cut, coffeehouses and shops were shattered by bombs.

The chaotic political situation reigning in Europe left Dollfuss no other resort than to approach Italy's fascist leader Benito Mussolini for help. Standing next to Il Duce, the slight, gentle-eyed chancellor appeared almost touchingly frail. However, he had no doubts about what he needed to do. He swiftly banned the National Socialist Party in Austria, in its place declaring the Vaterländische Front ("Fatherland Front") with himself as its leader; he designed a so-called *Krückenkreuz* ("crutch cross") as the Austrian counterpart to Germany's *Hakenkreuz,* or swastika; he organized mass rallies in open fields; and, on the occasion of a religious festival called Katholikentag ("Catholics' Day") in

FIGURE 10.1 Escape from the university, 1933.

September 1933, he gave a blustering tirade filled with rabble-rousing words: "The time of capitalist, liberalist economic order is over! The time of Marxist materialist mass seduction is past! The time of party politics is gone! We reject conformity and terror! Our goal is Austria as a social, Christian, German state built on a corporate basis and under powerful authoritarian leadership!"

The notion of a "corporate basis" for society had been promulgated by Othmar Spann (1878–1950), a professor of economics at the University of Vienna. Spann's book *The True State* claimed that the proper way to understand the structure of any society was by breaking it down into professions rather than into classes. Moreover, it informed the reader that the whole is more than the sum of its parts, from which it followed that social groups are more important than individuals.

Spann's ideas found an enthusiastic following among students who had a nationalistic and pro-Catholic bias. Some of them belonged to the

Students' Free Corps, which engaged quite regularly in bloody brawls with National Socialist groups, especially on Saturdays, when a ritual showdown called a *Bummel* ("rumble") would often take place in the arcade-lined inner courtyard of the university. Classes were abruptly cut off, lecture rooms were trashed, and hate-filled slogans were smeared all across the corridor walls. Again and again, the university was forced to close down because of the turmoil.

And yet the Vienna Circle, undaunted, continued with its regular Thursday evening meetings. Members were let in by Menger and Hahn, who, being professors, had keys to the university building, now often eerily deserted. Their meeting place was a sanctuary of silence, while outside, the streets resounded with ferocious tribal war chants and the ominous thump-thump-thump of heavy boots.

EXTRA!

Karl Menger wrote in his *Reminiscences:*

> In 1933, the year of Hitler's coming to power in Germany, there were periods when life in Vienna was almost intolerable. The newspapers published extra editions round the clock, and vendors ran shouting through the streets hawking the latest extras. Groups of young people, many wearing swastikas, marched down the sidewalks singing Nazi songs. Now and then, members of one of the rival paramilitary groups paraded along the wider avenues. I found it almost impossible to concentrate and rushed out hourly to buy the latest extra. On one of those days, I met Dr. and Mrs. Schlick in a streetcar. "It is impossible to concentrate," the professor said. "I read the extras from morning to night."

And Menger added: "It was sad to see Schlick's quiet serenity slowly disappear. In one of my conversations with him during that terrible period he said that in his opinion the rise of Hitler meant the *Untergang des deutschen Volkes*—the decline and fall (more precisely, the perdition) of the German people."

In Germany, the Nazis were burning books. Jewish shops were being boycotted, Jewish officials were being dismissed. Albert Einstein was among the first to seek asylum abroad: having been forewarned in time, he simply did not return home from one of his journeys. As a punitive measure, or *Strafausbürgerung,* he promptly lost his German citizenship.

Thus began an exodus of scientists on an unprecedented scale. When Prussia's minister for education asked the aging David Hilbert whether his institute in Göttingen was suffering from having lost all its "Jews and Jew-friends," Hilbert drily replied: "No, the institute is dead."

Hans Reichenbach and Richard von Mises were among the German professors who were forced to retire. Reichenbach had been the head of the Berlin Society for Scientific Philosophy and, together with Carnap, editor of the journal *Erkenntnis.* After his expulsion from Germany, he became a professor at the University of Istanbul.

Richard von Mises also moved to Istanbul. The Berlin Institute for Applied Mathematics, which von Mises had created, was taken over by Vienna-born Theodor Vahlen, a fanatical Nazi who undertook to build up a "German mathematics." In his Turkish exile, Richard von Mises wrote *A Small Textbook of Positivism.* The book was not small at all, and in its title was defiantly displayed the word *Positivism,* so loathed by German philosophers.

In Freiburg, Martin Heidegger, the notorious philosopher of Nothing, was named rector of the university. Not only did he publicly celebrate his joining of the National Socialist Party, he even took to lecturing while wearing a Nazi stormtrooper brown shirt. Heidegger declared, "Adolf Hitler, our great leader and chancellor, with his National Socialist revolution, has created a new German state, which will safeguard for its people the stability and continuity of its history. Heil Hitler!"

Meanwhile, in Vienna, Moritz Schlick wanted to take a firm stand in the struggle against National Socialism. "Led by an inner urge," as he put it, he wrote a carefully worded letter to Austrian chancellor Dollfuss, in which he congratulated the latter for his recent activities. An excerpt follows:

"Highly respected *Herr Bundeskanzler,* you have rightly recognized that the spirit that dominates Germany today, as a result of postwar sufferings, is not the true German spirit at all, as embodied by the greatest of that nation. The German people and the whole world are best served by firmly opposing the spread of this evil."

Schlick ended the letter with the expression of his most sincere devotion.

It was an unabashed show of solidarity. At that time, a fair number of liberals and conservatives, including Karl Kraus and Sigmund Freud, saw the Fatherland Front within the so-called corporate state as the last bastion against Hitler. Karl Kraus wrote that Dollfuss was "the little savior against the great threat," while Freud declared that "only Catholicism protects us from Nazism."

The National Socialists fully shared this view, and as a result took it as obvious that Engelbert Dollfuss had to be removed. A member of the Nazi party shot Austria's chancellor in front of the Parliament but succeeded only in wounding him. In court, the offender was diagnosed as "mentally lacking" and got off with a light sentence of five years in jail. As soon as Dollfuss was back up on his feet, he introduced martial law, including the death penalty.

## COMBINATORIAL ETHICS

Karl Menger, surrounded by terrifying turmoil no matter which way he looked, found it hard to focus on pure mathematics. All around him, society was being torn to shreds, and all parties were convinced of being in the right.

Increasingly, Menger's thoughts turned to ethics, which, after all, was supposed to be a field offering solutions for profound conflicts between opposing parties. That was the core of ethics. Without conflicts of interest there would be no need for morality.

However, there was a snag. Most members of the Vienna Circle agreed that it is impossible to speak in a scientifically valid way about values. In Carnap's view, values were devoid of cognitive meaning, according to the empiricist criterion of significance. And Wittgenstein had proclaimed that there is no value in the world—and that if there were, it itself would have no value. This, he declared, is why there can be no ethical propositions (*Tractatus* 6.24).

Karl Menger looked for a way out. He tried to develop an ethics that was value-free—a formal type of ethics relating to traditional ethics in much the same way as formal logic relates to traditional logic. Mathematics surely ought to be able to help in establishing this new type of ethics.

The idea of such a link was not new. Robert Musil had also held that "there exist certain connections between mathematical and moral thinking." Menger's attempt to apply exact thinking to ethics was not without precedent. Far earlier, in fact, Immanuel Kant had sought a formalized ethics based on pure reason. But in Menger's view, Kant's famous categorical imperative did not take sufficient account of the diversity of viewpoints in a society. Kant's command says (in an old-fashioned but oft-cited translation): "Act only according to that maxim whereby you can at the same time will that it should become a universal law"—or, in simpler terms, "Do X only if you like the idea of everyone doing X." This left numerous questions open. What to do, for instance, about those who "will" other laws? When are the various maxims of different people compatible with each other, and when not?

Decades later, Menger described his agitated state of mind at the time: "While the political situation in Austria during the winter of 1933–34 made it extremely hard to concentrate on pure mathematics, sociopolitical problems and questions of ethics imposed themselves on everyone almost every day. In my yearning to find a consistent, comprehensive worldview, I asked myself whether some answers might not arise by means of exact thought."

In the autumn of 1933, when university riots resulted in all lecture courses being suspended for six weeks, Menger withdrew to Prein, a village lying in the shadow of the Rax mountain, a favorite excursion spot for the Viennese. There, beneath an imposing cliff face, Menger wrote a short booklet called *Morality, Decision and Social Organization: Toward a Logic of Ethics*.

When philosophers mused about ethics, so Menger claimed, they considered their main tasks to be (1) searching for "the concept of morality," (2) gaining insight into "the essence of goodness," and (3) establishing a "list of duties" or uncovering "the principle of virtue." But these were not his priorities: "As for me, I will deal with none of these questions." Menger intended to leave all such blurry issues in the hands of philosophers. What he wanted to investigate, by contrast, was the formal compatibility of various moral or legal norms, eschewing all talk of values: "Morals will, so to speak, be identified with the groups of their [ . . . ] adherents."

This led to a combinatorial calculus. Menger painstakingly avoided all current ideological issues. For his examples, he used the contrast between smokers' opinions and nonsmokers' opinions, which at the time was as innocuous a disagreement as could be imagined. Menger was interested in how a society could deal with individuals who adopted different viewpoints, and he always refrained from judging the viewpoints themselves. Thus he was attempting, in a certain sense, to apply Rudolf Carnap's tolerance principle to ethics.

At first, Menger's book found little acclaim, not even among the Vienna Circle. Its stilted style (a mixture of letters and dialogues) and its strict avoidance of all value judgments were not in keeping with the public mood of the time.

Oswald Veblen, now back at the Institute for Advanced Study in Princeton, politely queried whether such questions were deep enough for a mathematician of Karl Menger's caliber. Less politely, Georg Nöbeling, who once had been Menger's favorite student, wrote that the mere act of raising such questions filled him with disgust. Nöbeling, much to the dismay of his former teacher, had returned to National Socialist Germany to accept a professorship in Frankfurt, and Menger viewed the step as an act of blatant treason.

It was the worst possible moment to approach ethical problems by means of "social logic." Applying that idea to economics, however, seemed much more reasonable. Menger had anticipated this when he wrote: "Groups similar to those considered in my notes [i.e., single-issue

FIGURE 10.2 Karl Menger and Oskar Morgenstern on the rocks.

groups] might also be formed according to extra-ethical criteria—especially aesthetic, political, or economic criteria. Groups of the last kind, incidentally, might be not irrelevant to theories of economic action."

This view was embraced with enthusiasm by Oskar Morgenstern (1902–1977), an economist of about Menger's age.

## MORGENSTERN (ARYAN)

Although Morgenstern was born in Prussia, he belonged to the Austrian school of economics, or more precisely, to its fourth generation. As the first generation had consisted of Carl Menger only, it was only natural that Morgenstern should hold the views of the latter's son in high esteem.

During his first semesters at the University of Vienna, Oskar Morgenstern came under the influence of the so-called Spann Circle, which had formed around Othmar Spann, the prophet of the "true state." Young Morgenstern also assiduously studied the German idealistic philosophers Johann Gottlieb Fichte, Friedrich Schelling, and Georg Friedrich Wilhelm Hegel. And as a result of his upbringing, Morgenstern held anti-Semitic prejudices. The lectures on economics given by Friedrich von Wieser (third generation) aroused his contempt. "Could he by any chance be a Jew or a half-breed, or a liberal?" Morgenstern mused in his diary. Actually, Wieser was not particularly liberal; in fact, he admired Mussolini. But first and foremost, Wieser behaved, as Morgenstern later noted, like an "aristocrat of the old school." Actually, though, it was Oskar Morgenstern who had aristocratic blood flowing in his veins. His mother was an illegitimate daughter of Germany's short-termed Emperor Friedrich III, a member of the Hohenzollern dynasty.

Soon enough, Oskar Morgenstern turned away from Othmar Spann's romantic fantasies and toward the theory of marginal utility. He became the assistant of Wieser's successor. Morgenstern had done poorly in mathematics at school, having had to repeat one class, but at the university, his studies shifted into high gear. After only six semesters he obtained his doctorate and moreover received a luxurious three-year travel grant from the Rockefeller Foundation.

Morgenstern was convinced of the importance of mathematical methods in economics. However, mathematics had counted for little in the Spann Circle and for hardly any more in the Austrian school of economics. This is why the young Dr. Morgenstern chose to visit England, the United States, France, and Italy: to learn mathematical economics. After the polyglot prodigy finally returned to Vienna, he applied for his *Habilitation.*

Othmar Spann, however, got wind of this plan and did his best to sabotage it, launching the rumor that Morgenstern had a Jewish background. The name, after all, seemed to give it away. Indeed, a Jewish friend of Morgenstern's had even once advised him, albeit as a joke, that he should sign all his articles "Morgenstern (Aryan)," so as to nip all misunderstandings in the bud.

In the end, Spann's malevolent intrigue did not succeed, but the episode did teach Oskar Morgenstern one important lesson: it cured him forever of his earlier anti-Semitic leanings.

In Vienna, Morgenstern joined the economics circles of Ludwig von Mises (third generation) and of Friedrich von Hayek (fourth). Ludwig von Mises (1881–1973) was the brother of the applied mathematician and philosopher Richard von Mises, but Ludwig enjoyed the reputation of being even more arrogant than his brother. The two siblings could not stand each other.

Friedrich von Hayek (1899–1992) had also belonged, for a short time, to Othmar Spann's circle, but he had left it behind before Morgenstern arrived. Hayek, a remote cousin of Wittgenstein's, now led the so-called Geist Circle, which was also frequented by Felix Kaufmann and Karl Menger, two members of the Schlick Circle. It would seem that, at that time, only the most resolute of hermits could abstain from all Viennese discussion circles.

Hayek and von Mises were both diehard liberals, and staunch opponents of a full socialization in the spirit of Otto Bauer and Otto Neurath. Ludwig von Mises even prided himself on having gloriously saved Austria from Bolshevism. "This event was all to my credit, and mine alone," he told the future readers of his memoirs. "However," he added in a sudden attack of modesty, he had unfortunately not been able to prevent the Great Depression—though he had delayed its arrival by ten years. These heroic feats were performed by the dextrous von Mises as he sat in his quiet office in the finance department of the Viennese Board of Trade. He had no foothold at the university.

Friedrich von Hayek was the director of the newly founded Institute for Business Cycle Research. Impressed by the expertise that Morgenstern had acquired abroad, he hired the young man, and when Hayek left Vienna for the London School of Economics in 1931, Morgenstern became his successor. That same year, the Creditanstalt, Austria's largest bank, went bankrupt. The crash triggered a huge currency crisis, and soon a swamp of corruption and financial scandals came to light, with the unemployment rate reaching dizzying heights. Morgenstern witnessed all this firsthand at his Institute for Business Cycle Research.

One of the main challenges facing Morgenstern's institute was to create a trade barometer that would facilitate the making of economic forecasts. However, such a notion seemed to verge on paradox: indeed, as was well known in expert circles, Oskar Morgenstern firmly held the opinion that economic forecasts were fundamentally impossible. Indeed, this view had been the central claim of his *Habilitation* thesis.

Morgenstern's argument was simple. Any intelligent economic agent would of course react to any forecast. And yet, that reaction would have to have been correctly anticipated when the forecast was made—but the agent would take this into account, which also had to be anticipated, and so on, in an endless tail-chasing loop. Clearly, an economic prediction is different from a weather prediction. Predictions can never affect the weather. The atmosphere does not react to what people say—but business does. Whoever tries to model the self-modifying process is led straight into a vicious circle, argued Morgenstern. It is similar to playing a chess game, with each player trying to outguess the other, fully knowing that the other, on their turn, will be doing the same.

On more than one occasion Oskar Morgenstern remarked, in a modest aside, that his revelation of the fundamental unpredictability of economics was a counterpart, in that discipline, to Gödel's renowned incompleteness theorem in mathematics. Economic equilibrium is *not* compatible with perfect foresight. Some years later, however, Morgenstern would learn that already in 1930, in a paper on parlor games, John von Neumann had found an escape from the vicious circle of mutual anticipation: forecasts about the *probability* of this or that event can be compatible with an equilibrium from which none of the parties would have any incentive to deviate.

When he heard about John von Neumann's finding, Oskar Morgenstern was impressed, as it seemed to confirm his convictions about the usefulness of mathematics in economics. He wrote an essay called "Logistics and the Social Sciences," devoting its first three pages to the axiomatic method that David Hilbert had so masterfully applied to geometry. Such a rigorous technique was missing in economics, said Oskar Morgenstern. His new mission in life, as he now saw it, was to find a way to approach economic problems with "truly exact thinking and truly exact methods," as would be expected from any adherent to the scientific worldview. But he bitterly regretted having wasted so much of his youth struggling to absorb the outpourings of German philosophers such as Fichte, Schelling, and Hegel. Luckily, though, such aberrations belonged only to his distant past. Now, he visited the meetings of the Vienna Circle, and once he lectured to its members on "Perfect Foresight and Social Equilibrium."

### Another Mathematics Lesson

Karl Menger proved to be an ideal sounding board for Morgenstern's ideas: "Yesterday noon, lunch with Karl Menger in the German House. After a lengthy interval since our last meeting, we debated for 2½ hours. He has carefully read my essay on foresight, agrees with it, and wants me to keep on pursuing this interesting line of thought."

As far back as 1922, Karl Menger had written a paper titled "On the Role of Uncertainty in National Economics." That youthful work, however, had ended up in a drawer, since the editor of the *Zeitschrift für Nationalökonomie* had strongly advised Menger not to submit it for publication.

At the time, the editor's assistant was upset by his boss's recommendation, seeing it as proof of the latter's ineptitude. That assistant was Oskar Morgenstern—but now he was the journal's editor! Times had changed. In his new role, Morgenstern took great pleasure in sending Menger's paper off to the printer, and as an extra token of his support, he invited Menger to give a talk at the Society for National Economics. Menger used the occasion to criticize the semirigorous misuse of mathematical concepts that was so rife among economists. As Joseph Schumpeter later put it, "He read them the mathematical riot act."

In other ways too, Morgenstern enthusiastically supported the use of exact methods. His institute sponsored a series of lectures called *Mathematics for Economists,* given by Karl Menger, with two of Menger's former students, Abraham Wald and Franz Alt, taking care of the exercise classes. This provided these two young unemployed scientists with much-needed income. As a result, they now started developing an interest in issues of economics, and both wrote papers that later became classics of mathematical economics. After the *Anschluss,* the two had to flee from Austria, but luckily both were able to find positions in the United States thanks to their publications in economics, which were far more useful to them than any geometry papers would have been.

In 1936, Franz Alt wrote a short paper called "On the Measurability of the Utility Function," in which he showed that the likes and dislikes of economic agents can always be encoded as real numbers and arranged on a line, and thus can be compared in size. For a discipline that aspired to exactitude, this was welcome news. Today, *utility theory* is a field of its own in economics.

Abraham Wald's contribution turned out to be even more seminal. Many years earlier, Léon Walras (the French counterpart to Carl Menger, the father) had proposed fundamental equations for the prices and quantities of goods in a market economy. The Viennese banker Karl Schlesinger, who took mathematics lessons from Abraham Wald, pointed out problems with these equations. He gave a talk on this in the Mathematical Colloquium of Karl Menger, the son.

Schlesinger argued that if there is an oversupply of some good, then that good will cost nothing, no matter how indispensable it is. Air, for instance, is free. When one takes this idea into account, then Walras's equations have to be replaced by a system of equations and inequalities.

This is where Abraham Wald stepped in. In 1937 he proved the existence of a well-defined price equilibrium for such a system. Wald's paper motivated John von Neumann to publish his own theory of economic equilibriums. All of these results appeared in Menger's journal *Ergebnisse eines Mathematischen Kolloquiums,* and from them soon emerged a brand-new branch of economics. From the 1950s onward, economics was dominated by the *general theory of equilibrium,* which became a fertile breeding ground for Nobel Prizes.

In probability theory, too, Abraham Wald did pioneering work. Scientists had long known how to compute probabilities, but since the word *probable* is riddled with hidden ambiguities, a truly firm foundation was lacking. To establish such a rigorous foundation was a challenge that David Hilbert had already proposed in 1900, as one of his twenty-three celebrated problems.

Richard von Mises had tried his hand at this challenge. He focused on serial phenomena—for instance, the repeated tossing of a coin. The result is a random sequence of heads and tails—or more abstractly, a random sequence of 1s and 0s. Now, Mises asked, if such a series is given to us, how can we check whether it is truly random? Obviously, the 1s and 0s have to show up roughly equally often. But the periodic sequence 0, 1, 0, 1, 0, 1, 0, 1, 0, 1, . . . has that property, and obviously it is not in the least random. Its regularity is revealed by the fact that the *subsequence* consisting of all the even-numbered terms consists solely of 1s, and its complementary subsequence consists solely of 0s. Hardly random! This led Mises to the idea of looking at subsequences. Shouldn't *every* subsequence of a truly random sequence have roughly equally many 0s and 1s? This, though, as it turned out, was asking for too much: no such sequence exists.

However, Abraham Wald was able to provide an elegant definition of random sequences, capturing the concept of irregularity in mathematically exact terms. Roughly speaking, it said this: no gambling system will allow you to win against a truly random sequence. Wald had been led to this idea while listening to a lecture given by Karl Popper in Menger's Mathematical Colloquium. Obviously, the Colloquium was not forbidden ground for young Popper in the same way the Vienna Circle was.

Abraham Wald gave mathematics lessons to Oskar Morgenstern, who wrote: "Yet one more mathematics lesson. We've already reached derivatives. Wald thinks that by next year I'll have made enough progress to understand almost everything in mathematical economics. That sounds good to me!"

Morgenstern decided to use some of the methods in Menger's book *Morality, Decision and Social Organization* to try to understand economic behavior more deeply. What happens if an agent takes into account how other agents react? Morgenstern was hoping to write a paper on the topic, with the title "Maxims of Behavior." At first, his progress was slow. But then he was joined by John von Neumann, and their collaboration led to the founding of *game theory*, a fertile new branch of mathematics dealing with conflicts of interest.

Game theory in turn had repercussions on ethics. Indeed, by the mid-1950s, Richard Braithwaite (1900–1990), once a colleague of Wittgenstein's at Cambridge, pointed out that game theory provides a tool for studying moral philosophy. It allows one to deal, in an exact manner, with questions such as fair sharing, rewards and punishments, and self-interest and the common good. One of the key roots of this area of philosophy can thus be traced all the way back to Karl Menger's aspiration to find a formal way of studying ethics.

## THE MEANING OF LIFE

Like Karl Menger, Moritz Schlick was skeptical about Kant's famous categorical imperative. An imperative is a command, but who issues the command? How can an imperative work if there is no "imperator"—no emperor, no commander? According to Kant, it is *duty* that commands. But what is duty? Is duty defined as "that which commands"? If so, this is circular, and thus not helpful. Or is duty simply another name for God? But is God not dead? And in any case, why should one obey a command? Obedience may rank high among the so-called Prussian virtues, but in most other cultures, obedience is not particularly prized. Kant, to be sure, was a Prussian; but then, so was Schlick, who from boyhood on had hated rigid rules that enforced dutifulness.

Friedrich Schiller had already poked gentle fun at Kant's stern style of morality. He wrote: "I like to be of help to my friends, but alas, I do so with pleasure; thus it often worries me that I am not virtuous."

According to Kant, "Morality, lying at the core of our sense of free will, opposes our inherent desires, and thus gives rise to feelings that may be called pain." In other words, there is no such thing as virtue unless it comes through long, arduous striving and painful self-discipline. Schlick, however, described this view of morality as "a shopkeeper's determination to recognize virtue only in hardship and effort," and in contrast, he claimed that "instinctive virtue is the most beautiful and sublime." He felt far closer to Nietzsche than to Kant.

Like Karl Menger, Schlick had to come to terms with the fact that ethics was a touchy topic inside the Vienna Circle. In his *Overcoming of Metaphysics,* Carnap had gone along with Wittgenstein's assertion that there exist no propositions ("*Sätze*") in ethics (6.42). Carnap's reasoning is straightforward enough: value judgments cannot be empirically verified. If one accepts the idea that the meaning of a proposition resides in its method of verification, then value judgments must be meaningless—or at least "meaningless as scientific statements," as Carnap more cautiously put it later in life. Even later in life and even more cautiously, he put it this way: "According to the empiricist criterion of meaning, value judgments are without cognitive relevance."

In Schlick's eyes, all of this went too far. His first book, the youthful *Wisdom of Life,* had dealt with ethics, after all. From then on, he had returned time and time again to the manuscript of his *New Epicurus* to add more ideas, and he regularly lectured on matters of ethics. He finally wrote a book on the topic, called *Questions of Ethics,* when he was "in deep solitude on the rocky coast of the Adriatic Sea," and it came out as a volume in the Vienna Circle's series of *Writings on the Scientific Worldview.*

According to Schlick, the task of philosophy was simply to elucidate the meaning of propositions, rather than to assert them (that being the province of science). On matters of ethics, however, Schlick definitely wanted to assert propositions. By his own lights, then, ethics would belong to science rather than to philosophy. Indeed, ethics could be seen as a branch of psychology, as it deals with human behavior.

In order to forestall any misunderstandings, Schlick sent Ludwig Wittgenstein a copy of *Questions of Ethics,* "less as an invitation to read it (for I know you have better things to do) than as evidence that I have no intention to hide it from you. If you do chance to take a look at it, your judgment will probably be, or so I suspect, that it has nothing at all to do with ethics—and I wouldn't even take this as a negative critique." But Wittgenstein himself had had no qualms giving a talk about ethics in Cambridge, regardless of what his sound bite 6.42 says, and he never held Schlick's book against him.

There are no such things as absolute moral values, said Schlick. Ultimately, in his view, the only justification of a moral principle is to see it as an intrinsic part of human nature. Indeed, the human tendency to conform to moral principles is as natural as the tendency to pick up one's native language as a child. But what leads us to believe in moral principles? It seems clear that what is deemed "good" is determined by society, most likely because it seems useful. Such usefulness, however, is not defined by an explicit utilitarian calculus that aims at ensuring "the maximal amount of happiness for the maximal number of people." Rather, moral principles are unconsciously internalized in people. Morality is simply part of the social instinct of each individual, and it is rooted in the universal experiences of pleasure and pain, of happiness and suffering. And the internalized moral principles of many individuals, taken all together, add up to form the society's collective morality, which is transmitted to the next generation in turn.

Social instincts, in other words, are just as natural to us as are our most primitive bodily urges. Teaching, punishment, and reward may of course influence human behavioral tendencies, but social instincts are even stronger and longer-lasting than is any control imposed from outside. To act morally constitutes its own kind of pleasure.

In *Questions of Ethics,* Schlick wrote: "Whoever has understood, as we do, that feelings of pleasure are the only basis for values will see without further ado that the concepts of value and happiness are identical."

In another passage he wrote: "I am always amazed by the superficiality of the observations and arguments that are supposed to prove that happiness and morality have nothing to do with each other, or even that virtue is harmful to happiness."

Schlick had studied not just physics but also psychology and was even well-versed in evolutionary biology. He had enthusiastically reviewed the book *The Dancing Bees* by Karl von Frisch and had written a glowing letter of praise to an unknown young zoologist and physician named Konrad Lorenz; nota bene, this was more than three decades before these two researchers shared the Nobel Prize for their pioneering work on animal behavior. In his letter to Lorenz, Schlick expressed his "utmost intellectual delight" and added: "Thank God that there are still some psychologists that a philosopher can read with unmitigated pleasure!"

Thus it was not frivolous speculation but evolutionary psychology that led Schlick to give an affirmative answer to his own rhetorical question: "Is a kindhearted person not also a serene one? . . . Here we find . . . a most remarkable hint. Indeed, if we use the well-tried method of assessing inner feelings by looking at their outer appearance, it turns out that the same charming interplay of facial muscles expresses both benevolence and pleasure. People smile not only when they are happy, but also when they feel sympathy."

And this led Schlick to formulate a succinct moral principle: "*Sei glücksbereit!*" or, rendered literally in English, "Be ready for happiness!" The spirit of the remark, however, is perhaps more accurately caught by the less compact phrase "Always be ready to give happiness a chance!"

Morality, Schlick was convinced, is not tied to self-denial: "It does not come dressed in a nun's habit." Quite the contrary: "Moral behavior springs from pleasure and pain; if one acts nobly, it is because one enjoys doing so. . . . Values are not dictated from above, but lie within; it is human nature to be good."

Thus in his musings on ethics, Schlick replaced duty with benevolence. For some, however, this sounded more like blarney than like a scientific worldview. Karl Menger rather condescendingly wrote: "[*Questions of Ethics*] is a book breathing its author's spirit of gentleness and kindness, while the influence of analytic thinking is merely implicit—namely, in the author's avoidance of some of the worst traditional verbiage."

However, Schlick's thoughts on ethics must have seemed anything but harmless to the Catholic clerics, who controlled a large fraction of Austria's public opinion. Many philosophers, too, were troubled by Schlick's views: Had hedonism not been refuted long ago? And didn't the word *happiness* belong more in a love nest than in a scholar's study? Admittedly, the authors of the American Declaration of Independence had listed "the pursuit of happiness" among the unalienable rights, but in German philosophy, happiness was rarely mentioned, Nietzsche being the great exception. Pleasure is deep, went the *Song of Zarathustra.* But for pious minds, pleasure smacks of frivolity. How could anyone dare to claim that morality is based on feelings of pleasure, and then aim, on top of that, to propagate such a twisted doctrine at universities? Weren't the youths of today already loose enough, what with their bobbed hairdos and their wild dances?

To make matters even worse, Schlick had the cheek to write an essay called "The Meaning of Life." He even gave a simple answer to the question. The meaning of life does not reside in a higher purpose, but can be expressed in just one word, said Moritz Schlick: "The meaning of life is *youth.*"

This seemed paradoxical even to Max Planck, Schlick's old teacher. For the first time ever, Planck felt himself obliged to contradict his favorite student. Isn't youth a stage of immaturity, something yet unfinished, merely a preparation for life? Then how can it contain the meaning of life? And isn't youthful folly downright proverbial? Can meaning emerge from silliness?

But for Schlick, youth was not necessarily tied to a particular age. It was characterized by being open to happiness and ready for play. "Only in play can we grasp the meaning of life," said Schlick. Only in play are our actions freed from the dictates of necessity. The Greek gods had played all day long. Play is its own justification, not amenable to any further queries.

"Man only plays when he is a man in the fullest meaning of that word, and he is only completely a man when he plays." This may sound sexist to modern ears, but Schiller used the term *Mensch,* meaning "human."

## THE AGREEABLE AND THE BEAUTIFUL

It is but a small step from ethics to aesthetics—and not even that, according to Ludwig Wittgenstein ("Ethics and aesthetics are one" [6.421]). Philosophy these days may have lost some of its zest for aesthetics, but in the old days, Hume and Kant reasoned at length on the beautiful and the agreeable.

"The Agreeable and the Beautiful in Philosophy" was the title of a PhD thesis that Schlick had suggested to a student named Sylvia Borowicka, the daughter of a well-to-do Viennese family. This led to a fateful chain of events.

It happened that one Johann Nelböck (1903–1954), another of Schlick's students, was obsessed with Sylvia Borowicka. It also happened that life had never been kind to Nelböck. He came from a hamlet named Krandeln in Upper Austria and had grown up in humble conditions. To walk to school from his family's home took him an hour—in good weather. In winter, darkness and snow often rendered the track well-nigh impassable, and yet the young Nelböck never gave up, sometimes having to literally plow his way to school. By the time he finally set foot in the University of Vienna—already a remarkable achievement for someone from such a backwater—he was twenty years old.

But being older did not hold him back, nor did being from a backwater; in fact, nothing held him back. He went as far as getting a doctorate in philosophy, with a thesis titled "The Meaning of Logic in Empiricism and Positivism." Schlick approved it, although with a mediocre mark—a mere "sufficient." This was the lowest possible passing grade for a PhD. Clearly, Johann Nelböck was not someone whom Schlick could imagine joining the Vienna Circle.

As far as is known, the relation between Nelböck and Borowicka never became intimate. But one day the young woman told her admirer that Professor Schlick was interested in her, and that she reciprocated that interest. We have no way of knowing if any of what she said was true, but hearing those words drove Nelböck crazy with jealousy.

Soon after he had gotten his doctorate, the young philosopher from Krandeln took to haunting the hallways of the university and loudly shouting that Schlick had engaged in "immoral games" with Borowicka. He waved a gun in the air and on several occasions claimed he was going to shoot the offender and commit suicide afterward. At night, he would often phone Schlick at home.

The professor, thoroughly alarmed, informed the police, and Dr. Nelböck was arrested and charged with illegal possession of a firearm, for starters. Where had he obtained his revolver? From Sylvia Borowicka, as it turned out. The girl said she had purloined it from her father. The case grew ever more bizarre, and in the end, both Nelböck and Borowicka were admitted to a psychiatry clinic for a thorough mental examination.

Sylvia Borowicka was medically adjudicated to be "a nervous girl with a somewhat strange character," but nothing to worry about. The consulting psychiatrist in fact recommended that she be permitted to finish up her doctoral studies. And this she was able to do, although with a new supervisor: Schlick's colleague Robert Reininger took over, and approved Sylvia's thesis on the agreeable and the beautiful.

Nelböck, on the other hand, proved to be a far graver case. He was diagnosed as a schizoid psychopath and transferred to the mental hospital Am Steinhof, where he was kept under observation for several months. "A psychopath with bizarre and megalomaniacal ideas, and homicidal and suicidal impulses. Examinee reports having wanted to murder Professor Schlick," said the report.

After some months, though, Nelböck seemed to be recovering. He began making himself useful in the asylum as a clerk. Apart from his odd choice of reading material—vast amounts of philosophy—nothing about him seemed out of line any longer. And so he wound up being discharged. He stated that he wanted to prepare for a teaching position.

However, the moment that Johann Nelböck had turned his back on the psychiatry clinic, he started stalking Moritz Schlick anew. He resumed his nightly phone calls. Schlick switched to an unlisted number, but to no avail—the threatening calls kept coming. He finally had no choice but to disconnect his phone line. Occasionally Nelböck would lie in waiting for Schlick in front of his house. Whenever Schlick spotted the haggard silhouette of his nemesis standing at the streetcar stop, he would quickly hail a taxi and hop in. Nelböck would then rush up and bang his fists against the windows of the cab, his face contorted with rage. The frightening affair soon became known around the university, and some students organized security patrols for Schlick.

Schlick never underestimated the threat. Whenever there was talk of Nelböck, he referred to him as "my murderer." One time, Schlick's daughter, a schoolgirl and an ardent pacifist, reported at home that all the boys in her class were required to undergo paramilitary training. She was upset, but her father merely asked whether in her opinion he should not be allowed to carry a gun to defend himself against his murderer.

There was a brief period of respite from this torment, when Schlick, together with his family, traveled to the United States to spend another guest semester, this time at the University of California at Berkeley. Their stay offered them relief in every respect: nothing but sun and peace and friendly colleagues. But when the family returned to Vienna, Schlick once again found himself engulfed in his nightmare.

It all seemed inconceivable: here was one of the most famous professors in Vienna, a highly respected friend of Albert Einstein and Ludwig Wittgenstein, the philosopher of happiness and the embodiment of reason, and he was being persecuted by a psychopath with homicidal tendencies.

A renewed complaint to the police led to a second indictment: Nelböck was again interned in the psychiatric clinic, and the university rector's office notified the Viennese school board of this news. That marked the end of any chances that Nelböck might ever obtain a teaching job.

In Nelböck's view, there was no doubt at all that it was Schlick who had pulled the strings and caused this catastrophic blow to him. Schlick—only Schlick—had destroyed his professional prospects. But when, after some time, Nelböck's paranoia once again appeared to subside, he was released, just as before. This time, however, by police order, he was barred from staying in Vienna.

Nelböck thus returned to his native hamlet of Krandeln, estranged, pale, and addicted to nicotine. Everything cut him off him from the villagers, starting with his horn-rimmed glasses. Hadn't he aspired to become something better? And now he was living entirely out of his father's pocket. How could there be a Herr Doktor who was not a physician? Not a blessed soul in Krandeln had ever heard of anything of the sort! And all the village children giggled behind his back.

What had all the drudgery of Nelböck's endless studies been good for? Heads were shaking. If only he had gone off to a faraway city to become a priest or a teacher, people would have understood. But a bookworm scholar? Nobody in Krandeln had ever heard of such a thing. It made no sense at all.

In short, everything seemed to be pulling Johann Nelböck back to Vienna.

# The Circle's End

*Vienna, 1934: Civil war in Austria. Social Democrats crushed. Neurath escapes into Dutch exile. Ernst Mach Society dissolved by police. Vienna Circle barely tolerated in Vienna. Hahn's chair abolished after mathematician's premature death. Nazi putsch fails. Vienna Circle grinds down. Edgy debates about protocol sentences. Schlick and Neurath exchange scornful rebukes. Circle splits over Wittgenstein. Gödel falls prey to mental problems. His work picked up by Tarski and Turing. 1936: Stalker shoots Schlick; press blames victim. 1937: Vienna's thinkers think ahead and disband: Menger to United States, Waismann to United Kingdom, Popper to far-off New Zealand.*

## TO RUSSIA WITH LOVE

On a frosty day in February 1934, a telegram arrived in Moscow with the terse message "Carnap expects you." It was addressed to Otto Neurath, c/o IZOSTAT on Kuznetski Most. This street leads to the notorious Lubyanka, where the People's Commissariat for State Security had its offices; its torture chambers were also located there. Some jokers said it was Moscow's tallest building: you could see Siberia from the basement.

The telegram was in fact a coded message. It came from Vienna, and it meant: "Do not return. The police are on the lookout for you."

Otto Neurath was not surprised to receive such a telegram. Before leaving for Moscow, he had settled on that exact text with his partner, friend, and muse, Marie Reidemeister, who had remained in Vienna to take care of Neurath's wife, Olga, and his socialist museum. "Carnap expects you" sounded harmless enough. Neither the Viennese censors nor Joseph Stalin's secret police would see anything to object to in it.

The political situation in Austria had worsened with each passing month. The ruling regime of the Ständestaat, or corporate state, took hold of the levers of power ever more firmly by passing new emergency measures on a daily basis. The Social Democrats were paralyzed. Their head, Otto Bauer, shared the views of Julius Deutsch, the founder and chair of the paramilitary leftist Schutzbund, which by now had been declared illegal: as soon as the government committed an act "raising a tempest of outrage and passion sweeping through the whole working class," then the moment for action would have arrived, but not before. The party felt too weak to make the first move. It had always been handicapped by a sense of responsibility. A well-known expert on the art of insurrection named Leon Trotsky had diagnosed this a few years earlier: "If you agree with Karl Marx's slogan that 'Revolution is the locomotive of history,' then Austro-Marxism is its brake."

And thus the Austro-Marxists held back their rank and file and anxiously waited for the moment of truth to arrive. Otto Neurath sarcastically joked: "We are known for our great cunning because we have armed the workers, but we don't let them shoot."

The Social Democratic leaders were convinced that sooner or later there would come an occasion for a sudden uprising. But Chancellor Dollfuss was not willing to give them any such chance. The Fatherland Front's leader stuck patiently to his slow and measured strategy. One of these days the Sozis would surely lose their nerves.

The long-awaited moment arrived on February 12, 1934. In Linz, a group of Schutzbund members defied a search warrant and machine-gunned the police raiders. This gave the signal for a spontaneous revolt of the working class. The leadership of the Social Democrats urgently advised against it, but in vain. Overwhelmed by the events, Otto Bauer at once rushed to call for a general strike, and in great haste the Sozis set up their headquarters.

But the government was far better prepared. Just one day earlier, the minister for security, Emil Fey, head of the fascist Heimwehr, had publicly declared: "Tomorrow we will get down to business and will finish off the job."

The Dollfuss regime imposed martial law, and the army aimed its guns at some of the city's largest apartment buildings. It had not escaped the experts that those huge dwellings were ideally suited to serve as citadels of the proletariat. After some shelling, even the most obdurate resistance subsided.

Three days and three hundred casualties later, the civil war was over. The top leaders of the Schutzbund—Bauer and Deutsch—escaped across the border. The Social Democratic Party was officially disbanded. It had held 40 percent of the votes; every tenth Austrian had even been a card-carrying member. But now there was to be no more voting. From this point on, the working class would be represented only by a puny office located in some foreign country, and by the Socialist International. The latter's chair had been held for more than ten years by Friedrich Adler, who lived in Switzerland and later moved to Belgium.

During the February fighting, the police repeatedly tried to arrest Otto Neurath in his Viennese apartment. Through Marie Reidemeister, they found out that he was abroad. The fact that he happened to be in Moscow, capital of the proletarian revolution, did not placate the forces of the law.

However, Otto Neurath's stay in Moscow had not been contrived by a conspiracy of the Communist International. He was in Moscow in a purely professional capacity, and not for the first time either. As director of Vienna's Museum for Social and Economic Affairs, Neurath was taking part in launching an Institute for Pictorial Statistics for the state publisher of the Soviet Union. The leaders of Red Vienna knew of his doings and approved of them. Although they were staunch anticommunists, they had pledged solidarity to the proletariat, whether under Stalin's rule or not. A proper education in social affairs could not hurt anybody. So let the Soviet worker learn some statistics!

But now, Red Vienna had been demolished by Engelbert Dollfuss. Its Lord Mayor Karl Seitz was under arrest. The Fatherland Front took control of the city's administration. Police sealed the rooms of Neurath's Museum for Social and Economic Affairs.

Neurath, however, had made plans for this eventuality. He had no intention of remaining in Moscow, as, say, just one more guest of the Hotel Lux, where hundreds of German emigrants were housed, courtesy of the Soviet regime. No, Neurath had prepared a fallback position for himself in Holland. There, in the previous year, he had founded the Mundaneum, which was the Dutch branch of his Viennese museum.

And so he now made his way from the Soviet Union to the Netherlands, necessarily in a roundabout manner: the first stop was Prague ("Carnap expects you"), then Poland, then Denmark. He took care never to set foot on German soil: after all, in the Gestapo there were surely a few who had not forgotten the days of the Munich Soviet Republic. The Germans would certainly not let Otto Neurath slip through their fingers a second time. Moreover, the Austrian ministers who years before had helped to free him were now in deep trouble themselves: Renner was languishing in jail, and Bauer was in exile.

It was not easy for Otto Neurath to construct a new life for himself in the Netherlands. On more than one occasion, everything would have fallen apart had he not gotten a last-minute loan of a few hundred dollars. But it was not for nothing that friends called him "the great locomotive": Neurath thrived on crises. Soon Olga and Marie came by train from Vienna to join him, and the cozy trio moved into an apartment in The Hague.

A short while later, Neurath's graphic artist Gerd Arntz joined them, too, with his family. It even proved possible to transfer a large part of the material from the museums in Vienna and Moscow. Within a short time, the Mundaneum started hosting exhibitions and churning out posters and brochures.

From that point on, however, Neurath carefully refrained from using Marxist rhetoric.

## "Up with Dollfuss! Down with the Unity of Science!"

One casualty of the civil war of February 1934 was the Ernst Mach Society: it was dissolved by police order. The official explanatory note stated that "it has been brought to our attention" that the society had been acting on behalf of the Social Democratic Party. As said party had been outlawed, the group clearly had to be disbanded.

Less than a year earlier, the Dollfuss regime had ordered the dissolution of the freethinkers' association, much to the satisfaction of the Catholic Church. The fact that the original proponents of the Ernst Mach Society had all belonged to that dubious clique certainly had also "been brought to our attention."

Moritz Schlick, as chair of the Ernst Mach Society, was summoned to the police headquarters of his town district and apprised of the decision. He immediately objected and tried to get it reversed. His argument: the society had always been absolutely apolitical, and its work had always been "true to the spirit of the man after whom it was named." It had engaged only in purely scientific activities, just as its statutes declared. The whole enterprise had no connection whatsoever with atheism.

Schlick firmly denied any "leanings toward the Social-Democratic Party." He was well known to be apolitical to the bone. "Not for a minute would I remain chair of an association that engaged in any sort of party politics."

The aim of the society was solely to promote the scientific worldview. This was the exact opposite of ideological propaganda! And it was merely by "pure chance" that one of the officials of the society happened to reside in a communal apartment building. (This was Otto Neurath, of course.) "Moreover, the fact that Otto Bauer had once delivered a talk under the aegis of the Ernst Mach Society was similarly due to just one of those chance events." Hence, the decision to dissolve the Ernst Mach Society was based on fallacious grounds.

"It is somewhat tragicomical," Schlick explained to the police directorate, "that I should be cast as leader of an association with anti-governmental tendencies. Indeed, I am among those university professors who feel the most profound and most sincere sympathies toward the present government, and I even expressed this sympathy in the summer semester of 1933, in a spontaneously written letter addressed to Chancellor Dollfuss."

Moritz Schlick's protests were eloquent, but in vain; they fell on deaf ears. Disheartened, he wrote to Carnap: "So the Ernst Mach Society has really been disbanded now, obviously before my protest had reached the relevant circles."

Schlick, who had never belonged to any political party, was now required to join the Fatherland Front. This action was expected from every state employee. Anyone who refused to do so had to face dire consequences.

Professor Gomperz had refused, and before he knew it, he had been dismissed. It would have been counterproductive to point out to the authorities that he had been instrumental in bringing Ernst Mach to Vienna. That "distinction" would not have earned him any kudos from Dollfuss. And so, at age sixty-three, Heinrich Gomperz was forced out of his native land, and eventually he became a professor at the University of Southern California in Los Angeles.

In contrast to Gomperz, Schlick was compliant, but this, too, was of no use. All it did was further upset Neurath and Carnap, who, from their positions abroad, were keeping a weary watch on happenings in Vienna. "Up with Dollfuss! Down with the unity of science!" sarcastically railed Neurath.

It is most unlikely that Engelbert Dollfuss had ever heard of the unity of science. His sole goal, after all, was the unity of power. But Dollfuss did know a thing or two about positivism, and he was dead set against it. The most famous proponent of "legal positivism," Hans Kelsen, had framed the constitution of the young Austrian republic. But Kelsen's constitution was now ancient history, having been replaced by the so-called May Constitution. This deft act of legerdemain took place in May 1934, when the banished Austrian parliament finally managed to reconvene, if only for a few hours. The delegates from the left were not invited, however—no need to bother them! The briefly reconstituted House approved the law for replacing the constitution, and instantly thereafter, the parliament sank back into oblivion again. The proper formalities had all been duly respected.

The new constitution began with the words "In the name of God the Almighty, the origin of all law," thus making it crystal clear that the law of the land did not come from mere mortals. And to fully acknowledge the "corporate basis" of the state, two corporations—one of state officials and one of farmers—were established with great fanfare. Other corporations could wait.

The Nazis, however, chose not to wait. In July 1934, the Austrian National Socialists mounted a putsch. Men from the SS-Standarte 89 wearing stolen army uniforms stormed the chancellery and the headquarters of the national broadcasting service. Their amateurish coup flopped miserably within a few hours, but before it had run its course, Engelbert Dollfuss, bogeyman number one of the Nazis, had been shot in his office, and this time fatally. He was left to bleed to death. After that, various further uprisings cost several hundred lives, much as had the repression of the left a few months earlier.

Kurt Schuschnigg became the new leader of the Fatherland Front. He looked like the perfect personification of a stiff headmaster. As chancellor, he carried on with the politics of his predecessor, convinced that Mussolini was his only hope in foreign affairs, and that domestic politics was best served by clerical-fascist folklore.

## "A POTENT INTELLECTUAL EXPLOSIVE"

It was only natural that the scientific worldview found it hard to breathe in an authoritarian state that sought its support in Catholic pilgrimages and the reproduction of medieval pageants.

Although Schlick did not politically oppose the corporate state, his enlightened philosophy was utterly alien to it. Ernest Nagel, a keen-eyed young American philosopher who attended some of Schlick's lectures in 1935, wrote: "It occurred to me that although I was in a city foundering economically, at a time when social reaction was in the saddle, the views presented so persuasively from the Katheder were a potent intellectual explosive. I wondered how much longer such doctrines would be tolerated in Vienna."

Not much longer, this was clear, but the dénouement turned out to be far more dire than could have been imagined at the time.

Nagel, by the way, was just one of many young scientists who came to Vienna to get a taste of the scientific worldview, and he went away deeply enriched. He became well known as an American exponent of logical positivism in the Vienna Circle style, writing a seminal book called *The Structure of Science,* and he also co-authored a famous book on Gödel's incompleteness theorems.

First among the visitors had been Frank P. Ramsey, who visited not only Wittgenstein on several occasions, but also Moritz Schlick. While passing through town, Ramsey even took a quick taste of psychoanalysis with Theodor Reik, a disciple of Freud. Everything moved ultra-swiftly with Frank Ramsey: he was barely twenty-three when he became Director of Mathematical Studies at King's College in Cambridge, and he died tragically young, aged only twenty-six, of hepatitis.

Young Alfred J. Ayer arrived from Oxford. He stayed for half a year. After his return to England, he published a highly successful book called *Language, Truth, and Logic,* which presented the basic ideas of the Vienna Circle with masterful clarity. The American philosopher Willard van Orman Quine also showed up, shortly after receiving his doctorate at Harvard. As a consequence of meeting Herbert Feigl, Quine had decided to visit Vienna as a postdoc. After his return from "the city of my dreams," as he described it, Quine became one of the most influential American logicians and philosophers.

An intense exchange of visits and ideas between Poland's logicians and the Vienna Circle started with a lecturing tour of Poland by Karl Menger. And from Scandinavia came many visitors, such as the Norwegian Arne Naess (1912–2009), who later became a pioneer of experimental philosophy, and the Finnish philosophers Eino Kaila (1890–1958) and Henrik von Wright (1916–2003).

The Vienna Circle seemed to be appreciated everywhere more than in Austria—everywhere, that is, with the exception of Germany, where the Nazis made short shrift of positivism. The visits from Berlin—by Hans Reichenbach and Carl Hempel, for instance—had come to an abrupt end.

By this point, all the weight of the Circle rested on Moritz Schlick's shoulders. Carnap and Neurath were abroad, and the stalwart Hans Hahn, with his loud voice and clear words, had unexpectedly died in his fifty-fifth year. During one of his last lectures in the summer term of 1934, he had suddenly had to break off, overwhelmed by stomach cramps. It was cancer. An emergency operation was attempted, but Hahn did not survive it. He died the day before the Nazi putsch. After that, there was no longer anyone alive in Vienna who had belonged to the erstwhile *Urkreis*. An epoch was drawing to a close.

As Ernest Nagel wrote: "With Hans Hahn's premature death, Rudolf Carnap's departure for Prague, Otto Neurath's exile in The Hague, and Kurt Gödel's visits to Princeton, the Circle had been deprived of some of its most original and strongest members."

Shortly before his death, Hans Hahn had resumed his correspondence with Paul Ehrenfest, his close friend from student days, and one of the "inseparable four." In the meantime, Ehrenfest had become close to Einstein and was now a professor of theoretical physics in Leiden.

"As for myself," Hahn wrote to his friend, "there is little to report; the external conditions are unfavorable, since salaries have been cut drastically (with particularly loving attention given to university professors); my main interest is currently in philosophy (logical empiricism with anti-metaphysical leanings)."

Not long after that, Hahn was shocked by the unexpected news of the sudden death of his friend. Paul Ehrenfest had been suffering from serious depressions for many years, just as his former teacher Ludwig Boltzmann had. Though he was a highly original physicist, he felt that he had lost touch with quantum physics, which was moving ahead quickly. Once, in a demoralized letter to his students, Ehrenfest candidly poured out his despair: "Every new issue of the *Zeitschrift für Physik* or the *Physical Review* immerses me in blind panic. My boys, I know absolutely nothing."

Hitler's rise to power horrified Ehrenfest, and to cap it all off, his son Vassili was mentally handicapped. It was all just too much for him to bear. Finally, at his wits' end, Ehrenfest decided to kill himself, and to take poor little Vassili with him. He left behind letters to his many friends, trying to justify his deed. He wrote that he had viewed it as his duty.

## VACANCIES

The dauntless professor Hans Hahn had been a royal pain in the neck for the Austrian government. He had never disguised his Social Democratic convictions. After Hahn's death, the ministry for education abolished his position. After all, when a professor of mathematics finds time for philosophy, and even for agnosticism and anti-metaphysics, this clearly proves that his professional duties leave him with far too much idle time. Ergo, Hahn's chair was obsolete, a "superfluous entity"—a perfect time to use Occam's razor, so to speak. Hence, away with the chair!

Karl Menger was the one most directly affected by this measure. He would have been Hahn's natural successor. Now such a prospect was gone. But the young associate professor still held out some hope: in the coming year, another university chair would become vacant, since the eminent mathematician Wilhelm Wirtinger would reach retirement age in 1935.

However, when the time came, the mathematics faculty's choice for that chair was not Karl Menger but the far less known Karl Mayrhofer, an amanuensis of Wirtinger. One thing they were sure he would not do was dabble in philosophy.

Extremely upset, Moritz Schlick once again wrote a letter to the ministry. He voiced profound opposition to the faculty's vote. Despite being a philosopher and not a mathematician himself, he knew perfectly well that two candidates for the vacant chair were far better qualified than Mayrhofer: namely Karl Menger, who enjoyed a worldwide reputation in his field, and Emil Artin, an algebraist who had solved Hilbert's seventeenth problem. Artin, incidentally, was a born Viennese and a disciple of Gustav Herglotz, who had been one of Hahn's "inseparable four." Also, it was no secret that Artin, currently a full professor in Hamburg, was eager to leave Germany: his wife was Jewish.

But Moritz Schlick's intervention had little influence on the tangled networks hidden in the subterranean depths of his university. It was Mayrhofer who got the job. After the *Anschluss* took place, it came out that he had long been an illegal member of the National Socialist Party.

Karl Menger and Philipp Frank had both stressed, in their obituaries for Hans Hahn, that their great friend had been "the true founder of the Vienna Circle." Within the most recent decade, that group had reached international renown. Ernest Nagel wrote in the *Journal of Philosophy* that "such a catholic interest" in logical analysis could hardly be found in any other place on Earth. It was a truly unique circumstance that the disciplines of mathematics, physics, law, medicine, and sociology—and of course philosophy—were all equally well represented in the Circle's weekly gatherings.

And yet, despite all these words of high praise, the highly "catholic" (though of course not Catholic!) Vienna Circle had already come to a standstill.


OTTO'S PROTOCOL

"There will be no meetings of the Vienna Circle this winter," Schlick wrote to an American colleague in the fall of 1933. "Some of our elder members have become too dogmatic and could discredit the whole group; I am trying now to build up a new circle of younger people who are still free from prejudices."

The regular Thursday evening meetings resumed only in the fall of 1934. "We naturally feel very intensely and painfully the fact that Hahn is no longer with us," Schlick confided to Carnap. On the other hand,

Schlick did not seem too upset at no longer regularly running into Otto Neurath, who clearly was one of "our elder members who have become too dogmatic."

The feud between Neurath and Schlick was of long standing. The two had never felt they were on the same wavelength. For several years, Hahn had been able to mediate, but during the infamous debate on "protocol sentences," their dispute suddenly boiled over and turned acrimonious.

It all started quite harmlessly, with Carnap's paper "The Physical Language as Universal Language of Science." This paper claimed that all scientific theories were ultimately based on so-called protocol sentences, which described facts "in all their plainness" and which "needed no further confirmation."

How to recognize protocol sentences? They were propositions, wrote Carnap, "for which no other protocol sentences are needed." Schlick gently objected that this definition seemed circular. Neurath took over and, in a written essay, tried to supply a more precise definition. A protocol sentence is not itself an assertion, but a report on an assertion. It has to contain the exact time of day, and the name of the person reporting. Thus, for instance:

"Otto's protocol at 3:17 PM:

[Otto's language thinking at 3:16 PM was:

(At 3:15 PM a table was observed by Otto in the room.)]"

Now this was not so remarkable in terms of "plainness." But Neurath insisted: protocol sentences were never to use subjective words like *here* and *now*, nor the pronoun *I* or the adjective *own*, for this would lead them straight into the fatal traps of idealistic philosophy.

Scientific knowledge, Neurath explained, must be communicable. Science has no room for subjective expressions. Rather, unified science ought to use a "universal slang," so to speak (by which he probably meant *jargon* or *interlingua*). Neurath added that every child could acquire this "slang" by training. Since there exist no sentences that can be asserted without validation, there exist no basic protocol sentences either. And in typical Neurath style, he added that the only people who believe otherwise belong to long-discredited "school philosophies." It is a principle of science that every claim is subject to revision. This principle must also hold for protocol sentences; a revision might well prove necessary, even in such a basic utterance—for instance, if one's watch has supplied the wrong time of day, that will invalidate the time stamp on the sentence, and hence it will need to be revised.

Neurath closed his essay, as was his style, with a few quick reflections on others' views: Wittgenstein's attempt to portray philosophy as a much-needed "staircase of clarifications" must be considered to be a failure, and the emergence of tight and collegial collaboration in support of the unity of science was underscored by the rapid progress of the Vienna Circle. Neurath concluded by saying that his remarks would doubtless provoke Carnap into correcting them in numerous spots.

Which is exactly what Carnap did in the next pages of the same issue of the journal. He wrote that there exist several methods for constructing a scientific language. He now believed that an approach attributable to Karl Popper was best, for which reason it would be highly desirable that the latter's results were published. (This gentle hint was aimed at Schlick.)

According to Popper, *any* statement can serve as a protocol sentence (or "basic sentence," as Popper preferred to say) under the proper conditions. On the other hand, no statement was ever absolutely certain— that is, so certain that it would never need any additional corroboration. At this point, Carnap candidly confessed that his previous views on this point had been wrong. Moreover, it seemed to him that Popper's approach "reduced the danger that younger persons would unwittingly stray into metaphysics while searching for protocol sentences."

"The next task," Carnap concluded hopefully, "will be to develop the theory of science in a constructive collaboration."

In fact, the protocol debate spread like a wildfire. Carl Hempel contributed to it, as did Karl Popper, Edgar Zilsel, and others as well. Moritz Schlick confided to Carnap that he viewed the whole debate as being "beside the point and of little interest." But nonetheless, even he also eventually entered the fray.

TRACES OF METAPHYSICS AND POETIC CONFIRMATIONS

Schlick wrote a short paper called "The Foundations of Knowledge" during his spring holidays in Amalfi, "while sitting relaxedly on a balcony overlooking the blue bay of Salerno." Feeling refreshed and in an excellent mood, he gave free rein to his thoughts, "in exactly the fashion that they popped into my mind," as he reported to Carnap.

It is quite true, said Schlick in his essay, that protocol sentences are of the same type as any other scientific statement: they are hypotheses, clearly nothing but hypotheses ("as some of our authors have pointed out almost triumphantly," he added, making a jab at Karl Popper). Indeed, as soon as an observation is written down or otherwise asserted, it becomes a thing of the past. The sentence recording the observation might conceivably have been based on an error, or might have been made in a dream, or during a drunken spree, or under hypnosis (etc.), and thus may at some point need to be revised. But what Schlick calls a *Konstatierung* (sometimes misleadingly translated as "confirmation" or "affirmation") is final and definitive at the moment it is made, and remains forever beyond appeal. A *Konstatierung* is always of the form: "Here and now this and that."

True *Konstatierungen* (acts of noticing) cannot be written down: this would turn them into protocol sentences, and thus hypotheses. No, Schlick had in mind something quite different. Scientific hypotheses lead to specific predictions: we expect something. Whenever we confirm or refute such an expectation via our own observations, we are *noticing* something. Science leads us to such acts of noticing, but science is not based on them.

Schlick wrote of his nebulous *Konstatierungen* that "they do not lie at the foundations of science, but knowledge touches them like a flame, reaching each one for only an instant and then immediately devouring it . . . These moments of fulfillment and combustion are essential. All the light of knowledge stems from them. And this is the light into whose origins philosophers inquire when they are seeking the foundation of all knowledge."

Philosophers from ancient Greece could not have put it any better—and Schlick's balcony in Amalfi certainly offered an enchanting view of the sparkling Mediterranean. From foggy Holland, however, Neurath replied derisively. "Some may appreciate such lyricism, but whoever takes up the cause of radical physicalism in the service of science [ . . . ] will not wish to claim to be a philosopher belonging to this school."

In his essay "Radical Physicalism and the 'Real World,'" Neurath launched a polemic against Schlick. He began with the cheerful remark: "A representative of the Vienna Circle once quipped that we are all far better at pointing out residual traces of metaphysics in our neighbors than in ourselves." And thus Neurath was proposing to "point out residual traces of metaphysics" in Schlick's thinking.

This he did at great length and with glee. At the end of his essay, Neurath summed up his diagnosis: "It seems as if we have encountered here the last traces of a coherent metaphysics." However, it was unclear whether Neurath, by using the pejorative phrase "the last traces of a coherent metaphysics," meant to suggest that Schlick would eventually overcome them.

At first, Schlick did not intend to deal with these "rather silly remarks of Neurath" in detail. He confided to his friend Carnap, "As you can imagine, I was taken a bit by surprise. Of course I am not going to reply directly." On the other hand, he did not wish to withdraw completely from the discussion, although as time passed, the whole thing struck him as "more and more ridiculous."

Therefore, in his next essay he wrote: "I was a little surprised when I was accused of being a metaphysician and a poet. Finding it impossible, however, to take this indictment seriously, I was neither shocked by the one nor flattered by the other, and did not intend to take up the discussion." But now he did, "in a humorous vein" and without mentioning Neurath by name—thus in a doubly hurtful manner.

Neurath had claimed that statements can be compared only with other statements, and not with facts. Schlick asked "in all innocence": Why should a statement in a tourist guidebook—say, that a certain cathedral has two spires—not be comparable with reality?

FIGURE 11.1 Schlick sees what he sees.

Neurath wanted to link the truth of a statement to its consistency with other statements. But Schlick saw in this nothing but a return to the coherence theory of truth, which was long since dead and buried. He himself opted for the correspondence theory, as intuitively understood by unsophisticated people. The truth of a sentence has nothing to do with whether it agrees with other sentences (as the coherence theory claims), but on its correspondence with reality—and whoever wished to use this claim to accuse him of metaphysics was free to do so, wrote Schlick; but come what may, he was not going to give up his *Konstatierungen*. "I see what I see!"—and he would surely still see it even if all the scientists on Earth were to claim otherwise. He continued with a barb slung at Neurath (but still not naming him): "If anyone were to tell me that the real reason I believe in the truth of science is that it has been adopted 'by the scientists in my culture circle,' all I could do is smile. Yes, I do indeed place great faith in scientists—but that is only because whenever I have been able to test the claims of these fine folks, I have always found them to be trustworthy."

Carnap tried to save Neurath from the charge of having fallen for coherence theory. Surely Neurath did not believe in such a crackpot idea? "And yet I must confess," Carnap added, "that some of his wordings can be read that way."

But was it merely a question of "wordings"? Was the slowly widening intellectual gap between Carnap and Schlick based on nothing but different ways of using words? Schlick tried to mollify his friend by suggesting that the split could be caused by "your mathematician's mind-set as opposed to my physicist's mind-set." Mathematical physicists, after all, must check the mutual consistency of their equations, while experimental physicists must check whether theoretical claims correspond to reality. Or, to revisit the words of old Mach, "The aim of science is the fitting of facts to thoughts, and the fitting of thoughts to each other."

Schlick's conciliatory gesture toward Carnap was of little avail. The debate about protocol sentences had clearly demonstrated that the old philosophical problems, or "pseudo-problems" if one preferred, could not be made to vanish overnight. Obviously, the widely praised collective of thinkers of the Vienna Circle was just as beset by mutual misunderstandings as ever the much-maligned "school philosophies" had been.

Karl Menger's valedictory comment on the subject of protocol sentences was: "This discussion was just one of the many that I followed silently."

## INDISCRETIONS

There was little momentum left any longer from the erstwhile "turning point in philosophy." Even Carnap and Schlick's long friendship had occasional dark moments. This happened in particular whenever the topic of Wittgenstein was broached. The preface of Carnap's *The Logical Syntax of Language*, for instance, required lengthy negotiations.

Schlick admonished Carnap: "The trouble arises entirely from your insistence on ascribing to Wittgenstein opinions different from those he actually has. . . . His actual views and statements are in every respect far more free. Haven't Waismann and I stressed that point for a long time?"

On another occasion, Schlick defended himself somewhat ingenuously with the words: "You cannot claim that I faithfully cling to Wittgenstein's position, since he himself has changed it considerably." This was indeed the case. Waismann had been kept busy around the clock by the necessity of updating his planned book on Wittgenstein's philosophy.

Shortly before the Nazi coup in July 1934, Schlick wrote to Carnap: "It has recently been very hot and exciting in Vienna. This time, Wittgenstein arrived late from England, and I haven't been able to talk with him very much."

This was when Schlick learned that Wittgenstein had suddenly decided to write his own book—this after Friedrich Waismann had devoted years of his life to writing a book explaining Wittgenstein's new philosophy. The Marxist student of philosophy Heinrich Neider conveyed the hot-from-the-oven news to Neurath in Holland: "Wittgenstein wants to prevent the publication of Waismann's book and buy the rights from him! Then he will be able to write the book himself. Oh my, oh my . . . How petty is this great man!"

Neurath found his darkest suspicions confirmed. In blocking Waismann's book, Wittgenstein would be using his money in an ugly way. "This masterstroke planned by the master himself is really sickening! I would have lost my patience long ago! What does the prophet say about this?"

"The prophet" (meaning Schlick, of course) said nothing at all, and ere long, "the master," in his reliably unreliable manner, changed his mind once again. Thus Waismann kept toiling away on his cursed book, and he even gingerly resumed his occasional reports to the Circle on his conversations with Wittgenstein. But not a word thereof was to be spread to the outside world—not even to former members of the group. Schlick made sure that neither Neurath nor Carnap was informed about what transpired during the meetings of the Vienna Circle.

Neurath wrote to Carnap: "Only dark reports on the Schlick Circle reach my ears. Everyone is bound by oath to keep silent on everything that is whispered from the holy books."

Naturally, not everyone respected the oath. Heinrich Neider, a fervent admirer of Neurath, kept the latter up to date until one day he remembered: "To my dismay, it just now occurred to me that I forgot to inform you that the strictest silence has to be maintained concerning the readings from the 'holy book.' I thus must most urgently beg of you not to make use of my indiscretions. However, under the seal of strictest secrecy, reports will continue to flow to you."

Physicist Wolfgang Pauli gently teased Schlick that the Vienna Circle had become "a religion." Less gently, Gustav Bergmann jeered that Schlick's group had been converted into a "holy sanctuary for Wittgenstein's philosophy," and all this without Wittgenstein ever appearing in person, since over the years, Waismann had managed to become his master's mouthpiece.

In fact, this was the only claim to success that Friedrich Waismann had to his name. He was nearing forty, and despite twenty years of valiant efforts, his doctoral studies were not yet completed. His Wittgenstein bible, advertised again and again as being "in print," seemed more unfinished than ever. And on top of all this, his modest position of assistant librarian was abruptly chopped. The government had no money to spend on such trifles.

Everything seemed to be conspiring against Waismann. He was the chief target of the student association's relentless campaign against the "Jew-polluted environment" reigning in the institute of philosophy. And his lectures on mathematics at the adult education center, the Volksheim in Ottakring, had been suspended by the Fatherland Front.

Schlick tried to help out, by throwing his international reputation into the ring. But even this had little effect on the ministry. Moreover, the fact that Schlick was himself a member of the Fatherland Front counted for nothing—at this point, every civil servant was required to be a member.

In desperation, the unemployed Waismann was forced to face his terror of examinations: plucking up all his courage, he finally took his doctoral defense. To his great relief, he passed, but despite now having a PhD, he was still jobless.

Porträt des Inhabers.

Eigenhändige Unterschrift des Inhabers:

*Rauel Rose*

Wien, den *16 . X . 1924*

FIGURE 11.2 Rose Rand.

Waismann was not the only problem child among Schlick's students. Rose Rand was poor as a churchmouse. She managed just barely to make ends meet by translating scientific texts from Polish to German. She also faithfully kept the minutes of the Vienna Circle's meetings, but this she did without any remuneration.

For a brief moment it looked as if she would get paid for her dedication to the Circle after all. Carnap wrote to her from Prague, saying that he and Neurath "would be very grateful if you could provide us with a full transcript of the meetings of the Circle during the winter semester. I will then be able to send you some money through Neider. We are particularly interested in original texts from Wittgenstein. Please highlight any verbatim quotes."

Soon, however, Carnap had to report to Neurath: "Rose Rand just informed me that Schlick has refused to give his approval, and of course this only happened after she had completed the task."

To mitigate at least a bit of Rose Rand's rough plight, Moritz Schlick found her a job with psychiatry professor Otto Pötzl—the one who once, as a fresh young MD, had diagnosed the budding writer Robert

Musil as suffering from "severe neurasthenia." At this point, Pötzl had succeeded his former boss Julius Wagner-Jauregg as head of the clinic for psychiatry and neurology. Schlick had known and appreciated Pötzl for several years: indeed, on two occasions, the latter had examined Johann Nelböck and recommended him for detention.

Rose Rand, while hard at work in Pötzl's university clinic, managed also to write a philosophical paper: "The Concepts of 'Reality' and 'Unreality' in the Minds of the Mentally Ill" was based on interviews conducted with patients. It remained unpublished, however, and the Vienna Circle did not take up her essentially experimental approach, although it clearly held more promise than all their prolonged debates on protocol sentences.

## "THE GREATEST LOGICIAN SINCE ARISTOTLE"

Not long thereafter, Moritz Schlick had another reason for turning to Otto Pötzl once again.

In the spring of 1934, Kurt Gödel had returned, just as planned, from his guest semester at the Institute for Advanced Study in Princeton. Soon afterward, however, it became clear that the young logician was suffering from mental problems and urgently needed medical help. Schlick wrote to Pötzl, asking "for your kindness in permitting me to bring my esteemed colleague, *Privatdozent* Dr. Kurt Gödel, to your special attention."

In his letter, Schlick stated that it was impossible to praise Gödel's intellectual faculties too highly. Gödel was a pure genius. "He is a mathematician of the highest caliber, and his findings are universally recognized as epochmaking. Einstein had no qualms in describing Gödel as the greatest logician since Aristotle, and it is beyond any doubt that Gödel, young though he is, is the world's greatest authority on foundational questions in logic."

The "greatest logician since Aristotle" was not yet thirty years old, but he was constantly plagued by fears of persecution and poisoning. Pötzl did his best, but it is never easy to treat a case of profound paranoia.

There exist countless anecdotes about psychiatrists, and probably more are set in Vienna than anywhere else. After all, Sigmund Freud and Alfred Adler did their main work here, as did Julius Wagner-Jauregg and Richard von Krafft-Ebing. There is even a canonical collection of these stories, in a chapter of the book *Aunt Jolesch, or the Decline of the West in Anecdotes,* by the Viennese writer Friedrich Torberg. In that chapter, Professor Pötzl occupies a place of honor. One of the stories describes how Pötzl took special care of one patient suffering from paranoia (it surely cannot have been Gödel or Nelböck).

The treatment was going well and the patient seemed almost cured. But then, suddenly, he suffered a relapse, to his doctor's dismay. "There are people here who are trying to kill me," the patient kept muttering, stubbornly averting his gaze. "Come now, come, now," said Pötzl reassuringly, "no one here is planning any such thing." "But they are!" said the patient, "I have proof. They are trying to poison me." "Nobody wants to poison you, my dear friend." But the patient persisted, and so did Pötzl. With unflappable gentleness, he rolled out all the standard arguments that had worked with other patients. But in this case, they had no effect. All his best strategies failed him. "Everyone persecutes me!" insisted the patient. "I know it. Everyone. Even *you,* professor." "What?! *Me*?!" exclaimed Pötzl, suddenly losing his cool. "Have you gone mad?"

To return to Kurt Gödel: Professor Pötzl efficiently committed the greatest logician since Aristotle to the sanatorium in the village of Purkersdorf. This art nouveau building is an architectural jewel designed by Josef Hoffmann at the turn of the twentieth century. It is not too far from Vienna's western suburbs and is quite close to Otto Wagner's beautiful St. Leopold church, on the grounds of the psychiatric clinic Am Steinhof.

After some time, Kurt Gödel was released, but he was by no means cured. In the following years, he kept commuting between the mathematical seminar in Vienna, the Institute for Advanced Study in Princeton, and various psychiatric clinics such as Purkersdorf, Rekawinkel, and Aflenz. Again and again, he was afflicted by his strange "states." He no longer frequented the gatherings of the Vienna Circle, but he did attend those of the Vienna Mathematical Colloquium as often as he could.

It was already beyond any doubt that Gödel's work was epoch-making. In the 1930s, mathematical logic bloomed into one of the leading sciences, comparable in its importance to quantum physics. Brilliant young researchers such as Alfred Tarski from Poland and Alan Turing from Britain were drawn to it, and added their own marvelous insights to it.

Alfred Tarski (1901–1983) was a frequent visitor to Vienna, where he had endless discussions with Gödel, Popper, and Carnap. It was in this period that Tarski developed a formal definition of the concept of truth. His idea mirrored the commonsensical correspondence theory of truth. For example, the statement "Fido is barking" is true if and only if Fido is indeed barking. In this example, a higher-level statement—namely, "The following is true"—is applied to a lower-level statement—namely, "Fido is barking." Formally, this act linking two distinct linguistic levels requires a metalanguage, like those that Rudolf Carnap was developing at the time.

In later years, when Tarski was living in the United States, he became, to his chagrin, not the greatest logician of his day—that enviable title indisputably belonged to Kurt Gödel—but "the greatest *sane* logician," as he liked to quip.

## FROM GÖDEL'S ABSTRACT MUSINGS TO THE BIRTH OF COMPUTER SCIENCE

The impact of Alan Turing was even greater than Tarski's, and it shapes every facet of today's world. In the mid-1930s, however, Turing's abstract ruminations seemed light-years away from any possible application.

In 1931, Gödel's incompleteness theorem had revealed the seemingly crazy fact that in any axiomatic system that is rich enough to include number theory, there must exist true sentences—in fact, an infinite number of them!—that have no formal proof. This is highly disconcerting, but one might still hold out the hope that there could exist a mechanical procedure for determining whether any given statement of number theory is true or false. Such a truth-determining mechanism might conceivably totally bypass the concept of formal provability. Maybe there was some telltale sign lurking deep in all those patterns of symbols that expressed mathematical truths, but lacking in all other patterns. Maybe there was some way to detect the presence or absence of that sign. Could this somehow be done?

David Hilbert, in his famous 1900 lecture in Paris, had formulated this crucial question touching on the most central core of the nature of mathematics, and he gave it the name *Entscheidungsproblem,* or "decision problem." Alan Turing thought and thought about this question of Hilbert's; it simply would not let go of him.

Turing had of course deeply absorbed Kurt Gödel's ideas, and he brilliantly adapted them to the world of computing machines (which at the time did not yet exist) to explore the deep question that Hilbert had posed in 1900. To do this, however, Turing first had to answer the question: Just what *is* a general computational procedure, or *algorithm*? To this end, he devised a set of rudimentary machines, later called *Turing machines,* that could do all sorts of computations. For instance, one type of Turing machine could add two given numbers (but it could do nothing else). Another type could multiply two given numbers (and nothing else). Yet another type could determine whether a given number was prime—and so forth and so on. These hypothetical machines, if they were actually built, would not be practical at all—in fact, they would be pitifully ponderous and unimaginably slow—but they would be guaranteed to work, and that's all that mattered, in this abstract context.

Some Turing machines would never stop churning, since they might be given a task that never ends, such as "find the largest prime number." Since there is no such beast, a Turing machine built to carry out that task would never halt. Turing showed that the question "Will such-and-such a machine eventually come to a halt?" was equivalent to Hilbert's *Entscheidungsproblem,* and with that insight, he had effectively converted questions about mathematical truth into questions about computing machinery. This was an amazing contribution.

However, Turing's greatest conceptual breakthrough came when he showed that there is a special kind of Turing machine that would be able, in principle, to imitate the activities of any particular Turing machine. He called such a general machine a *universal automaton,* although today it is most deservedly called a *universal Turing machine,* and here we will call it a *UTM* for short.

The idea is that if you feed a UTM a description of machine X, as well as specific numerical input that you might feed to X (for instance, suppose X is the primeness-checking machine, and you set X to work on the number 641), then the UTM will start churning away, imitating machine X working on its numerical input, and eventually the UTM will give you exactly the same result that X would have given you (in this case, "yes," since 641 is a prime number), although the UTM will work far more slowly than X itself would work.

Now Alan Turing was a most imaginative fellow, and one wonderful day, while he was lying in a meadow, catching his breath after a solitary long-distance run, he hit on a brilliant idea, inspired by Gödel's work. By the time he rose to his feet to walk home, Alan Turing had solved Hilbert's *Entscheidungsproblem.*

Turing's idea—a beautifully Gödelian one—was to feed a universal Turing machine a description of *itself.* Thus instead of imitating some *other* machine, the UTM would imitate itself. This deeply twisty idea grazed paradox in a most provocative fashion: after all, if a machine is imitating itself, then it is perforce imitating itself imitating itself, which means that it is imitating itself imitating itself imitating itself, and so on and so forth, ad infinitum.

By looking carefully at the implications of this kind of infinite regress (like mirrors reflecting each other, or wheels within wheels, or snakes swallowing their own tails . . . ), Turing discovered that there are fundamental limitations on all possible Turing machines, and in particular that it is not possible to build a machine of any sort that can always tell whether a given Turing machine will halt. This deep insight was equivalent to showing that there is no solution to David Hilbert's *Entscheidungsproblem.* This was a result of almost indescribable significance, in mathematics, in logic, and in the theory of computation. Its implications reverberate everywhere in today's world of computers.

A few years later, after World War II had broken out, Alan Turing was working in Bletchley Park on deciphering the Enigma code of the German Wehrmacht. This was a deep computational challenge, and before long it brought him straight back to his earlier ideas about universal Turing machines. This in turn led him to the challenge of designing and building an actual, physical programmable computer (not just a theoretical abstraction).

At the same time, but on the other side of the Atlantic, the Gödel-inspired mathematician John von Neumann was spearheading a parallel development. Thus was born the era of electronic computers. And as our story shows, the digital world that holds the entire world in sway today sprang out of extremely abstruse investigations into mathematical logic carried out by a quiet and self-effacing (and sadly paranoid) member of the Vienna Circle, way back in the early 1930s.

### "It was Nelböck!"

Not every philosopher waxed enthusiastic about mathematical logic. Some even thought that it missed by a mile the true essence of logic. Such was the opinion of young Leo Gabriel (1902–1987), a philosopher who had written his doctoral thesis, called "The Concept of God," under Heinrich Gomperz. Gabriel sought a logic that was entirely different from formal logic: his "integral logic" was going to grasp "the truth of the whole." It would be revolutionary, if it worked.

In philosophy, "the whole" is even fishier than "the nothing." It was used, however, as a password by thinkers who opposed analytical philosophy, and Leo Gabriel was one of them. He was deeply miffed when Moritz Schlick pointed to the vacuity of phrases such as the venerable Aristotelian cliché "The whole is more than the sum of its parts." What, asked Schlick, is meant by the vague phrase "the sum of its parts"? Leo Gabriel hated this type of pedantic hair-splitting. Although he had earned his teacher's diploma with Schlick, he saw nothing in the Vienna Circle except for positivism, whose deadly dull spirit totally contradicted his noble idea of "the whole."

Soon Gabriel became a teacher of history and philosophy at a secondary school. As a deeply religious Christian, he occasionally wrote for the Catholic daily *Die Reichspost*. After the 1934 civil war, he took charge of the philosophy section in the Volksheim in Ottakring. This center for adult education had previously been Social Democratic terrain, just like Otto Neurath's Museum for Social and Economic Affairs. The Dollfuss regime, rather than closing these institutions down, chose to fill them, in a not-so-friendly takeover, with new blood and a so-called "new spirit" whose stated goal was "the propagation of a Christian attitude rather than an ethically neutral scientism."

Ever since their student days, Leo Gabriel had been a friend of Johann Nelböck, the dreaded nightmare figure in Moritz Schlick's life. Nelböck, in defiance of police orders, had returned to Vienna, as he had felt utterly out of place in his home village of Krandeln. He was now living on the drab Westbahnstrasse, subletting a small room that opened onto a backyard. He gave lessons and tutorials to philosophy students. He was one of many destitute self-employed scholars stranded in Vienna. His parents sent him five to ten schillings per month—less than what a worker on the dole would receive for a week.

Like his friend Dr. Leo Gabriel, Dr. Johann Nelböck wrote now and then for the *Reichspost*. One time, he published in it a review of Albert Einstein's 1934 book *Mein Weltbild*. Without wasting a moment, Nelböck launched into a vicious tirade against the physicist's ideas. He drew attention to "the basic deficit" in Einstein's worldview: that deficit was, needless to say, "the lack of a deeper penetration into the matter of a strictly scientific foundation and method."

Neilböck railed against this "lack of a deeper penetration" with great vehemence, all the more so because "positivistic circles close to Einstein reject the existence of the objective (in particular in the fields of morality and law), and view it as a mental construct devoid of meaning."

When Nelböck wrote these words, he had in mind, of course, his archenemy Moritz Schlick, with whom he still had to settle accounts. Far be it from Johann Nelböck to let himself be deflected from the straight and narrow! And Moritz Schlick understood this ominous fact only too well.

In his *Reminiscences*, Karl Menger reports a poignant scene. During an official reception, he noted that one of the men standing near the president waved "rather intimately" to Schlick, and the latter responded in the same way. "Oh, so you have friends high up in the government?" Menger asked teasingly. Schlick's expression grew dark, and he replied in a grave tone, "That's not a friend. It's a security agent who used to be my personal bodyguard."

This was the first time that Karl Menger had heard that his friend Moritz Schlick had been stalked for years by a psychopath threatening to kill him. But since these threats had never been followed up by any action, the police would always stop sending the bodyguard after a while, on the grounds that nothing worrisome had happened so far. The point was obvious and made perfect sense: Professor Schlick had never been harmed, so what on earth was the use of giving him a bodyguard? Thus over time, it became more and more difficult for Schlick to renew his request for protection.

Menger writes: "And, as though it had happened only yesterday, I remember the way that Schlick added, with a forced smile: 'I fear they are beginning to think that *I* am the one who is mad.'"

In the spring of 1936, Nelböck entertained high hopes of getting a teaching job, not at a secondary school—that had been ruled out forever—but at the Volksheim in Ottakring. His good friend Leo Gabriel had put his name forward. According to Gabriel's plan, Nelböck would give a lecture course on positivism. After all, having studied with archpositivist Schlick, he had a great deal to say on that topic.

But then the chairman of the Volksheim learned that Nelböck had twice been committed to a psychiatric clinic. After conducting an official inquest, he had to tell the candidate that he could no longer be taken into consideration for the job. The chairman expressed great regret at this negative outcome but added that some of the students might have objected to the course anyway, since positivism was so completely out of tune with the times.

For Nelböck, it was self-evident that the positivist Professor Moritz Schlick was behind this back-stabbing betrayal, and his friend Leo Gabriel confirmed his suspicion that Schlick had thrown a monkey wrench into the works in order to secure the Volksheim job for his old crony Friedrich Waismann.

Schlick again! Well, of course, Nelböck had never expected anything else. Schlick was behind it all. Schlick was carefully tracking his every move. This horrid behavior had to be stopped!

Over a year earlier, Nelböck had again acquired a revolver and some ammunition. A while later, though, he had reconsidered his plan and had tossed all his ammunition into the waters of the Danube. But now he was going back to his original plan, so he went out and purchased ten brand-new cartridges. He was smoking a great deal, and that night, all night long, his landlady could hear him pacing up and down in his little room.

Moritz Schlick's last lecture for the summer term of 1936 was scheduled for June 22. Nelböck decided early in the morning that this was the day on which he would kill Schlick and would commit suicide afterward. Stowing his revolver in his jacket pocket, he left his apartment at around eight a.m. His landlady asked him whether he would be back at noon. "No," replied Johann Nelböck.

The young philosopher from Krandeln made his way to the university district of Vienna, and waited on the so-called Lawyers' Staircase. Schlick would take the opposite turn, heading to the Philosophers' Staircase. Nelböck plunked himself down on a nearby bench and removed the safety catch.

Schlick took the D streetcar. A female student of his got on at the same stop. The two exchanged a few words. The weather was glorious. When they entered the university building, the student ran ahead to make sure she could find a seat for herself in the classroom.

Nelböck caught up with Moritz Schlick on the Philosophers' Staircase and fired four shots. Three of them were lethal. Then the murderer,

FIGURE 11.3  Murder on the Philosophers' Stairs.

rather than turning his gun on himself, calmly waited to be arrested. He seemed quite emotionless when the police hauled him away. Later, he said that he had forgotten all about his plan of suicide.

The news of the crime spread like wildfire throughout the building. "The Nazis did it!" was the first report. "No, it was Nelböck!" was the next. And the murderer had shouted, just before firing: "Take that, you miserable cur!" No, no—that wasn't it at all! He had shouted, right after firing: "You lecherous vermin!" Wildfire has many darting fingers.

When Nelböck was questioned, he claimed full responsibility for his act. He had reached the conviction that Schlick would oppress him for as long as he lived. Now that the affair was over and done with, he, Johann Nelböck, no longer cared about anything.

Austria happened to be in a state of emergency, and for that reason, Nelböck might well have been executed right away, but doubts were raised as to whether the murderer was fully accountable for his action.

This, then, would require a serious trial, and at a regular court. And during the inquest, it came out that Moritz Schlick had in no way been involved in the negative outcome of Nelböck's application for a position in the Volksheim.

Leo Gabriel was summoned as a witness but did not show up in court. He asserted that he was obliged to stay in Innsbruck during the trial. Thus his role in the fatal deed remained forever murky. The connection with student Sylvia Borowicka was also downplayed in the trial, by common consent. In front of the judges, Nelböck stressed philosophical rather than personal reasons for his action. He stated: "For me, Schlick's behavior expressed the utter lack of scruples of his so-called scientific worldview."

The judge summed things up as follows: In Nelböck's eyes, Schlick had "robbed him of his love, of his creed, and of his means of living." However, in the eyes of the law, Johann Nelböck was fully responsible. He was found guilty of murder and of illegal ownership of a firearm and sentenced to ten years in prison.

Eighteen months later, he was a free man again.

## CHRISTIAN PROFESSORS FOR A "BETTER FUTURE"

Both the crime itself and the subsequent trial had huge repercussions in the press. It is not every day that one philosopher shoots another. The crime chronicles gleefully gloated that "the sad sensation of the case" was that "conflicts of a philosophical nature could become motives for murder."

Nelböck was described as "a man of a truly philosophical physique, with hardly any muscles and bad posture" and with a thin-lipped face whose sole character trait emanated from his thick "philosopher's spectacles." Apparently those who do philosophy have a poor public image, or at least they did then.

The first wave of emotion had hardly subsided when sympathy for the culprit began to be voiced, together with harsh attacks on the character of his victim, laced with anti-Semitic barbs. This elicited semiofficial replies whose intentions were better than the taunts but hardly less depressing. It was pointed out that Schlick was of Aryan descent, that he had never left the Protestant Church, and that it was not in the least true that he had employed two female Jewish assistants, as was rumored everywhere—he had merely hired one Jewish librarian, and this only for a limited time.

There were also some touching obituaries. The young writer Hilde Spiel, who had studied with Schlick and had completed her doctoral thesis under Karl Bühler on "The Representative Function of Film," wrote in the daily paper *Die Neue Freie Presse:* "It is not often that a scholar becomes, as deeply as did Moritz Schlick, a model of excellence for his students. . . . There is no one who did not learn from him both lucid thinking and pure moral sentiments."

This warm viewpoint was definitely not shared, however, by a lead article in *Better Future,* one of the best-known journals in fascist Austria. It was titled "Admonition for Soul-Searching" and was signed by an anonymous "Professor Austriacus." It described with great sympathy the profound "clash of worldviews" that had taken place between the lonely young Dr. Nelböck and Moritz Schlick, that "idol of Vienna's Jewish circles." Professor Austriacus stressed a "truly scary aspect" of the case—namely, that Dr. Nelböck had not been born as a psychopath but had only become one "under the influence of the radically devastating philosophy professed by Dr. Schlick."

The eloquent if anonymous author bewailed the fact that under Schlick's leadership, the so-called Viennese Circle had carried out much activity "to the detriment of Austria's reputation as a Christian state." Even after the dissolution of the affiliated Ernst Mach Society, its doctrines, although "prohibited because they corrupted the people and the culture," were somehow still able to be openly taught at Austria's foremost university. Who could ever count the students whose innocent souls had been forever corrupted by the teachings of Moritz Schlick?

The author also reported that poor Johann Nelböck had been terribly upset whenever his tormentor Schlick proclaimed his nihilistic philosophy in class. After all, one can quite easily imagine the turmoil raging in the souls of young students when they hear, delivered from a high pulpit, words that negate everything that they have held sacred so far.

The article ended with these pious sentiments: "The philosophy chairs at the University of Vienna in Christian-German Austria must henceforth be filled by Christian philosophers! And it is to be hoped that the terrible murder case at the University of Vienna will hasten a satisfactory and total solution of the Jewish question." Thus wrote Professor Austriacus, looking forward to better days.

As Philipp Frank wrote to Albert Einstein, this seemed a strange conclusion, since neither Schlick nor Nelböck was Jewish. "But a cat always lands on its feet." And indeed, soon afterward, a certain Franz Sievering wrote in the pages of *Better Future* that there had never been any intention to brand Schlick as a Jew. "We claimed merely that he was a *friend* of the Jews, and that he had somehow even become an idol within Viennese Jewish circles."

As usual, the daily *Reichspost* found in the sad event a clear political lesson: "There is no bridge from Schlick's philosophical doctrine to the spirit of the new Austria." Everyone with a sense of responsibility ought to know where prompt action was needed. "The sanctuaries of science must open up to the new Austria."

The minister of education Hans Pertner understood this important message very well. He wrote that the need for "science's freedom" should never lead to the negation of truth or the teaching of erroneous views. He added that he was confident, and would remain so. Indeed: "The fact that the thorough rejection of materialism and liberalism has not yet given rise to any drastic change cannot shake our faith in our final victory." It all sounds quite logical.

In the daily *Linzer Volksblatt,* one Bernhard Birk wrote about the problematic activities of Moritz Schlick: "For a full fourteen years, young, tender flowers of humanity were forced to drink from the poisonous vial of positivism as if it were the very water of life. The effect must have been horrible." Robust souls would simply throw up, said Birk. However, "there also exist delicately organized minds, fragile porcelain from the roots of the *Volk,* patriotic children of the Austrian soil, people who yearn for the beautiful and the noble. To pour the doctrine of positivism into these wide-open minds is like pouring chloric or nitric acid down their throats."

## HALFWAY TO THE MOON

The remaining members of the Vienna Circle were well aware of how precarious their situation had become. The public reaction to the news of Schlick's murder would have been proof enough, had it been needed. Although they still met occasionally—for instance, in the flat of Friedrich Waismann or of Edgar Zilsel—this was mere rearguard action. For all intents and purposes, the once-bright days of the Vienna Circle were all over.

Nauseated by the political climate in Austria, Karl Menger decided to emigrate with his young family to the United States. He had been elected vice chair of the International Congress of Mathematics in Helsinki in 1936, and he profited from the occasion to make tentative inquiries into jobs abroad. In 1937, he accepted a position at Notre Dame University in Indiana, "far removed," as he wrote to his friends in Vienna, "from all the centers of production of mental and physical poison gas."

At first, Menger asked only for a leave of absence from the University of Vienna. From afar, he valiantly strove to keep alive his Viennese Mathematical Colloquium. On the last day of the year 1937, he wrote to his former student Franz Alt, who had remained in Vienna:

> I am deeply saddened to be able to do so very little for the splendid circle of Viennese mathematicians so dear to me. I believe you should all get together from time to time, and especially should see to it that Gödel takes part in the colloquium. It would be of the greatest benefit not only to all the other participants but also to Gödel himself, though he may not realize it. Heaven knows what he might become entangled in if he does not talk to you and his other friends in Vienna now and then. If necessary, be pushy, on my say-so.

But Franz Alt could do little for Kurt Gödel, and the Mathematical Colloquium fell apart, just as the Vienna Circle had already done.

Not only Karl Menger but also Karl Popper and Friedrich Waismann managed to leave Austria before the *Anschluss*. The situation was looking more and more hopeless to Jews. Official Austria was against anti-Semitism but could not stem its tide.

Although his *The Logic of Scientific Discovery* had attracted considerable attention, Karl Popper could not hope to find any academic position in Austria. Even the idea of getting his *Habilitation* seemed out of reach. And thus he simply worked as a teacher, like his wife, Hennie, at a secondary school in Vienna.

In Popper's eyes, Austria's political situation was doomed. Thanks to Italy's ill-fated Abyssinian campaign, Mussolini had become politically isolated. Il Duce's only remaining option was to move closer to Nazi Germany. Therefore, he could no longer act as Austria's protector against Hitler. The Austrian state had lost its last external support.

As part of the treaty of July 1936, two years after the July putsch, the Austrian chancellor Kurt Schuschnigg promised Hitler that he would include sympathizers of the National Socialists in his cabinet. In reply, Hitler lifted the *Tausend-Mark-Sperre*. The borders were thus opened up again for German tourists and German journals. They mostly had good things to report about life in the Third Reich. More and more Austrians felt envious of their German neighbors. It would certainly be nice to belong to a great power again—although of course things might turn a bit worse for the Jews.

Carnap wrote to Popper that he understood that "the ground is burning under your feet." Already in 1935 Karl Popper had asked for a leave of absence so that he could explore his options in the United Kingdom. Popper's English was poor, but that would soon change, and quickly. He jumped on every possible opportunity to give talks, which were arranged by a growing network of contacts. Aside from Carnap and Bühler, this included many Viennese now living in Britain: for instance, the art historian Ernst Gombrich, who had just arrived there; the physicist Erwin Schrödinger, who was hoping to return to Austria;

and the economist Friedrich von Hayek, by then well established in London. Popper also got to know Alfred J. Ayer, G. E. Moore, Bertrand Russell, and Niels Bohr, and Albert Einstein wrote a letter of recommendation for him.

However, the most effective help for Karl Popper came from an old member of the Vienna Circle, namely Felix Kaufmann. Kaufmann had studied law and philosophy in Vienna and belonged to the so-called legal positivists, who had gathered around Hans Kelsen and had frequented the economists' circles linked to Friedrich von Hayek and Ludwig von Mises. Kaufmann had not been a central figure of the Vienna Circle, but this could hardly be expected from someone belonging to so many other groups. He had written on the concept of infinity and had authored a book called *Methodology in the Social Sciences.*

Thanks to his cushy job with the Anglo-Iranian Oil Company, Felix Kaufmann had excellent contacts in Britain. Now he used them to arrange a fellowship for Karl Popper at the University of Cambridge. This fellowship was intended solely for Austrians, and more specifically for Austrians who were in need. In a strict sense, Popper was not "in need." After all, the Austrian state did not (at least in theory) take an anti-Semitic line; moreover, Popper even had a tenured job at his high school. This was quite a privileged situation, especially in light of the rampant unemployment in Austria. But Popper abruptly decided to quit his job, thus renouncing his reliable steady income. Now he could claim to be truly in need, even in a strict sense. Though a foolhardy move, this worked, and he got the fellowship.

Shortly thereafter, on Christmas Eve of 1936, Popper received another offer, again through Felix Kaufmann's connections. This was for a position at Canterbury College, in Christchurch, New Zealand. Christchurch was as remote from Vienna as was geographically possible ("halfway to the moon," as Popper chuckled), but it nonetheless offered a real job and job security. So the risk-taking young philosopher accepted.

The Cambridge fellowship went instead to Friedrich Waismann. Now this was someone who beyond any doubt had been in dire need, ever since the ministry had canceled his position as an assistant librarian and his courses at the institute for adult education had been suspended. And now, all at once, he was able to move to Cambridge, of all places—Cambridge, that beautiful haven of tranquillity and intellectualism in which his idol Ludwig Wittgenstein lived.

However, Ludwig Wittgenstein's Trinity fellowship came to an end at exactly that moment, in 1936. For some time, this ex-heir of a fabulous fortune had planned to settle in the Soviet Union. The University of Kazan had even offered him a chair in philosophy. However, after an extended visit to the Soviet Union, Wittgenstein dropped this project. "One could live here," he wrote while still in the Soviet Union, "but only on condition of always remembering that free speech is not possible."

Joseph Stalin decided whereof one could speak and whereof one must be silent. Wittgenstein found out that this was not a great thing after all—at least not for himself. Thus he kept on commuting between Vienna and Cambridge, with occasional visits to the fjord in Norway where his humble hut was still standing. Soon after G. E. Moore retired, Wittgenstein was appointed to fill his chair.

A quarter of a century earlier, G. E. Moore had taken dictation on logic from Ludwig Wittgenstein. That had been in the year just preceding World War I. Now World War II was just around the corner.

# Circling the Globe

*Vienna, 1938–1945: Hitler rules Vienna. Schlick's killer freed. Ethnic cleansing at University. Gödel finds wife, loses visa, achieves breakthrough on Hilbert's Problem Number One. Hapless logician found fit for Wehrmacht, departs for United States via Siberia. Never travels again, but claims time travel feasible. In Holland, Otto and Marie jump onto hijacked lifeboat, float over mines, are rescued by destroyer. Interned in England, Otto holds forth on durable tennis courts. From Isle of Man to All Souls. In Oxford, "Man with a load of happiness" meets sudden death.*

## A WITCHES' SABBATH

In March 1938, the Austrian chancellor Kurt Schuschnigg decided to hold a referendum. Did the people want a free and independent Austria, yes or no? Hitler was not ready to wait for their answer. He ordered his troops to move into Austria. The invasion went like clockwork. Resistance seemed pointless. In fact, the arriving columns of the Wehrmacht were wildly cheered on by delirious crowds. Those who did not cheer were not shown on the newsreels.

The German playwright Carl Zuckmayer (1896–1977) described the *Anschluss* as "truly a witches' Sabbath of the mob." Like many others, he had left Nazi Germany in 1933 to find refuge in Austria. Now he was forced to pack again: "That night hell broke loose. The underworld opened its gates and vomited forth the lowest, filthiest, most horrible demons it contained. The city transformed itself into a nightmare painting by Hieronymus Bosch; phantoms and devils seemed to have crawled out of sewers and swamps. The air was filled with an incessant screeching, and horrible, piercing, hysterical cries from the throats of men and women who continued screaming day and night. People's faces vanished and were replaced by contorted masks: some of fear, some of cunning, some of wild, hate-filled triumph."

Karl Schlesinger killed himself the day Hitler entered Vienna. He was the Viennese banker who had introduced the ideas of economist Léon Walras to the Mathematical Colloquium and thereby inspired Abraham Wald to develop his equilibrium theory. Schlesinger's suicide was only one among many. The newspapers were ordered to stop reporting on that depressing topic.

From the University of Notre Dame, in Indiana, Karl Menger wrote to the University of Vienna: "Cabled today: Accepted position abroad giving up Viennese professorship. Letter follows. I hereby confirm in writing that I have accepted a position abroad and resigned from my Viennese professorship. Karl Menger."

Immediately after the *Anschluss,* or "annexation," a ruthless "purification," or *Säuberung,* of the civil service got underway. In the Viennese Faculty of Philosophy, 14 out of 45 full professors, 11 out of 22 associate professors, 13 out of 32 emeriti, and 56 out of 159 lecturers were registered as so-called *Abgänge,* or "departures." In the Faculty of Medicine, the losses were even more drastic. Never had there been such an acute brain drain. Austria, or what was now called the Ostmark, meaning the "Eastern Borderland," was amputating its head.

The Law for the Re-establishment of the Professional Civil Service provided the legal leverage for the "transfer into retirement." Civil servants were required to declare an oath of allegiance to Adolf Hitler, the Führer. This sacred act and its various privileges were reserved solely for Aryans, of course. All others simply had to look out for themselves.

FIGURE 12.1  The Hitler salute in front of and within the university.

In his inaugural address, the newly commissioned rector, Kurt Knoll, criticized the aloofness toward National Socialism that many professors had formerly displayed. Ominously, he added: "Everything has changed now. The material lesson experienced by the professors during the Führer's presence in Vienna will not fail to produce its effects."

One of the professors who felt such effects was Otto Pötzl, the head of the psychiatric clinic. Though not Jewish, Pötzl had long been a member of Sigmund Freud's Psychoanalytic Society. Thus he was a *Judenfreund,* obviously. As for Freud himself, now well over eighty, he had to pack his bags, of course. He left for London. Otto Pötzl merely lost his job, on the personal recommendation of the leader of the National Socialist Lecturers' Union.

But Pötzl was not going to give up his clinic, and his entire life's work, to an incompetent Nazi toady; he therefore applied for party membership. So many Austrians were making this same move, and all at the same time, that the Nazi Party had to declare a moratorium. But Pötzl's maneuver succeeded. Within a few weeks, he was reinstated as head of the clinic. He could be sure, of course, that the leader of the National Socialist Lecturers' Union would keep a close watch on him. Nothing escaped the new rulers.

Students now had to smartly stand up whenever a lecturer entered or left a classroom, and had to salute the lecturer with a resounding "Heil

Hitler!" This "German greeting," however, sometimes fell short of what was desired. In May 1938, the leader of the Nazi student union felt obliged to send out a circular requiring that the student body should act "in a sharp and united manner" against any signs of laxness in the Heil-Hitlering.

Starting in the summer semester of 1938, a numerical restriction, or *numerus clausus,* was imposed on Jewish students, and for good measure on female students as well. Anyone who was enrolling for the first time in the university had to confirm, "according to their best knowledge and conscience," that they were not of Jewish descent. The few remaining Jewish students were forbidden to use the university library. The university kept close tabs on them. Without being asked, it submitted a list of all their names and addresses to the Gestapo (that is, the Geheime Staatspolizei, or Secret State Police).

In the winter semester, after the violent pogroms of the infamous Kristallnacht, or "Night of Broken Glass," which took place on November 9–10, 1938, Jewish students were barred completely from entering the university. The ministry explained that this measure was taken "to avoid discomfort." Shortly thereafter, the university could boast that it was now totally *judenrein,* meaning "cleansed of Jews." Those Jews who still had books to return to the library had to deliver them on the steps in front of the main building.

## OBITUARY

The Viennese mathematician Walter Rudin, who experienced the *Anschluss* as a seventeen-year-old high school student, wrote later: "In one respect, we were better off than the German Jews. There the screws were tightened gradually, and for the first couple of years there was hope that it would all blow over, that a different government would be formed, that things would get back to normal. As a result, many German Jews procrastinated until it was too late. In Austria, though, it became absolutely clear within a couple of days that the only option was to get out."

Playwright Zuckmayer wrote in his memoirs *A Part of Myself:* "What was unleashed upon Vienna had nothing to do with the seizure of power in Germany, which proceeded under the guise of legality and was met by parts of the population with alienation, skepticism, and an unsuspecting nationalistic idealism. What was unleashed upon Vienna was a torrent of envy, jealousy, bitterness, and blind, malignant craving for revenge. All better instincts were silenced."

During the months following the *Anschluss,* as the harassment rapidly escalated, some of the few remaining members of the Vienna Circle succeeded in escaping. Among them were Rose Rand, Gustav Bergmann, and Edgar Zilsel. Rand, who had passed all her examinations some time earlier, received her doctoral degree in a so-called promotion for non-Aryans, a hurried affair. At the same time, she was barred from any future professional employment in Germany. In order to emigrate to England, Dr. Rand had to document that she had taken cooking classes. Her paper "The Concepts of 'Reality' and 'Unreality' in the Minds of the Mentally Ill," which had already been accepted for publication, was now abruptly rejected ex post facto. Once she arrived in England, Rose Rand found refuge as an auxiliary nurse at the St. Albans Hospital for mentally retarded children, where conditions were ghastly.

Rudolf Carnap had long ago moved from Prague to the United States, and now Philipp Frank followed suit. Some members of the Viennese Mathematical Colloquium, such as Abraham Wald and Franz Alt, also were able to escape in time, before the Shoah was definitively set in motion.

International public opinion was outraged at Hitler's *Anschluss.* But with the sole exception of Mexico, no government in the world lodged a protest against it. This is why today there is a square in Vienna named Mexikoplatz.

One of the protagonists in Leo Perutz's unfinished novel *One May Night in Vienna* said with a deep sigh: "What the foreign press is writing about Vienna sounds like the obituary for a famous film star, to whom the world is indebted for many hours of artistic delight. And now fate has taken her away, but the great film producers will manage without her."

FIGURE 12.2 Carnap bids farewell to Europe.

Adolf Eichmann, who decades later was executed for crimes against humanity, was just now beginning to make his name by setting up the Central Agency for Jewish Emigration in Vienna, soon to become a model all over the Reich. Adding to the irony was the fact that it was located in the luxurious Palais Rothschild, and was financed by the draconian Reichsfluchtsteuer, which taxed to the bone all those who were fleeing the Reich. The queues of people waiting for emigration permits quickly grew longer and longer.

## MERCY FOR NELBÖCK

For Johann Nelböck, too, the *Anschluss* brought about a change of residence. He had initially been an inmate of the jail in Stein, on the Danube, for about one year. However, a certain Professor Sauter felt that the moment had come to release Moritz Schlick's murderer. Little is known about Sauter, except that he probably was also the "Professor Austriacus" who had authored the infamous article casting the blame on Schlick for having mentally derailed Nelböck. Sauter appealed to the minister of justice to grant Nelböck a pardon, his reasoning running roughly as follows.

The murder had been an act of desperation due to political distress. It was well known that the shady Moritz Schlick had been "a representative of Jewry in the philosophical faculty," and an early member of the Fatherland Front (now a highly stigmatized movement). By contrast, the upright Nelböck had always been "strongly motivated by national sentiments and an outspoken anti-Semitism." It had not been possible for Nelböck to profess his true motivations at the trial. Indeed, such convictions had been unacceptable during the Systemzeit, or "Era of the System" (this was the peculiar new name for the five-year period during which the old Ständestaat, or "corporate state," had reigned in Austria).

Professor Sauter's reasoning convinced the new authorities, and as a result, Nelböck was released in November 1938. For the first time in his life, he obtained a regular job: a position at the geological department of the Mineral Oil Administration.

A couple of years later, and well before the end of his probationary period, Nelböck requested that his name be stricken from the penal record, so as to "fully return to the fold of the *Volk*." He wrote that he could pride himself in having rendered "a service to National Socialism by removing a teacher who professed Jewish doctrines that were injurious and alien to the *Volk*." He deemed it a hardship to have to still suffer as a martyr for the National Socialist cause, now that its worldview had prevailed.

The Viennese authorities endorsed Nelböck's plea, but the private secretariat of the Führer held that such a step would be premature. Unsurprisingly, the chief attorney of the state concurred with this view: he maintained that "it remained a fact" that Nelböck's actions "had been guided primarily by personal motivations," and not by the intention to deliver the community from a *Volksschädling,* or "social pest." Moreover, it seemed "to be not without danger for the legal order" to pardon individuals who believed that it was within their right to remove anyone that they personally saw as undesirable. Thus for the time being, Nelböck's name remained on the penal registry.

The Dancer and Her Umbrella

Kurt Gödel was not racially persecuted and had little to fear from the Gestapo. He had always been circumspect in his political utterances. Gödel had lost his right to lecture after the *Anschluss,* but this was a consequence of a general measure: the Third Reich had scrapped the title *Privatdozent.* This fact irked Gödel, but it spared him the embarrassment of having to pledge allegiance to the Führer.

On a Saturday evening in 1938, in the apartment of Edgar Zilsel, a few weeks before the Wehrmacht entered Vienna, Gödel gave a lecture to the few remaining members of the Schlick Circle. It was probably the last meeting ever of the small group. Gödel's highly technical talk on consistency was understandable only to mathematicians, of whom there were none in the audience. By now Gödel had no scientific contacts left in Vienna.

But Gödel was not completely isolated. In 1929, the year of his PhD, he had met Adele Nimbursky (1899–1981) and fallen in love. She had been a dancer in nightclubs (especially Der Nachtfalter, or "The Moth"), and was divorced and several years older than Kurt. Her father was a photographer. Adele lived with her parents and worked in their apartment as a masseuse, or physiotherapist. "Her type: Viennese washerwoman," as Oskar Morgenstern later described Gödel's paramour in his diary. "Loquacious, uneducated, full of resolve—and has probably saved his life." The brilliant Austrian logician Georg Kreisel recalled that Adele had "a real flair for *le mot juste.*" All in all, she seems to have been quick-witted, though not in the least mathematically inclined.

For many years, even Kurt's closest colleagues knew nothing about his love affair. His mother and brother knew of it, but both were icily opposed to Kurt's liaison with a woman who, in their opinion, was entirely unsuitable for him. But Adele took doting care of him. During his various stints in Princeton in the 1930s, she faithfully waited in Vienna for her beloved and incomprehensible genius to return. And she did not always have to wait for long. One time, a nervous breakdown forced Gödel to cut short his stay in America. On another occasion, he

FIGURE 12.3   Adele Nimbursky becomes Gödel's lifeline.

had not even managed to board the steamer for the States. When he was obliged to withdraw to a sanatorium, Adele discreetly visited him.

Increasingly, though, Kurt was beset by terrible, uncontrollable fears of being poisoned. When such "states" were particularly intense, he could eat food only if Adele herself had prepared it, and even this could happen only after she had first tasted it, right in front of his eyes, and from off his plate, and with his own spoon.

After the *Anschluss,* Adele protected her adored Kurt with steadfast resolve. One evening, as they were strolling through the streets of Vienna, a few young Nazi thugs came up to them, shouted nasty words, and then swept Gödel's spectacles off his face and hurled them to the ground, because they took him for a Jew. Adele, instinctively grabbing her umbrella in a most menacing fashion, chased them away.

Attacking Jews needed no explanation in those dark days. In fact, a frequently told joke ran this way: "Have you heard? Hitler is killing the cyclists and the Jews." "Why the cyclists?" "Why the Jews?"

In any case, after almost ten years of friendship, the time seemed ripe for Kurt and Adele to exchange vows—even if Gödel's mama had always hoped for a better better half for her dear son. She finally relented. Gödel himself was more than happy to let Adele take care of all the preparations. Their new apartment was located in a stately house on the outskirts of town, in one of the best neighborhoods, only a few steps away from the famous vineyards of Grinzing.

The wedding took place in September 1938. Two weeks later, Gödel traveled to Princeton once more, to spend another guest semester at the institute. Luckily, he had no problems with the paperwork: for years, he had held a visa permitting him multiple entries into the United States. Despite the ominous political tensions racking Europe, he was in an upbeat and confident mood. These were the days of the treaty of Munich. War had been avoided once again. And Gödel's Princeton trip came off without a hitch. Adele, meanwhile, remained in Vienna. The plan was that she would join him on his next journey.

## BIGGER AND LITTLER INFINITIES

This time, Gödel was bringing sensational new results across the ocean. Since his last visit to Princeton, he had switched from logic to set theory. To those who study the foundations of mathematics, set theory is often seen as the underpinning of all other branches of mathematics, such as geometry or analysis, and it is therefore of considerable philosophical weight.

A great deal had changed in set theory since the discovery of Russell's paradox at the dawn of the new century. People had grown extremely cautious. Sets containing themselves were taboo. "The set of all ideas" or "the set of all sets" had turned into mathematical nonentities. The new axioms had strict rules for defining sets.

One such rule was the so-called *axiom of choice*. Given a family of non-empty sets, you were allowed to pick one element from each set and thereby form a new set. That this can always be done is by no means as obvious as might seem at first glance, and if one uses it one can arrive at what seem extremely weird results, such as the famous Banach-Tarski theorem, published in 1924, which says that you can take a sphere and disassemble it into five parts that can then be reassembled into two spheres of exactly the same size as the original one. The "parts" are not standard parts that could be carved with a knife, or even physically realized at all, but surrealistically strange subsets of the sphere that depend on the axiom of choice to be dreamt up. Quasi-paradoxical results like this made the axiom of choice seem deeply suspicious to some logicians, although to others it seemed as natural as could be.

The first person to postulate this axiom had been Ernst Zermelo, a former student of David Hilbert, a former thorn in the side of Ludwig Boltzmann, and a former collaborator of Hans Hahn. Now Kurt Gödel had succeeded in showing that Zermelo's axiom, despite its weird consequences, was compatible with the other axioms of set theory. Adding it to the other axioms would not give rise to any contradiction. Gödel's proof of this result was considered a sensational tour de force by those in the know.

Even more remarkable were Gödel's findings concerning Cantor's *continuum hypothesis*. Georg Cantor (1845–1918) had been the first person to transcend the deep historical ambivalence about the infinite that mathematicians had always manifested. Taking the bull by the horns, Cantor dared to look infinity straight in the eyes. And to his amazement, when he gazed into this forbidden territory, he found that there was a vast universe of infinities of different sizes. No one had ever imagined anything of the sort.

The set N of all natural numbers 1, 2, 3, . . . constitutes the smallest infinity. You might think that the set E of all even numbers 2, 4, 6, . . . , since it is obviously a subset of N, would be only "half as big" as N, but quite counterintuitively, E turns out to be no smaller than N—or more precisely, "no less infinite" than N. Indeed, every natural number in N can be matched with its exact double in E—thus 1 maps onto 2; 7 maps onto 14; 300 maps onto 600; and so forth, ad infinitum. This way, one sees that every element in N has exactly one partner in the "smaller" set E, and vice versa. Thus these two infinite sets E and N, even though the former is a subset of the latter, are declared to have the same magnitude.

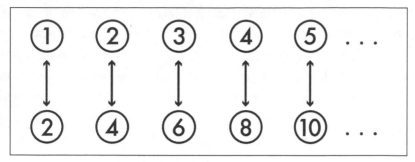

FIGURE 12.4 Every number has a partner in a smaller set.

As this shows, the notions of "bigger" and "smaller" don't apply to infinite magnitudes in the same way as they do to finite ones.

In contrast to E, the set Q of all *rational* numbers—that is, fractions such as 7/2 or 1/13—seems to be bigger than N. In fact, one would tend to guess that Q is "much bigger" than N, the set of all natural numbers: after all, between any two successive natural numbers, such as 3 and 4, there are infinitely many rational numbers. (There are even infinitely many rationals between any two given rational numbers!) But Cantor had shown that despite this apparent huge numerosity, one could still form a list of all rational numbers, meaning that one can make a one-to-one mapping between the elements of Q and the elements of N, just as we did above between N and E. This shows that, in the jargon of set theory, the set Q is *countable*. Thus, surprisingly, the set N of all natural numbers and the set Q of all rational numbers have the same magnitude. Could it be, then, that *all* infinite sets are countable?

Using a deeply ingenious proof, which foreshadowed Gödel's wonderful proof of incompleteness several decades later, Cantor showed that this is not the case. In fact, he showed that the set R, which consists of all the points on the real line (the so-called *continuum*), is *uncountable*. That is to say, the continuum R has a greater magnitude than the set of natural numbers N and the set of all rationals Q. Going beyond this first remarkable result, Cantor proved that there are not just *two* different "sizes" of infinity, but *infinitely many* sizes, and he proved many theorems about these new "numbers."

Despite Cantor's amazingly rich and fertile imagination, he left one major question about infinite sets unanswered. This question asks: What is the "smallest" infinity that is greater than N? In other words, what is the smallest of all the uncountable infinities? Cantor surmised that it was R, the continuum, but he was unable to prove his guess. Cantor's famous *continuum hypothesis* thus asserts that the smallest infinity, other than the first infinity (that of N), is the infinity of the continuum.

To provide a proof of this claim had been Problem Number One in Hilbert's famous list from 1900. And in 1938, Kurt Gödel succeeded in solving one half of Hilbert's problem, so to speak: he proved that the continuum hypothesis is compatible with the other axioms of set theory; that is, it does not contradict them. This does not imply that it must be true, of course. Gödel's proof was exceedingly complex, technically far more demanding than his proof of his more famous incompleteness theorem of 1931. But just as the incompleteness theorem had done, this new Gödelian result once again opened up a whole new mathematical realm.

Already by that time, it was conjectured that the *negation* of the continuum hypothesis is also compatible with the other axioms of set theory. In other words, the continuum hypothesis would be undecidable, given the other axioms of set theory. Gödel tried for years to prove this, too, but he never succeeded. A proof was finally found in 1963 by the young American mathematician Paul Cohen (1934–2007).

From Cohen's and Gödel's results, most mathematicians concluded that the situation in set theory is precisely analogous to that in geometry. That is, just as Euclid's parallel axiom is independent of the other axioms of geometry, so Cantor's continuum hypothesis is independent of the other axioms of set theory. Just as one can do either Euclidean geometry (accepting the parallel axiom) or non-Euclidean geometry (accepting its negation), so one can do either "Cantorian" set theory (accepting the continuum hypothesis) or "non-Cantorian" set theory (accepting its negation). You are free to decide which way you would like to go. This vision of different possible set theories agrees nicely with Carnap's tolerance principle!

Gödel himself, however, had a different viewpoint. He did not believe that there were different set theories. Although he was perfectly happy with the idea that there are different geometries, each one perfectly self-consistent, and none of them being "the truth," he believed that there is only *one* set theory—the *true* one. Deep down, Gödel was convinced that the "correct" axioms of set theory had not been discovered yet. This presumes that somewhere, in the realm of ideas, there is the "true" mathematics. In other words, we human beings do not *invent* mathematical truth: we *discover* it, much as Columbus discovered a new continent, or as physicists discovered that atoms exist.

Only a true-blue Platonist can hold such a view; in the eyes of thinkers such as Carnap, Wittgenstein, Hahn, and Menger, Platonism seemed to have been long transcended and fully discredited. But Kurt Gödel did not see things that way at all. He had been committed to a Platonic view of mathematics for many years, though he hardly ever talked about it—just as he had kept mum about his liaison with Adele. And he never let go of this view of mathematics.

In 1939, Gödel took part in the yearly meeting of the American Mathematical Society, and there he gave a talk on the continuum hypothesis. As a result, he was invited to deliver a plenary lecture at the next congress of the International Mathematical Society, which was scheduled for 1940. These extremely prestigious congresses were the equivalent of the Olympic Games. It soon became clear, however, that current political developments would preclude not only the mathematics congress but also the Olympic Games.

In March 1939, during Gödel's stay in the United States, Hitler's troops invaded Prague, in flagrant violation of the Munich treaty. Even the most naïve of observers could now clearly see that war was just around the corner.

After his guest semester at the Institute for Advanced Study in Princeton was over, Gödel spent a few months with his friend Karl Menger at Notre Dame University in Indiana. But then, despite Menger's most urgent pleadings, he traveled back home to Vienna. He intended to return in the fall to Princeton, this time with his wife. On June 14, 1939, he boarded the *Bremen* in New York. Little did he suspect that he had thereby embarked on a journey that would wind up circling the globe.

## "The Coffee Is Wretched"

When Kurt Gödel had reached Vienna in the summer of 1939, he was drawn into a bureaucratic maelstrom. First off, his Austrian passport was canceled; after all, Austria no longer existed! Even the name *Austria* was shunned. As a German citizen, Gödel received a brand-new passport, but it no longer included the multiple-entry visa for the United States, as that had been invalidated along with the old passport. The American consulate, hopelessly overtaxed, was unable to transfer the visa to the new travel document. Thus Gödel had to join the queue, along with hundreds of thousands of others, of those who were hoping to obtain permission to enter the United States.

The guileless logician could not ask for special consideration, as he was not being politically persecuted. Nor could he apply for a privileged "professor's visa," since one of its requirements was that the applicant must have given lecture courses in the previous two years. Gödel had done no such thing.

Moreover, his *Habilitation* had been suspended. To make up for this loss, he was welcome to apply for the status of *Dozent Neuer Ordnung,* or "lecturer of the New Order," which had recently been introduced by the Nazi administration, but the odds of getting it were biased heavily against him.

Gödel no longer had any supporters at the university. No one in Vienna had the slightest inkling of his great accomplishments. His work, with its high degree of abstraction and formality, seemed the opposite of what was currently being hailed as "German mathematics." Of course, no one was able to explain just what German mathematics actually was, but at least one thing seemed abundantly clear: the ballistics of bombs was *in* and the logic of axioms was *out*. Even David Hilbert, who in collaboration with the Swiss mathematician Paul Bernays (1888–1977) was currently finishing up a major two-volume work called *Principles of Mathematics,* had been relegated to the sidelines. When Hilbert died in 1943, his passing attracted far more attention in the United States than in Germany.

During the hectic weeks of the summer of 1939, Kurt Gödel's problems with the Viennese bureaucracy reached the point of utter unmanageability. In a letter to his friend Karl Menger, he bemoaned his "endless errands." Red tape had hit record heights.

In a fragment of a novel, Leo Perutz described the plight facing those wishing to leave the Reich: "Obscure offices that no one had ever heard of before would suddenly emerge from hiding, would make their demands imperiously known, and would insist on being satisfied, or at least noticed and consulted."

Kurt and Adele Gödel had already given notice that they would vacate their flat in Grinzing, so now a new apartment had to be found. And the foreign-currency bureau was raising all sorts of unanswerable questions about the money that Gödel had earned in America.

Official inquiries brought to light the fact that Gödel had not properly taken leave from the University of Vienna before his departure in the fall of 1938. The ministry and the university began to engage in an agitated correspondence about the unknown whereabouts of Kurt Gödel. He was actually back in Vienna, but nobody seemed to have noticed. Moreover, an official started complaining that Dr. Gödel's certificate of Aryan descent contained sixteen documents, but the marriage certificate of his parents was still missing. Gödel was instructed to submit it with the shortest possible delay. Moreover, the official hinted that the marriage certificates of his grandparents might also be needed.

The leader of the National Socialist Lecturers' Union reported that Gödel enjoyed a high reputation in scientific circles, but that his teacher Hans Hahn had been a Jewish professor, and that "generally, mathematics in the Vienna of the System Era had been deeply Jew-polluted." Gödel himself was considered to be apolitical, the report said, "but he cannot be expected to be up to the difficult political situations he will encounter in the United States as a representative of New Germany."

In September 1939, the war broke out, with Gödel still trapped in Vienna. He was called up for duty by the German Wehrmacht. Somehow, the military medical examiners found the frail thirty-three-year-old intellectual "fit for garrison duty." His conscription was only a matter of time. The situation seemed hopeless. But all at once came a magical deus ex machina. In far-off Princeton, Gödel's great admirer John von Neumann intervened. "Gödel is in a class by himself," he wrote to the director of the Institute for Advanced Study. Something had to be done for him, and there was no time to lose. And in another letter: "Gödel is absolutely irreplaceable. He is the only mathematician about whom I dare to make this assertion."

At the end of this letter, von Neumann appended a masterful analysis of the confused situation and pointed to a way out of the impasse. Thanks to Johnny's clear thinking in a moment of highest urgency, the bureaucratic hurdles of the State Department were overcome, and in early January 1940, the Gödels finally obtained their longed-for entry visa to the United States.

Because of the ever-present threat of U-Boots, crossing the Atlantic was out of the question. Thus the Gödels would have to take the long way around—all the way across Siberia and then the Pacific Ocean. Given the political situation, this was hardly a cakewalk. In fact, it was a highly risky tightrope walk. Poland had already been chopped up and divided between Hitler's Germany and Stalin's Soviet Union. (Those two powers, soon to be each other's bitterest enemies, were still acting lovey-dovey with each other.) Lithuania had not yet been occupied by Stalin. As for Japan, well, it had conquered Manchuria, but it was not yet at war with America. All things considered, then, the trip seemed . . . feasible. With luck.

The Trans-Siberian railway rolled endlessly on and on through the deep tundra and the frozen winter nights. And once they had come out the other end of the Soviet Union and reached Yokohama, the Gödels somehow managed to miss their America-bound steamer. But finally—a few weeks later than planned—they sailed into port in San Francisco. "It is the most beautiful town I have ever seen," a much-relieved Gödel

FIGURE 12.5 Oskar (with hands crossed) and Kurt (with fingers enlaced).

wrote to his brother Rudolf. The happy ending was near. Forty-six days after their departure from Vienna, Kurt and Adele at last reached the safe haven of Princeton.

His friend Oskar Morgenstern noted in his diary: "Gödel arrived from Vienna. Via Siberia. This time with wife. When asked about Vienna: 'The coffee is wretched.'"

As for Oskar Morgenstern, he had happened to be in America during the days of the *Anschluss*. There, he learned that his name had been blacklisted by the Nazis. Whether Aryan or not, he had made himself thoroughly unpopular with those newly in power. Indeed, a major portion of the final budget speech delivered by the last Austrian minister of finance had been penned by Morgenstern. As the director of the Viennese Institute for Business Cycle Research, he had repeatedly and quite correctly stressed the economic viability of Austria as an autonomous entity. This viewpoint hardly jibed with the Nazi party line; in fact, to express such a view could well be the kiss of death. Clearly, Morgenstern had neither the desire nor the intention to end up in the concentration camp at Dachau, as had half of the last Austrian cabinet. He therefore did not go back to Europe and eventually succeeded in landing a professorship in Princeton.

Way back in Vienna, however, the heavy bureaucratic wheels continued to turn long after Gödel's departure. A ministry official kept on harping on the outdated news that in 1938 Kurt Gödel had taken leave without proper authorization, and inquired into the details of his return to Austria, "which obviously must have taken place"—little suspecting that by then, Gödel had already come home and gone away again, this time forever.

And then, ironically enough, against all odds, Gödel was named in absentia a lecturer of the New Order. According to the decree, this grand honor included "special protection by the Führer." And this status could not fail to impress: Hitler had just taken Paris.

Gödel never collected his doctoral diploma. Even today, it is still waiting in the university archives, together with a slip all ready for Gödel to sign. For years, the Nazi authorities kept inquiring into the whereabouts of Dozent Gödel and the reasons for his delay in picking up his diploma. The replies of his brother Rudolf turned increasingly curt: "As already reported several times, my brother is currently in the US." And Rudolf added that it was the German consulate that had warned his younger brother against crossing the Atlantic.

In the spring of 1941, Gödel was ordered to visit the German consulate in New York. There, he was informed that the German Reich would like to "repatriate" him, and as soon as possible. Gödel parried with the remark that he would find no salaried position there, and that his health, moreover, was frail. However, the whole affair soon became moot, for in June of 1941, Hitler's troops attacked the Soviet Union. From then on, there was no way to get anyone back to Germany—neither via the Atlantic nor via Siberia.

Half a year later, Adolf Hitler declared war on the United States. Gödel thereby became overnight an "enemy alien" in his newly adopted land. Even his solitary walks at night in the quiet zone of Princeton where he and Adele lived started to arouse suspicions among their neighbors. Luckily, the Institute for Advanced Study was able to allay their fears. In the following year, Gödel was summoned to a US Army examination board, but once again the stalwart institute managed to save him from that peril, too.

## Travels Through Time with Kurt Gödel

During the 1940s, Kurt Gödel and Albert Einstein became close friends, despite their considerable difference in age. This was a friendship of mythical proportions.

"Why did Einstein enjoy talking with me?" Gödel later wrote in a letter, and he surmised that one of the reasons was "that my views were often opposed to his, and that I made no secret of it."

"Gödel was the only one of our colleagues who walked and talked on equal terms with Einstein," said the physicist Freeman Dyson, then one of the institute's rookies, and one of the great mathematical minds of the twentieth century. And an assistant of Einstein's confirmed: "The one man who has certainly been by far Einstein's best friend over the past years was Kurt Gödel, the great logician. They were very different in almost every personal way [ . . . ] but they shared one fundamental quality: both went directly and wholeheartedly to the questions at the very center of things." Einstein himself liked to joke: "I go to my office just to have the privilege of being able to walk home with Kurt Gödel." Or perhaps it was not even a joke at all.

This "greatest intellectual friendship since Plato and Socrates," as it has been termed, was not just a friendship, but also gave rise to an amazing scientific discovery. Gödel had been asked by Paul Schilpp, the editor of a volume called *Albert Einstein, Philosopher–Scientist*, to contribute a philosophical essay about Kant and the theory of relativity.

The topic was by no means original. It had been investigated by Schlick and Carnap, among others. But Gödel, with his characteristic thoroughness, went more deeply into the matter. What had been originally commissioned as a philosophical essay turned into a mathematical theory, and in the process, Gödel discovered a truly remarkable new class of solutions to Einstein's field equations of general relativity.

His work implied that in principle, general relativity allows for rotating universes. Such a universe does not rotate around an axis; it rotates with respect to every local inertial system. It implies that general

FIGURE 12.6  Albert (with
dark suspenders) and Kurt
(with white jacket).

relativity need not obey what Einstein had called "Mach's principle,"
which he had depended on to motivate his theory. This came as a great
surprise. In particular, there would be no absolute time and no global
simultaneity in such a world, in contrast to the usual cosmological solutions. But this was not the end of the story.

Indeed, Gödel showed that in a rotating universe it is possible, in
principle, to travel into the past. It had been known for some time that
it is possible to travel into the future. Physicists had gotten accustomed
to the idea. But a trip into the past is far more paradoxical, since it plays
havoc with causality. For instance, a time traveler could meet with a
younger self, "and do something to that person," as Gödel wrote, with
a faintly sinister undertone. On the upside of things—and this was also
pointed out by Gödel—when a time traveler lands, time still flows in
the usual direction, rather than backward. That is certainly a relief.

Einstein and most other theoretical physicists concluded from
Gödel's paradoxical results that travels into the past were ruled out by

some as-of-yet-unknown physical principle. Gödel, on the other hand, concluded that our notion of time is deeply and fundamentally deluded.

At that time, Bertrand Russell visited Einstein in Princeton, and in the latter's house he found both Kurt Gödel and Wolfgang Pauli. Quite a foursome! Or as Pauli himself might have put it, "*Gar nicht so blöd!*" In his autobiography, Russell wrote that all three emigrants had "the German penchant for metaphysics" and that Gödel in particular "unveiled himself as an unmitigated Platonist."

Apparently, Gödel no longer saw any reason to hide his true convictions.

## MODERN MAN IN THE MAKING

The first member of the Vienna Circle to die in exile was Olga Neurath, in 1937. Despite her blindness, she had returned from The Hague to Vienna on several occasions for short visits. Like her brother Hans Hahn, she died at age fifty-five from the aftereffects of cancer surgery. Otto Neurath was thus widowed for the second time, although as a consolation, he still had Mietze Reidemeister, his faithful muse, as it were, at his side.

Neurath's son Paul had been arrested after the *Anschluss* by the Gestapo, while he was trying to escape into Czechoslovakia. After bouts of internment in the concentration camps of Dachau and Buchenwald, he succeeded in emigrating to Sweden, and later to the United States. There, the thirty-year-old lawyer enrolled as a student again, this time in sociology. He would never forget his brutal experiences in the concentration camps, and his book *The Society of Terror* became Paul Neurath's best-known work.

During his exile in the Netherlands, Otto Neurath had managed, within the span of just a few years, to build up an impressive Unity of Science movement. He was particularly eager to carry out his megaproject of an *Encyclopedia of the Unity of Science,* which he had once briefly discussed with Albert Einstein, during the latter's visit to Vienna in 1921. Einstein had said it was a good idea but failed to follow up.

Neurath planned for his encyclopedia to have twenty-six volumes, each of which would contain ten monographs. It would be issued in English, German, and French and would cover natural science, law, medicine, and the social sciences. Although the foundations had already been solidly laid, the project had had to be put on hold because of World War II.

Each year, an international Congress for the Unity of Science had been held, and each time it attracted hundreds of participants: in 1934, it was held in Prague; in 1935, in Paris; in 1936, in Copenhagen; in 1937, in Paris again; in 1938, in Cambridge, England; and in 1939, in Cambridge, Massachusetts. Each congress was organized by the Unity of Science Institute, a branch of Neurath's Mundaneum. Celebrities such as Bertrand Russell and Niels Bohr gave the keynote talks.

Pictorial statistics, too, had progressed with giant steps. Since the Viennese Method no longer had anything to do with Vienna, Marie Reidemeister coined the new name *Isotype* (standing for *International System of Typographic Picture Education*). Neurath planned to write a book on it: *From Hieroglyphs to Isotype.* Moreover, he combined his signature pictorial style, featuring "rows of little men," with Basic English, a minimal dialect of English that makes do with only 850 words. This language had been crafted by the English philosopher Charles K. Ogden in order to facilitate international communication, and it met with more success than Esperanto had (although that's not saying too much).

As a young man, Ogden had translated the *Tractatus Logico-Philosophicus* into English. Later, he had coauthored a book with the catchy title *The Meaning of Meaning.* Ogden's linguistic vehicle named Basic English was different from the "universal slang" that Otto Neurath envisioned as the ideal language of science, but it struck a related chord: nebulously hazy words such as *reality, transcendence, appearance,* and so forth simply did not exist in Basic English. Neurath's fanciful notion of an *index verborum prohibitorum* ("list of forbidden words") became unnecessary. After all, it would be difficult, if not downright

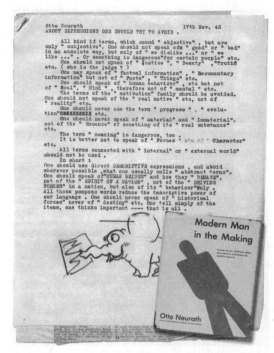

FIGURE 12.7 Otto Neurath avoids words and writes a best seller.

impossible, to carry out metaphysics in Basic English, or to translate Martin Heidegger into it. Now that was clearly a fine thing!

Emboldened by the success of its connection with Basic English, the Mundaneum expanded into cooperative ventures with British and American publishers. In 1939, Neurath's book *Modern Man in the Making* turned into a surprise best seller, despite the political thunderclouds threatening its optimistic message. It employed a tight mesh of texts and pictures to describe the dawning world of globalized exchange, international migration, and limitless progress.


## OTTO NEURATH IMPROVISES

The next international congress on the Unity of Science had been planned for May 1940—this time in Oslo. But Hitler had gotten there first; moreover, starting on Whitsunday, May 10, his troops overran

France in a spectacular *Blitzkrieg*. At the same time they also invaded the neutral countries of Belgium, Holland, and Luxemburg. Doing so gave more punch to the *Panzers*. On May 14, the Dutch Army was forced to capitulate.

Caught totally off guard in Holland's capital, The Hague, Otto Neurath and Marie Reidemeister heard the news on the radio. Being "enemy aliens," they had heretofore been required to stick to their rooms, but now they were within reach of Hitler's terrifying Stosstruppen and had no choice but to decamp head over heels. However, the unflappable Neurath—he who thrived on crises—did not lose his aplomb. "We will improvise," he declared. By roundabout ways and without any luggage, Otto and Marie reached the nearby harbor of Scheveningen, finding it packed to the gills with masses of desperate fugitives.

A huge smoky pall loomed over distant Rotterdam. The sharp sound of explosions filled the harbor. All the local fishermen refused to put out to sea, even if someone offered them a small fortune to do so. But Otto Neurath was unshaken: "If we cannot find a boat, we will use a piece of wood."

But the couple did not have to build a vessel on the high seas in the fashion of the mariners in Neurath's famous parable. They eventually found a boat that would take them. It was critically overloaded already and almost capsized when the bulky philosopher leaped into it from the pier. He was the next-to-last passenger, coming right on Marie's heels. The last one, a psychiatrist, also jumped, but ended up in the water. He was hauled aboard as the vessel turned toward the west and chugged off into the setting sun, sluggish and deep in the water, carrying fifty passengers instead of the prescribed fifteen.

The little boat was named *Zeemanshoop,* or "Sailor's Hope," and its journey has become something of a legend in the Netherlands: indeed, the *Zeemanshoop* carried to England the first so-called *Engelandvaarders,* young Hollanders who were determined to continue the fight against Hitler. But most of the passengers were Jews from Holland and Germany—some rich, some poor, all of them clearly aware that this was their last chance to escape from the Nazi storm.

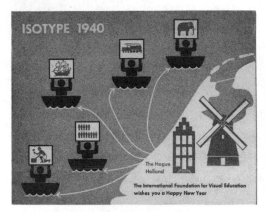

FIGURE 11.0 Wish and reality. New Year's greetings for 1940, and the hijacked lifeboat *Zeemanshoop*, with Marie Reidemeister and Otto Neurath. (This picture was taken on May 15, 1940, by Lt. Peter Kershaw [RNVR] from the deck of HMS *Venomous*.)

## HARRY HACK TAKES TO SEA

Otto Neurath was pleased to learn that the skipper asked for no fare and was even more delighted to hear that his name was "Harry Hack"—just the right name, he felt, for an adventure yarn. In fact, Neurath had jumped right into a Hitchcock film. Young Hack was no seaman at all, but a student who, with a couple of friends, had hijacked the boat by breaking the padlock of the hatch to the engine compartment. Somehow he had managed to get the engine running, although on one cylinder only. The *Sailor's Hope* was a lifeboat that belonged to the Dutch Navy: stealing it to save lives was certainly justifiable.

The sea remained blissfully calm all night. Here and there, strangers began conversing in subdued voices. The engine went on strike several times. There were no navigational charts in any of the lockers that they broke open, and in any case, nobody on board knew how to navigate. The compass light failed. The deck was so crowded—with women and children huddling together on suitcases, and men standing shoulder to shoulder against the rail—that whenever Harry Hack needed to reach the engine room from the cockpit, he had to climb overboard, and then he would edge his way along the outside fender. Some passengers lost hope and wanted to turn back. A timid voice suggested a vote. "You can swim if you like," said one of the student crew.

In the morning, a formation of bombers thundered overhead, seeking bigger fish than the *Sailor's Hope*. Later on, Harry Hack managed to nurse the second cylinder back to life. The boat became easier to steer, but no one had a clear idea about the proper tack. There were almost no provisions left—and as to fuel, there seemed no way to guess how long it would last. Harry Hack asked the passengers to watch out for floating mines, and the psychiatrist for signs of panic.

Toward evening, four smokestacks came into view, and eventually a British destroyer, HMS *Venomous,* took the fugitives aboard. "We were received with bananas, kindness, and tea," as Neurath remembered later. The Navy officers told Harry Hack that his *Zeemanshoop* had miraculously wound its way through several minefields: only the shallowness of his boat's draft had saved them.

Upon their arrival in Dover, Otto Neurath and Marie Reidemeister somehow got separated. Neurath, who had no passport, resourcefully pulled out from his pocket a review of *Modern Man in the Making*—it showed his picture. The bobbies nodded kindly.

But wars are not won with kindness, and all German citizens who were on British soil were interned as "enemy aliens." After all, Great Britain was fighting alone, with its back to the wall, and the islanders' fear that a fifth column of German saboteurs might spring up in their midst was tangible.

Thus Otto and Marie ended up in separate camps on the remote Isle of Man, halfway between England and Ireland. At least they were allowed to write to each other. Otto signed the letters to his darling Mietze with his trademark, an elephant. The two were even allowed to meet once a month, together with hundreds of other couples who were in a similar plight. However, Marie and Otto were granted this privilege only after promising to marry each other right after their release. The date of that release was uncertain and remained so for a long time.

Neurath, who had been imprisoned before, put up with his current fate without complaining. He felt "neither persecuted nor aggrieved, just simply interned," he wrote. It was in Otto Neurath's nature to endure all these vicissitudes quite philosophically, or rather, sociologically, since, after all, life in his camp supplied him with all manner of interesting observations.

"I have always been interested by the conditions in British jails," Neurath wrote to Felix Kaufmann, "and would gladly have paid for this information—but now I can get it for free!" Neurath gave lectures to his fellow inmates in the camp, with mouthwatering titles such as "How Do You Make the Tennis Court so Durable?" As Neurath himself liked to say, he spoke "fluent broken English."

A large portion of the camp's inmates were Jews, and many were released during the following months. Einstein, Russell, and others wrote letters on behalf of Neurath, and All Souls College offered him a position as lecturer at the University of Oxford. And so, after eight months of confinement, Otto Neurath and Marie Reidemeister were finally set free—and just as promised, they promptly married.

## A Strong Exit

For the first time in his life, Otto Neurath, now nearly sixty, experienced the joy of giving lectures at a university. Nobody in England cared about whether he had ever earned his *Habilitation*—in fact, the concept was unknown. From 1941 on, Neurath gave courses on social sciences and on logical empiricism. And before too long, he was even able to revive his Unity of Science movement.

He also carried on with his anti-metaphysics crusade. Was he not being proven right on a daily basis by current events? The crude ideology of the Nazis had always tended to side with idealistic philosophers, all the way from Plato to Martin Heidegger, and the blind obedience of Hitler's troops could well have distant roots in Immanuel Kant's ethics of duty.

With help from England's first female professor of philosophy, Susan Stebbing (1885–1943), the Neuraths were soon able to establish another Isotype institute in Oxford. This was their third. Experience and contacts paid off, and soon they were busy again with exhibitions and books.

Most importantly of all, they collaborated closely with Paul Rotha (1907–1984), a leading British pioneer of documentary film, whose talent offered brilliant possibilities for the pictorial work of the Neuraths. With the support of the British Ministry for Information, they produced more than a dozen short films and a full-length movie, *World of Plenty*, which became a box-office hit.

Neurath steered clear of the political bickering among Austrian emigrants. More than ever, the communists and the Social Democrats were at each other's throats. After the death of Otto Bauer, another acquaintance of Neurath's had become the informal leader of the scattered Social Democrats: Friedrich Adler, the former physicist. During the interwar years, Adler had been secretary of the Socialist Workers' International, which steered its own course between the Second and the Third (Moscow-dominated) International, and hence was jocularly called the "Second-and-a-Half International."

In one meeting after another, these groups tried to unite in their fight against fascism, but in vain; during one of Adler's endless tirades, a French politician whispered: "He shoots better than he talks." During the 1930s, Adler's office was first in London, then in Zurich, then in Brussels; in 1940, he fled from France to Spain with a forged passport, under the name of Herzl.

Now based in the United States and just as headstrong as ever, Adler made it clear that he was not going to contribute to a "legend of happy Austria." As in 1918, he still pleaded for a union with Germany. For that reason, during World War II, he torpedoed all efforts to set up a Free Austrian battalion.

Friedrich Adler was not prepared to support the "utopia of an Austrian nation," which he considered "as disgusting as it was reactionary." When the war was over, he avoided all contact with the new government in Vienna, dissolved the Socialist International, moved to Switzerland, and devoted his entire time and energy to writing a biography of his father, Viktor. He did not wish to have anything to do with day-to-day politics. But inevitably, when Friedrich Adler died in 1960, he found his final resting place in Vienna's Central Cemetery, in the monumental tomb of his father. Sigmund Freud would have smiled.

In contrast to Friedrich Adler, Otto Neurath was not plagued by even the slightest trace of bitterness or resignation. When the war finally ground to an end, peace found him as ready as ever, and indeed, grand new vistas opened up. The war economy seemed all set to develop almost by itself into a planned economy for peace, just as had happened after World War I. It would satisfy all needs by avoiding wasteful surplus.

In June 1945, the optimistic projects of a fair and social postwar order led to a landslide victory for Labor. Not only the United Kingdom but an entire continent was in need of reconstruction. A democratic "plan of all plans" seemed within reach. Déjà vu? Assuredly. But Neurath was stone deaf to disillusionments.

The newly reborn and renamed Austria (no longer Ostmark) was in urgent need of reeducation, as an antidote to the brainwashing years of Austro-fascist and National Socialist propaganda. Few observers suspected, in 1945, that Hitler himself had already provided for sweeping reeducation, during the grim final year of the war. Neurath threw himself with relish into the task of politically enlightening a whole people. Even his pet project, the *Encyclopedia*, came back to life.

But then, out of nowhere, like thunder booming in a clear blue sky, on December 22, 1945, a stroke stopped Neurath in his tracks. Death was instantaneous. Otto and his adored wife, Mietze, had been dining with friends. They had been talking about a recent newspaper article on Neurath, titled "Man with a Load of Happiness." On their return home, he read aloud to Mietze a few humorous lines from the letters he had written that day. Musing back over the past, he said that he could not have wished for a better life.

FIGURE 12.9  Otto and Marie (Mieze) Neurath, with unidentified cat.

Mietze, as she was cleaning up in the kitchen, heard a low sound that at first she took for a guffaw. Upon turning around, she saw that Otto had collapsed and lay lifeless on his desk.

It was an exit similar to that of Dr. Faust in Goethe's play. In his pact with Mephistopheles, Faust had pledged to give up his life as soon as he felt contented: "When I say to the Moment flying: Linger a while— thou art so fair."

As it happened, Faust had been the subject of Neurath's first paper ever—a fitting *ouverture* to an oversized life. Indeed, with his indomitable zeal and his boisterous fashion of dashing top-speed through life, Neurath clearly was a Viennese descendant of the restless medieval doctor.

# Fadeout

*Vienna, post-1945: Circle is dead. Vain attempts at reanima-
tion by librarian Kraft. Kraft is sued by Schlick's murderer.
Feyerabend concludes anything goes, is hailed as science's worst
enemy. In Cambridge Popper clashes with Wittgenstein, Wittgenstein rushes off. Legendary philosopher claims on deathbed:
Had wonderful life. Popper refutes own legend. In US Carnap
becomes living philosopher. Gödel sides with Plato, works on
theological worldview, finds fault in US Constitution, arouses
FBI's suspicions. Popper confesses: "I killed positivism."*

## Reparations in the Strict Sense

The bomb-torn walls of the University of Vienna were soon rebuilt. In
the summer semester of 1945, a makeshift schedule of lecture courses
was drawn up, but it was impossible to compensate for the loss of brain-
power. A recovery of the university's once-brilliant intellectual tradition
seemed so utterly out of reach that it was not even attempted.

When the state of Austria came back to life in 1945, one of the first
measures it took was to dismiss all civil servants who had been members of the Nazi party. A high percentage of professors thus had to go.
But in the following years, many of the ex-Nazis somehow managed to
claw their way back up and eventually reassumed their chairs, seemingly untainted.

It seems that the Austrian government never seriously contemplated inviting those who had been purged in 1938 to return home. According to the prevailing view, Austria had been Adolf Hitler's first victim. There was a murdered chancellor to prove this, and the Moscow Declaration of the Allied Powers had stated this as well. Those who had left Austria were a source of envy, not pity. After all, they had been spared years of hardship. Few of them, so the consensus went, would ever consider returning to the mournful Vienna of the postwar era; few of them were invited. Who would want to leave a comfortable life and come back to a crumbling town divided among four occupying armies? In their efforts at making reparations to victims, German universities proved much more active than their Austrian counterparts.

A typical episode centered on Karl Menger. Shortly after the *Anschluss*, Menger had sent in his resignation by cable from America. Nevertheless, eager university officials were not willing to be robbed of the pleasure of dismissing him on their own terms. In March 1938, they set in motion an investigation by the *Sippenamt*, or "kinship bureau," whose task it was to check proper Aryan descent. To their delight, Menger was found to be a *Mischling*, or half-breed. In a parallel investigation, a self-appointed *Sippenforscher*, or "kinship researcher," obsequiously asked the university for a photo (in profile, ideally) of the bust of Karl Menger's father, which stood in the arcade court of the main building. The diligent *Sippenforscher* added that taking such a photo would presumably require a special permit by the rector of the university. History did not record the reply. But in July 1938, Karl Menger was duly sent the decree proclaiming his dismissal, file number OZ.8146/1083.

In May 1946, however, the University of Vienna told the ministry that Menger could not count "in a strict sense" as someone who had been dismissed by the Nazi regime, because he himself had given notice after the *Anschluss*. In fairness, this argument was not convincing, even to the dean who had proposed it. But the upshot was that Menger, who would have been in mortal peril if he had stayed, was never invited to return to Vienna. The wound hurt deeply, and the loss for Menger's alma mater was beyond repair. By virtue of both his age and his personality, Karl Menger would have been uniquely suited to the mission of bringing the Vienna Circle back to life.

During the war, Menger had devoted himself to the task of teaching calculus to U.S. Navy officers, lecturing nineteen hours per week. Since he also had a family with three children to take care of, it was practically impossible for him to do any research during these years. After the war, he moved from Notre Dame into the ultramodern rooms of the newly founded Illinois Institute of Technology, built by Mies van der Rohe.

This was a far cry indeed from the ruins of Vienna. Back there, in his hometown, people seemed to have forgotten that he had ever existed. This was a bitter pill for Menger to swallow, and only in the 1960s did he more or less make his peace with Austria. But Vienna had never strayed far from his mind. When he died in 1985, he left behind copious notes on the Vienna Circle and the Viennese Mathematical Colloquium, which were published posthumously. A translation into German is still waiting.

## INNER EMIGRATION

Those who had managed in one way or another to live through the Tausendjährige Reich's "one thousand years" were also prone to bitterness, although not always justifiably.

Early on, Nazi philosopher Martin Heidegger was forced to step down as rector of the University of Freiburg. His endeavor to "totally revolutionize the whole German being" had ended in failure. Freiburg's union of lecturers ranked him at the top of the list of the "most expendable" professors in their midst. But even after resigning from the rectorship, Heidegger did not resign from the party. In fact, he remained a Nazi all the way till the end of the war, when he had a change of heart. Philosophers all around the world who revered his "Nothing"-filled writing style were soon able to forgive him his errors. Postwar life was not too rough for Martin Heidegger.

Otto Pötzl, formerly the Vienna Circle's psychiatrist in residence, lost his position as professor and head of the clinic after the war. This was the price he paid for having joined the National Socialist Party in order to keep his position after the *Anschluss*. Probably Pötzl could have found a way of reversing this decision, especially as he had a powerful ally in his coworker Viktor Frankl (1905–1997), who, as an Auschwitz survivor, wrote a famous book, *Man's Search for Meaning*, about survival in concentration camps. Frankl also founded the branch of psychotherapy called *logotherapy*. He testified that Otto Pötzl had tried as hard as possible to sabotage the Nazis' euthanasia program. Some Nazis had been given clemency for much less. But as Pötzl was close to retirement age, he ultimately settled for simply getting his full pension.

Among the members of the old Vienna Circle, there were two who had been able to live through the years of the Nazi regime in greater Germany: Viktor Kraft and Kurt Reidemeister.

Reidemeister's "unblemished Aryan descent" had never been disputed. But since he had had the nerve to complain about the political agitation being carried out by National Socialist students, he lost his chair in Königsberg soon after Hitler seized power in 1933. Outraged mathematicians rose up to protest on Reidemeister's behalf. At that point, this kind of action was still possible. The regime begrudgingly made a concession: Reidemeister was given a position at the University of Marburg.

There, he increasingly withdrew into his own world. Aside from continuing to work on knot theory, he devoted himself to philosophy. His book *Plato's Mathematics and Logic* was politically innocuous enough to be published in 1942. When the war was over, Reidemeister was elected president of the German Society of Mathematicians: after the years of "German Mathematics," it was now essential to entrust the new beginning to someone beyond reproach.

In 1955, Reidemeister was appointed to a chair in Göttingen, once such a glorious mecca of mathematics. But despite many belated tributes, his last years were sad. Little was left of the high-spirited young

*V. Kraft*

FIGURE 13.1 Viktor Kraft steps
out from the background.

German who so often had inspired uproarious laughter in the meetings
of the Viennese Mathematical Society.

The other original member of the Vienna Circle to have lived
through the Third Reich was Viktor Kraft (1880–1975). He belonged
to the founding generation of the Vienna Circle and had known Hans
Hahn, Philipp Frank, and Otto Neurath from well before World War I.
Like them, he had been a keen member of the Viennese Philosophical
Society and a zealous lecturer in Vienna's institutes for adult education.
Kraft had studied philosophy and history in Vienna and Berlin. Early
on, he landed a job as librarian at the University of Vienna—a secure
position that left him ample time to pursue his philosophical interests.

Kraft had taken part in the meetings of the Vienna Circle from the
start, although more as an interested bystander than as a focal figure.
He also belonged to the Gomperz Circle. Most importantly, he had
been among the first to recognize and champion Karl Popper's talent.
Young Popper liked to meet him when his library hours were over, often
escorting him back home. During these years, Viktor Kraft wrote little.

However, his book *The Principles of a Scientific Theory of Value* appeared in 1937 in the Circle's own series, *Writings for a Scientific Worldview.* In fact, it was the last volume ever published in that series.

After the *Anschluss,* Kraft lost both his job as a librarian and his right to teach because he firmly refused to split from his Jewish wife. Despite the loss of his job, Kraft was somehow able to weather the war years in Austria, and when the curtain fell on the Third Reich, his prospects brightened: he became chief state librarian and associate professor at the university.

In the late 1940s, Kraft published, in quick succession, four books that he had written during his period of enforced retirement: a new edition of his *Theory of Values;* a book called *Mathematics, Logic, and Empiricism;* another volume called *Introduction to Philosophy;* and most importantly of all, *The Vienna Circle: The Origins of Neopositivism.*

Kraft was probably better placed than anyone else alive to write such a book. On the one hand, he was the only surviving member who had witnessed the entire story from the *Urkreis* to the bitter end. It was in his apartment that some of the Circle's last meetings had taken place, in the confused months after Schlick's murder. Yet on the other hand, Kraft had always kept sufficient distance to be able to offer an objective account of the diverse points of view.

## THE "MOST RECENT HISTORY" ON TRIAL

Kraft's book on the Vienna Circle was subtitled *A Chapter on the Most Recent History of Philosophy.* And recent it was indeed, as was soon proved by a lurid episode.

In his book, Viktor Kraft had written that Schlick had been shot by "a former student, a paranoid psychopath." This phrase was eventually discovered by Johann Nelböck, who was still working in the Mineral Oil Department, now run by the Soviet occupiers. In 1947, Nelböck's conviction had been expunged from the penal record, as he had requested years earlier, leaving no blemish in his police dossier. Although Kraft, in his book, had never stated the name of Schlick's murderer,

Nelböck sued him for libel. "Paranoid psychopath" indeed! Nelböck was not going to take this lying down and have his livelihood jeopardized by another godforsaken positivist. He, Nelböck, would defend himself. The Vienna Circle had done him harm enough. He was going to take this to court. Kraft was going to regret his words.

After a lengthy trial, however, the court rejected Nelböck's suit. The judge ruled that "paranoid" was merely an assessment and did not amount to spreading detrimental facts. Naturally Nelböck did not relent. He was not going to swallow this poisonous verdict.

Viktor Kraft had good reason to be alarmed. The years had not chastened Nelböck, who was as firmly convinced as ever of the righteousness of his cause. Newspapers reopened the case and stressed the ominous parallels. They pointed out that the tragic events of less than fifteen years earlier should serve as an object lesson. "Already at that time," wrote the weekly *Wiener Wochenausgabe,* "the allegation that Dr. Hans Nelböck was mentally ill caused an unfortunate chain reaction ending with Dr. Nelböck pointing the deadly weapon at the philosophy professor Dr. Moritz Schlick."

The sagacious editors of the weekly knew exactly how to avoid the next link in the "chain reaction." The culprit, "who had certainly suffered enough for his crime," should under no circumstances "be made as miserable as possible for the rest of his life. Our fellow citizen Nelböck, who was once led astray for psychological reasons that can easily be understood," ought not to be barred from returning to human society "by being labeled, long after the party was over, a fool." After this choice callousness, the weekly's article ended with these sanctimonious words: "It is time, for the sake of humanity, to turn toward peace, atonement and quiet. Yes, here too. . . . "

Nelböck's lawsuit against Kraft was a contest between Schlick's murderer and Schlick's successor. Indeed, in the same year when his book on the Vienna Circle appeared (1950), Viktor Kraft was appointed to a chair in philosophy. One year later Leo Gabriel, Nelböck's former confidant and supporter, became Kraft's colleague at the institute for philosophy. It was a small world indeed.

Nobody expected the seventy-year-old Viktor Kraft to build up a new circle. Indeed, in postwar Austria, not much was expected from anybody. The future looked nothing but bleak. The University of Vienna had lost touch with international developments. The scientific achievements made between the two world wars seemed to belong to a remote era. Nevertheless, once Kraft stepped to the fore, he quickly became the focus of a handful of young thinkers rebelling against the moldiness of postwar Vienna and who, with their endless discussions, often defied the closing hours of the poorly lit coffeehouses.

Before long, a passionate young woman from Carinthia graced the group, and she soon became known as one of the foremost writers in the German language: Ingeborg Bachmann (1926–1973). Her poems, librettos, and radio plays reflected the tragic intensity of her life—a candle burning at both ends.

Ingeborg Bachmann wrote her doctoral thesis under the supervision of Viktor Kraft. It was on the subject of ex-rector Martin Heidegger, whose murky writings had so often rubbed the Vienna Circle the wrong way. Soon after finishing her thesis, Bachmann went on to write an exuberant essay about Ludwig Wittgenstein, and then she authored a radio play on the Vienna Circle. "In Vienna itself," it lamented, "the Vienna Circle is dead."

Another student of Viktor Kraft's was the feisty Viennese philosopher Paul Feyerabend (1924–1994), who had ably served in the German Wehrmacht, rising to the rank of lieutenant. After the war, Feyerabend, as a wounded veteran, returned to his studies, and in his PhD thesis, "On the Theory of Basic Sentences," he took up the contentious old theme of "protocol sentences," calling them by Popper's alternate name.

Feyerabend had found his way to philosophy and the Kraft Circle through the yearly summer meetings in Alpbach, a village in the Tyrolean mountains. In the postwar years, tiny Alpbach had become a significant intellectual meeting place, where such notables as Karl Popper, Rudolf Carnap, Herbert Feigl, Erwin Schrödinger, Friedrich Hayek, and Philipp Frank would come, bringing with them exciting whiffs of the wider world. And thus the brand-new Dr. Paul Feyerabend,

FIGURE 13.2 Ex-lieutenant Feyerabend claims that "Anything goes!"

inspired by his chance to hobnob with such famous figures, applied to the British Council for a fellowship, in the hopes of making a name for himself in England.

In 1954, at age fifty, Johann Nelböck unexpectedly died, in the midst of giving a lecture in the apartment of a retired Nazi philosopher named Dr. Lauss. In that same year Viktor Kraft retired, this time for good.

This meant that Kraft's position was now vacant. The search committee came up with a remarkable list of candidates: tied for first place were Friedrich Waismann (Oxford) and Carl Friedrich von Weizsäcker (Göttingen). The former, of course, we know as one of the most stalwart members of the Vienna Circle. The latter was an eminent German physicist and later a philosopher, whose intellectual accomplishments, however significant, had been deeply tainted by his wartime efforts to develop an atom bomb for Germany, which after the war he uncomfortably tried to deny. Next on the list came Bela Juhos (1901–1971), once a junior member of the Vienna Circle; and lastly a certain Erich Heintel (1912–2000), who described himself as a "substance metaphysician" who understood "Man in his wholeness as transcendentality in being." Now that, no doubt, was a most lofty philosophical insight, as its meaning floated so high up in the clouds as to be nearly invisible. Carnap would have loved citing Heintel's vision in his discussions of metaphysical language.

At this stage, some ministry officials saw the need for a slick bureaucratic sleight of hand. First of all, they demoted the vacant chair to an associate professorship. This clearly meant that the illustrious candidates from the great centers of Oxford and Göttingen could not possibly be interested in such a modest position—thus there was no need to even ask those two. So much for Waismann and Weizsäcker! Next, it was pointed out that Dr. Heintel had lectured in front of far larger audiences than Dr. Juhos had. That telling fact turned the trick (Otto and Marie Neurath could have made a cute pictorial representation of the argument with "rows of little men"), and so Erich Heintel was given the job; and what do you know, within a short time, his modest status was raised to that of full professor. *Et voilà!*

As a result of these deft maneuvers, Ernst Mach and Moritz Schlick wound up being succeeded in Vienna by a "substance metaphysician," a former member of the Nazi party, card number 9.018.395, who had been classified as "less involved" after the war. Professor Heintel would maintain his cathedra until well into the 1980s.

## Philosophy with a Poker

When Friedrich Waismann first arrived in Cambridge, he no doubt entertained hopes of bringing his book on Wittgenstein to a conclusion. But Wittgenstein himself had lost interest in their collaboration. He now wanted to write his *Philosophical Investigations* all by himself, and soon he refused to meet Waismann at all. "The man has grown fat," he declared. As a result, after a short while, the newly labeled "fat man" moved from Cambridge to Oxford as a lecturer in philosophy. His lecturing went well in the new venue, but tragedy soon overwhelmed his life. His wife committed suicide, and a few years later, so did their teenage son.

Friedrich Waismann's *The Principles of Linguistic Philosophy*, as well as his book on Wittgenstein (including the foreword written by Schlick almost forty years earlier, along with Waismann's records of their Viennese conversations), only appeared long after the death of all persons involved.

Shortly before the outbreak of World War II, Wittgenstein had been appointed professor in Cambridge as the successor of G. E. Moore; in addition, he became a British citizen. His German passport had, as he wrote, "burnt him in his pocket." Now—in the summer of 1939—as a British citizen, he could safely travel to Berlin. And this he did, so as to further the cause of his sisters. With Hitler's consent, they became *Ehrenarier,* or "honorary Aryans," a relabeling that made it possible for them to remain in the Ostmark, unmolested by Nazis. This business transaction—for in truth that was all it was—provided a substantial amount of foreign currency to the Third Reich; indeed, the assets of the Wittgenstein family had been invested to a large extent in Switzerland, out of the Nazis' reach. A large part of those Swiss funds was now used to pay for the certificate attesting that the Wittgenstein sisters were *reinblütig,* meaning "true-blooded" (or, in clearer language, "not Jew-blooded").

During the war, Wittgenstein served as a medical orderly in a London hospital and later as an assistant in a Newcastle laboratory. In 1944, he resumed his philosophy chair. One day in his classroom, a Viennese woman showed up— Rose Rand, PhD, who in earlier days had religiously maintained the minutes of the meetings of the Vienna Circle. Now just as then, Rose Rand was still as poor as a churchmouse, having barely survived for several years in England by toiling on an assembly line in the armaments industry. She humbly asked Wittgenstein if he might recommend her for a fellowship. Wittgenstein replied that he could do nothing for her, adding that in his opinion, there was "no disgrace in living from manual work."

More and more, Wittgenstein viewed his professorship as an impediment to his high philosophical aspirations. He felt hampered by the "wet and cold psychic climate" of Cambridge. Moreover, his relationship with Bertrand Russell had suffered. The seventy-two-year-old Lord Russell had returned to Trinity College in 1944. With his *History of Western Philosophy,* he landed at the top of the best-seller lists, and not long thereafter he was awarded the Nobel Prize in Literature. For his own reasons, Ludwig Wittgenstein strongly disapproved of the writing career of his former philosophical mentor.

Karl Popper, who worshiped Russell, had spent the war years in New Zealand. At the University of Christchurch, his teaching duties had been crushing. Nevertheless, during that time he had managed to write two books on the philosophy of history and politics, which would soon assure his fame: *The Poverty of Historicism* and *The Open Society and Its Enemies.* In them both, Popper resolutely attacked any and all forms of thinking that could lead to totalitarian ideologies, whether from the left or from the right. He considered these two books his contribution to the war effort, although in fact, his *Open Society* would have its full impact only in the following war—the cold one.

During his Viennese youth, Karl Popper had been given ample opportunity to observe totalitarian endeavors at first hand. On that topic, he truly knew whereof he was speaking. Moreover, it so happened that Popper's theory of how science works placed great weight on the interplay of hypothesis and refutation, on trial and error, and on the provisional nature of all knowledge. All of this machinery proved to be eminently transferable to the alternate domain of politics, where any democratic government that failed the cause of its people—meaning all governments, sooner or later—could be discarded in a fairly painless way, just as a scientific hypothesis can be discarded.

Pugnacious as ever, Popper named and shamed all "enemies of the open society"; these included not only Hegel, Engels, and Marx, the holy trinity of communism, but also Plato, with his proto-fascist philosopher-kings. Many philosophers felt offended on Plato's behalf; Otto Neurath not so, for he had already noted how much of Plato's *Republic* could be seen as a blueprint for a totalitarian state.

The economist Friedrich von Hayek liked Popper's staunch anti-Bolshevist stance. Hayek had always deeply opposed government interference in private affairs; he had even written a book, *The Road to Serfdom,* against it. Thus Popper's *Open Society* resonated deeply with

FIGURE 13.3 Wittgenstein feels unhappy in Cambridge.

his views. The influential Hayek, who also happened to be a cousin of Wittgenstein's, managed to find a position for Popper in the London School of Economics. This offer finally brought the combative philosopher back to England from the antipodes.

In the meantime, Popper had overcome his language problems, and in England he found a stage on which to strut that was commensurate with his ambitions. When, one day, he was invited to give a talk in Cambridge, he suddenly sensed that this was his chance for a showdown with Ludwig Wittgenstein. Way back in 1932, in the Gomperz Circle, he had only been shadowboxing with Ludwig, as his adversary had not been physically present. This time would be different.

Popper chose for his title: "Do Philosophical Problems Exist?" Just as the great "Do Atoms Exist?" debate between Mach and Boltzmann can be viewed as an overture to the Vienna Circle story, so the legendary "Do Philosophical Problems Exist?" clash between Wittgenstein and Popper can be seen as its final chord: a jarring dissonance. The angry argument took place on October 25, 1946, and lasted for roughly ten minutes.

For their entire lives, the two philosophers had managed to steer clear of each other. In a town such as Vienna, where everyone knows everyone, this in itself was no mean feat. In 1937, Popper, on his way to New Zealand, had given a talk in Cambridge. On that occasion,

Wittgenstein had decided not to show up, because he was nursing a cold. Or was it to give a cold shoulder? Wittgenstein could hardly have been unaware that once, in the Gomperz Circle, Popper had attacked him for being dogmatic.

Now Popper was visiting Cambridge again. Just before Popper's talk, and in the same room, Wittgenstein's highly exclusive seminar had taken place. As usual, Wittgenstein, sitting by the fireplace, had delivered a halting, groping monologue, occasionally grabbing hold of the poker used for pushing around the flaming logs. Now the public filed in for Popper's talk. And then entered the speaker, escorted by none other than Bertrand Russell. Altogether there were some thirty people in the ill-heated room, slightly more than the usual number.

By a twist of fate, Popper had been handed a good opening move on a silver platter. His talk had been announced under the erroneous title "Do Philosophical Puzzles Exist?" But of course his intention had been to speak about philosophical *problems*. The title change had been made without consulting him. To his ears, the word *puzzle* sounded far less serious than *problem*. And this was precisely his point—he claimed that there really do exist serious, genuine problems in philosophy. Thus the erroneous title was a free gift, conveniently leading him straight to the heart of the matter.

Wittgenstein interrupted the speaker, as was his wont, after the first few sentences. Could Popper please *name* some philosophical problems? Popper, of course, had a well-prepared reply. However, Wittgenstein interrupted him again, and then again, but Popper, catching the drift of the exchange, did not let him finish. Both philosophers spoke English with a strong Viennese accent, and right then their accent did not sound pleasant at all.

Things heated up quickly. Wittgenstein started to gesticulate. Since in his hand he was still holding the poker, he found it easy, as a way of underscoring his words, to gesture with it, much like a conductor waving a baton. When he urged Popper to give him an example of a moral proposition, the latter glared at the heavy iron rod that Wittgenstein was brandishing, and replied: "Do not threaten the guest lecturer with a poker!" Wittgenstein, caught off guard and at a loss for words, flung the poker onto the floor and stormed out, loudly slamming the door shut.

This, at least, is how Karl Popper related the story. What some of the witnesses remembered was different. Some said Popper's barb was uttered only after Wittgenstein had left the room. Some said Wittgenstein did not slam the door. All that is certain is that Wittgenstein was unable to endure Popper for fifteen minutes. And ever after that brief encounter, Popper enjoyed the reputation of being the only person who had interrupted Ludwig Wittgenstein as ruthlessly as Wittgenstein used to interrupt everyone else.

However, a great deal more lies behind the scenes of this dispute than merely the incompatibility of two Austrian prima donnas. A large part of Popper's philosophy can be viewed as a reinterpretation of, and often as an improvement of, the thinking of the Vienna Circle. Popper played these differences up while the Circle played them down, but essentially their thinking was on the same wavelength. In one respect, however, Popper's opinions clashed fundamentally with what Schlick, Carnap, and company hailed as their Turning Point in Philosophy. Popper was always firmly convinced that philosophical questions were much more than mere pseudo-problems caused by a misuse of language. He vehemently opposed Wittgenstein's view of philosophy as "a struggle against the bewitching of our reason by means of language." Popper rejected the notion that the great problems of philosophy are caused by "language taking a holiday."

As Popper noted in his autobiography, he was barely fifteen years old when he made it a personal rule never to argue about words and their "true" meaning. "I still believe," so he wrote, "that the surest way to perdition is to neglect real problems for the sake of disputes about words."

Given this outlook, though, Karl Popper was uncommonly gifted in twisting the words of his opponents and hoisting them with their own petard. Whether it was a question of puzzles or of problems, whether science was about falsification or verification—words actually played a central role in Popper's style of philosophizing, when push came to shove.

Within a week of his quarrel with Ludwig Wittgenstein, Popper wrote to Russell that he had been gratified to find himself on the same side as Russell. However, Popper expressed a twinge of disappointment that Russell had not played a more forceful role in the debate. This had not been necessary, Russell replied. Popper had proved perfectly able to hold his own.

And indeed this was the case. At this point, nothing could stop the ascent of Karl Popper's star.

In 1947, Viktor Kraft had unofficially asked Popper whether he might possibly be interested in assuming Moritz Schlick's former chair, but Popper replied in no uncertain terms that he had no intention of leaving England; he had found his place. And so he had. Twenty years later, he had become Sir Karl Popper; and when Lord Russell died shortly before turning 100, it was arguably Sir Karl who took his place as the most widely known philosopher of the day.

Ludwig Wittgenstein resigned from his Cambridge chair in 1947, with the goal of devoting himself entirely to his unending manuscript. He started spending much of his time in Ireland, and he visited the United States, and also returned occasionally to Vienna. For cinema fans, this period of Viennese history will be familiar as the dismal era of *The Third Man*.

Paul Feyerabend tried to invite Wittgenstein to visit the Kraft Circle. In his autobiography *Killing Time,* he described his attempt: "I went to the Wittgenstein family residence [the Palais Wittgenstein, which still stood at that time]. The entrance hall was large and dark, with black statues in recesses in many spots. 'What do you want?' asked a disembodied voice. I explained that I had come to see Herr Wittgenstein and to invite him to our circle. There was a long silence. Then the voice— the housekeeper, who spoke from a small and almost invisible window high up in the lobby—returned: 'Herr Wittgenstein has heard of you, but he cannot help you.'"

A few days later, however, Herr Wittgenstein changed his mind, and his gaunt figure showed up at the Kraft Circle's meeting, albeit an hour late. He sat down, listened to Paul Feyerabend for a minute or two, and then interrupted: "Stop at once! This will not do." He then took over from his fellow ex-lieutenant, and on the whole seemed to enjoy himself.

But the fact is that by then Ludwig Wittgenstein was suffering from a serious cancer, much as his father had. He had no desire to go into the hospital. Instead, he retired to the home of his physician, which was located in Storey's Way, Cambridge. In a letter, however, he once misspelled the address, writing "Storey's End." And that was in fact his last residence.

Wittgenstein asked his landlady to tell the friends who were due to visit him on the following day that he had had a wonderful life; and on that day, April 29, 1951, he breathed his last.

Two years later, his *Philosophical Investigations,* a major work about the philosophy of language, appeared in print. Wittgenstein had found the right structure, after all: the book was conceived like an album of landscape sketches, showing the same places over and over again, but from different viewpoints, all connected by a dense lattice of paths having neither beginnings nor endings.

"The meaning of a word is its use in language" became a mantra, as had the famous time-honored assertions of the *Tractatus*—the treatise against which, said the author, his *Investigations* leaned like a wanderer tilting into a headwind. He wanted the new *Philosophical Investigations* to be bound together with the *Tractatus,* "to put them into the right light." The epigraph Wittgenstein chose for the *Investigations* came from Johann Nestroy, that most Austrian of playwrights: "The main thing about progress is that it always looks larger than it actually is." This sentiment echoes the closing sentence in the preface of the *Tractatus:* that the book's value is to show how little is achieved by solving problems.

Despite this seeming modesty, the author of the *Tractatus* actually had thought that in that work he had carried philosophical thinking to its ultimate limits. By contrast, the author of the *Investigations* had realized that he would find no satisfactory end to these matters.

The years and decades following Wittgenstein's death saw a flood of publications from his *Nachlass,* or philosophical estate; it comprises some thirty thousand pages of notes. And in university libraries, whole shelves are filled with erudite tomes explaining Wittgenstein's thoughts—a task as thankless as that of explaining jokes.

"Peace in thinking," Wittgenstein wrote, "is the wished-for aim of those who philosophize." On that sentiment at least, Ludwig Boltzmann would have heartily agreed with Ludwig Wittgenstein.

Shortly before Wittgenstein's death, and without knowing about the *Investigations,* the young Ingeborg Bachmann had written on modern philosophy: "Where to position the lever today? Maybe with Ludwig Wittgenstein, who still has to be discovered—the greatest and at the same time the least known of all philosophers of our age." The poet described him as "one of the strangest and most legendary figures in all philosophy.

"Thus legend took over his life while he was still alive, a legend of voluntary deprivation, of trying to be saint-like. . . . Like Pascal, Wittgenstein always moved with his personal abyss—and within it."

Ingeborg Bachmann's fellow student, the enterprising Feyerabend, had planned to use his British fellowship to visit Wittgenstein and study with him. After Wittgenstein's death, he expeditiously turned to Karl Popper. And Paul Feyerabend slipped as smoothly into Popper's critical rationalism as he had glided into the logical empiricism of his PhD supervisor, Viktor Kraft.

Soon, however, he started to blaze his own pathway in the theory of how science works. At that time, the figure looming largest on the scene, other than Popper, was the American Thomas Kuhn (1922–1996). Kuhn's critical treatise *The Structure of Scientific Revolutions* had been published as a volume in Otto Neurath's *Encyclopedia of the Unity of Science.*

To Kuhn's mind, Popper's notion of scientific progress was too narrow and far too rational. After all, scientists who are eager to falsify their theories are not a dime a dozen. "Normal science," in the Kuhnian view, consists of the drudgery of accumulating knowledge along previously accepted and unquestioned theoretical lines. Only rarely does a collected fact contradict expectations—and when that happens, the new fact is typically viewed as being erroneous, at first. It is only when such anomalies have piled up sufficiently thickly that a Kuhnian "crisis" is at last perceived. When the crisis stage is reached, then new Kuhnian "paradigms" come to the fore and compete. Eventually one of them will gradually replace the previous mind-set, offering new theoretical lines for those who practice "normal science" to obediently follow. Each of the successive "paradigm shifts" transforms the world within whose constraints the "normal scientists" will carry out their fact-collecting drudgery.

In short, the Kuhnian vision is not one of confirmation or falsification. Rather, a number of factions in the scientific community compete for a while, and eventually one of them wins out. The fact that the other factions wound up losing the popularity poll does not mean that they were made up of nonscientists, though.

Thus to Kuhn, progress in science is due less to a theory's refutation than to a changing of the guard: roughly speaking, the old geezers merely fade into the sunset, while the young Turks try to make their names by tossing new paradigms into the ring. Thus, rather than a theory ever being falsified, its supporters merely go extinct. Science progresses by funerals, as the saying goes.

However, for Feyerabend, the Kuhnian view, though cynical, was not cynical enough. Soon the ex-lieutenant had turned into a severe pain in the neck for both Popper and Kuhn. The title of his main work was *Against Method,* and his slogan (borrowed from the great songwriter Cole Porter) was "Anything goes!" This became the battle cry of an anarchist crusade against compulsory "house rules" in science. Like a court jester, Dr. Feyerabend ridiculed anything smelling of academic pomposity. A 1987 article in the journal *Nature* described him as "the worst enemy of science." This was slightly over the top, but it clearly showed how far this Viennese apple had fallen from the Vienna Circle's tree.

## GÖDEL'S VIEWS "RATHER UNPOPULAR"

Not all the exiled members of the Vienna Circle flourished in America. Edgar Zilsel found no foothold in the New World and killed himself in 1944. The hapless Rose Rand, who moved to the United States after many miserable years in Britain, futilely tried to improve her prospects. She managed to keep herself alive by translating the works of Polish logicians, now into English instead of German, and from time to time Herbert Feigl and his wife, Maria Kasper Feigl, sent her some money "as a token of our old friendship." That was sweet, but Rose Rand died of cancer of the pancreas, lonely and old.

On the other hand, Gustav Bergmann and Herbert Feigl enjoyed distinguished careers as philosophers of science, and Richard von Mises, who had left Turkey, became a professor of aerodynamics at Harvard University. His *Theory of Flight,* which had started out as an instruction manual for the air corps of the Habsburg Empire, kept on evolving and survived through many incarnations, all the way until the age of supersonic jets. And of course Ernst Mach's last name became a household word. In 1947, the American test pilot Chuck Yeager reached Mach 1.06 in a Bell X-1; in 1953, he flew at Mach 2.44. In that same year, Richard von Mises died. His old friend and co-author Philipp Frank wrote an obituary for *Science.* Professor Frank, too, had wound up at Harvard. There he completed the first serious biography of Albert Einstein, his old friend whom he had once succeeded in Prague, an episode that dated all the way back to the closing days of the Habsburg Empire.

The phrase "from the Vienna Circle to Harvard Square" could serve as a motto encapsulating the history of the philosophy of science of the twentieth century, and indeed, Gerald Holton, an eminent historian of physics, used it as the title of one of his papers. Holton knew well whereof he spoke: having grown up in Vienna, he fled from the *Anschluss* when he was sixteen, and wound up teaching at Harvard.

The most successful member of the Vienna Circle was undoubtedly Rudolf Carnap. In America he found the ideal soil for his brand of philosophy. Soon he became the official voice of logical empiricism. Carnap taught as a professor first at the University of Chicago and later in Los Angeles, with guest appearances in Princeton and Harvard. At Harvard he worked closely with logicians Alfred Tarski and Willard van Orman Quine. Both had visited the Vienna Circle in the early 1930s, converging from different directions, and both were very much on Carnap's wavelength. And both compelled him, again and again, to retreat from extreme positions that had been confidently staked out by members of the Vienna Circle.

In particular, Quine's essay "Two Dogmas of Empiricism" cast doubt on two fundamental assumptions. First was the hallowed distinction, originally formulated by Immanuel Kant, between the analytic and the synthetic. The truth of a so-called *analytic* proposition derives, in theory, entirely from the meaning of its terms ("Bachelors are unmarried"). The truth of a so-called *synthetic* proposition, by contrast, depends on some extra pieces of knowledge ("Bachelors are envied"). Quine did his best to show that this distinction was far from clear-cut.

The second dogma Quine criticized was that meaningful statements could be reduced to the immediately given, which is to say, to sense-data. Quine's arguments convinced many philosophers, and Carnap had his hands full fighting rearguard actions. But at the same time, he kept extending his formal logic of scientific language and proceeded with his investigations into syntax and semantics. This was as it should be, for in many ways his original Viennese dream had at last come true: the new breed of philosophers—or at least those who called themselves "analytical philosophers"—could hardly be told apart from hardnosed scientists any longer.

Soon Rudolf Carnap was granted one of the highest distinctions a philosopher could hope for: a volume dedicated to him in the prestigious *Library of Living Philosophers.* The editor of the series, the American professor Paul Schilpp (1897–1993), who once had been an assistant to Moritz Schlick in Berkeley, had often regretted that it was no longer possible to ask Plato or Kant what they had meant by things they had said. Schilpp wanted to save future generations from experiencing such pangs of regret.

He hit on a three-step procedure: first ask some of the greatest philosophers of our age to describe their own work, then ask some of their colleagues to come up with critical essays, and finally allow the great philosophers to reply.

Kurt Gödel was among those asked by Schilpp to write a critical essay on Rudolf Carnap.

After the war, Gödel had been made a permanent member of the Institute for Advanced Study. From that moment on, his future was secure. Soon afterward, he became a citizen of the United States. It was his fourth citizenship, after the Czech, Austrian, and German ones.

With his usual thoroughness, Gödel prepared for the required interview at the Trenton office of the US Immigration and Naturalization Service. His two witnesses, Albert Einstein and Oskar Morgenstern, themselves naturalized citizens, were just barely able to keep their friend from explaining to the examiner how the US Constitution contained inconsistencies. The story has turned into an urban legend. Actually, the examiner, on hearing that Gödel was Austrian, asked him kindly what government they had over there. "A republic," replied Gödel, "which, through a flaw in the constitution, was changed into a dictatorship." "Too bad," said the examiner. "Luckily, such a thing could never happen here in the States." "Oh yes, it could," exclaimed Gödel, "and I can prove it!" This is when the examiner, a wise and experienced judge, noticed signs of distress in the two witnesses and mercifully cut the interview short.

Gödel may have had a point—he knew how easily Hans Kelsen's Austrian constitution had been overturned in the fateful year of 1933. Kelsen, an exponent of legal positivism, was now a professor at Harvard, just like von Mises and Frank. He had spent the 1930s in Cologne, Geneva, and Prague—in Cologne, incidentally, on the personal invitation of its energetic mayor Konrad Adenauer. And now Adenauer had risen to become chancellor of Germany.

In the early 1950s, Gödel was promoted from permanent member to professor at the institute. The step had long been overdue. "How can any of us be a professor," John von Neumann had asked, "when Gödel isn't?" The academic laurels piled up, such as honorary doctorates from Yale and Harvard ("for the discovery of the most significant mathematical truth of this century"), the Albert Einstein Award, and election to the National Academy of Science.

When Gödel was asked in 1951 to deliver the prestigious Gibbs Lecture to the American Mathematical Society, he decided at last to give vent to his long-held Platonic convictions: "I mean the view that mathematics describes a nonsensual reality which exists independently both of the acts and the dispositions of the human mind and is only perceived, and probably perceived very incompletely, by the human mind." Gödel admitted that "this view is rather unpopular among mathematicians."

Not only among mathematicians, but also among philosophers, he might have added. But Gödel was not overly impressed by the nitpicking of philosophers. In his view, as he told his friend Morgenstern, contemporary philosophy had at best reached the stage that mathematics had reached with the Babylonians.

This was a lamentable situation that Gödel now wished to change radically. It so happened that in all his time as a professor at the Institute for Advanced Study, from appointment to retirement, he gave no lectures, held no seminars, and published just one paper—a paper he had written a long time earlier. But when it came to his essay for the Carnap volume in the *Library of Living Philosophers,* he worked with feverish intensity.

Gödel had already written essays for two previous Schilpp volumes (this before being named a professor): they were dedicated to Bertrand Russell and to Albert Einstein. Neither Russell nor Einstein had had much to say in reply to his essays. Gödel knew how to close every loophole in an argument.

So now, in 1953, it was to be Carnap's turn. Gödel wrote to Schilpp that his essay would provide an answer to the question "Is mathematics a syntax of language?" His answer would be no, and therefore diametrically opposed to the views of Carnap and the Vienna Circle. Mathematics could *not* be reduced to conventions defining the mindless syntactic manipulation of meaningless symbols. Mathematics was instead about genuine objects in some Platonic world of ideas. However, the delivery of Gödel's manuscript got delayed.

In 1954, soon after Rudolf Carnap had accepted a position at the University of California that had been left vacant by Hans Reichenbach's death, Gödel wrote to the editor saying that his essay was more or less finished—he only wanted to add a few paragraphs.

In 1955, he wrote that the job would soon be completed.

In 1956: two more weeks and all would be finished.

In 1957, Gödel announced his decision to shorten his essay by two thirds—it would quickly be done.

In 1958, he did not reply at all to the increasingly worried queries of Schilpp.

And one year later, when the desperate editor announced the publication of the Carnap volume for the coming year, with or without Gödel's essay (after all, it said Living Philosophers, and time was ticking away!), Gödel confessed that he would not submit his essay after all, since it would be too late for Carnap to write anything in reply.

Gödel tried to explain his remarkable procrastinations with the argument that "in view of widespread prejudices, it would be more harmful than helpful to publish a half-finished work." This sounds like a lame excuse, but it wasn't. Gödel had really tried hard to deliver. After his death, no less than six different versions of his essay on Carnap were discovered among his papers.

If ever there was a battle cry that united Mach, Boltzmann, and all the philosophers of the Vienna Circle, it was: "Down with metaphysics!" Metaphysics was incomprehensible balderdash and ultimately responsible for all pseudo-problems in philosophy, and for human backwardness in general.

But Gödel had never joined in this chorus. On the contrary: rather than demonizing metaphysics, he aimed to do for it "what Newton had done for physics." This may sound like the wild ramblings of someone in a manic state, but once before Kurt Gödel had managed to do the impossible, when with his incompleteness theorem he provided a mathematical proof of a philosophical statement: "There exist mathematical truths that cannot be formally derived from the axioms."

However, there was to be no encore: Gödel failed to find a conclusive argument for his Platonic convictions.

## Gödel Becomes Rather Popular

Albert Einstein died in 1955, and John von Neumann a year and a half later. They had been Gödel's closest colleagues in Princeton. After they were gone, the remote refugee from Vienna withdrew increasingly into his own universe. He wrote careful replies to many letters but did not send them off. He communicated with the external world almost exclusively by telephone, a disembodied presence on the way to another world.

Some rare glimpses into Gödel's thinking at this stage in his life are given by his letters to his mother in Vienna. He wrote to her on every second Sunday, and these were letters that he actually did send off. Again and again, he announced his forthcoming visit; again and again, he found reasons to postpone it. In the end, he confessed that he suffered from nightmares about being trapped in Vienna again.

At some point, Gödel's mother understood that she would only get to see her son again if she were to travel to Princeton. This she did, even though by now she was already in her eighties; her visit was such a success that the elderly lady repeated it three times more.

In the letters Gödel sent to his mother, he could afford to be candid. Or at least so he thought. In truth, American military censors read all of his letters. They even apprised J. Edgar Hoover, the redoubtable head of the FBI, of this singular correspondence. Fortunately, Gödel never had any inkling of these nefarious doings. He already had more than his fair share of paranoia, as it was.

Gödel freely revealed his philosophical convictions to his mother, and in a series of letters he described what he called his "theological worldview." Thus in 1961 he wrote to her: "Of course a scientific foundation of the theological worldview is still far off, but I believe that it ought to be possible today already to rationally understand (without

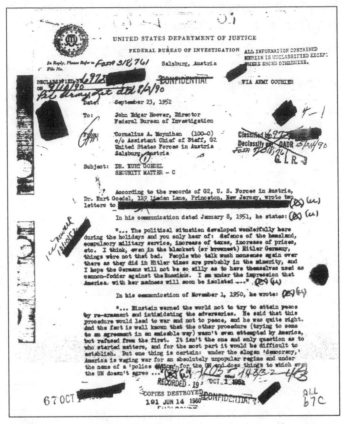

FIGURE 13.4  Mail for J. Edgar Hoover: the letters from Kurt
Gödel to his mother are brought to the attention of the FBI.

having to ground it on any religious belief) that the theological worldview is completely compatible with all known facts (including the circumstances prevailing on our Earth). . . . The world and everything
in it has a reason and meaning, and actually a good and indubitable
meaning. This immediately implies that our existence on Earth, since
it has by itself at best a very doubtful meaning, must be a means for
another existence."

Small wonder that Gödel had kept quiet during the discussions
of the Vienna Circle! His opinions were diametrically opposed to the
scientific worldview of the logical empiricists. Philosophically speak-

ing, Gödel was much closer to Leibniz than to Hilbert, Russell, or Einstein. As he wrote: "The objection that it is impossible that in another world we should remember experiences we had in this one is completely unjustified. Indeed, we could be born into another world with those latent memories. Moreover, we would of course have to assume that our reasoning power will be much better than it is here, so that we will know everything of importance with the same unfailing certainty as $2 \times 2 = 4$, where an error is objectively impossible."

He agreed when his mother wrote that there can be no beauty in a hopeless world, and in lines filled with a haunting uncertainty he urged her to take heart: "For we understand neither why this world exists, nor why it is constituted just as it is, nor why we are in it, nor why we were born in just these and no other circumstances. Why then should we fancy that we know one thing for sure, that there is no other world, and that we never were nor ever will be in another?"

From the copious notes discovered years after his death, it emerged that Gödel had been intensely interested in theology from his student days onward. Later in life, he formalized a scholastic proof for the existence of God by means of mathematical logic. Leibniz, too, had studied that argument: "A perfect being that unites all positive qualities in the highest degree is conceivable. This implies that it exists, since existence is a positive quality." Gödel was mesmerized by Leibniz and was convinced that a centuries-old conspiracy had suppressed the publication of his most important writings. "But who would have had an interest in destroying Leibniz's writings?" Menger asked incredulously. "Naturally, those people who would not want man to become more intelligent," replied Gödel.

Three times a day, Gödel took his temperature. He swallowed dozens of drugs. Shortly before he died, his sole remaining friend, Oskar Morgenstern, wrote: "Gödel clings to me—he has no one else, that much is clear—and thus he increases the burden that I have to carry."

In 1978, when Adele was transported home after a lengthy hospital stay, she found her husband in such a critical state that her ambulance

FIGURE 13.5 Summertime with a flamingo: Gödel in his garden.

carried him right away to the hospital. But it was too late already. Gödel died from starvation. He weighed only sixty-five pounds.

In the year after Gödel's death, Douglas Hofstadter's book *Gödel, Escher, Bach* rose quite high on the best-seller lists and wound up winning the Pulitzer Prize. The book turned Gödel into an icon of the computer age. But in truth Gödel did not belong to this age: he intruded like a visitor wandering in from another era, an interloper from the baroque world of Leibniz and Newton. For all his life, Gödel had been a stranger to the intellectual avant-garde of the twentieth century, in the Vienna Circle no less than at the Institute for Advanced Study.

Up to the end, his life was dominated by his hidden struggle with the Vienna Circle. His *Nachlass* proves this. In a questionnaire that he had filled out but had as usual failed to send off, some heavily underscored lines testify that he did *not in any way* belong to the intellectual climate of the twentieth century, least of all to that of the Vienna Circle. It had all been a huge mistake. And a handwritten note that Gödel hastily scribbled insists that Wittgenstein's views had "*no* inf" ("*no* influence") in the least on any of his work.

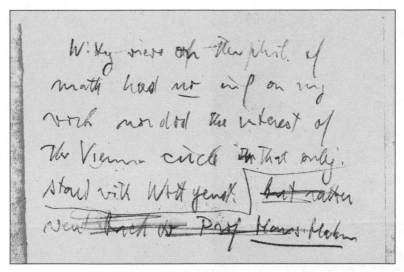

FIGURE 13.6 "no inf": Gödel leaves no margin for doubt.

## KARL POPPER'S CONFESSION

In 1974, it was Popper's turn: a volume in the *Library of Living Philosophers,* that hall of fame of philosophy, was now to be dedicated to him. The chapter on Popper and the Vienna Circle was entrusted to Viktor Kraft, by now ninety-four years old. It was a fitting choice. Kraft had been the first member of the Circle to take care of Karl Popper, during long walks in the Volksgarten, a park next to the university.

"Popper was never a member of the Vienna Circle, never took part in its meetings, and yet cannot be thought of as outside of it," wrote Kraft. "As Popper stands in a close, inextricable relationship with the development of the Circle, so the Circle was also of essential significance for his own development."

Popper devoted many pages of his reply to Kraft discussing what he termed the "Popper legend," which he compared with the Socrates legend. (Not that he wished to compare himself to Socrates, he added modestly: indeed, "nothing could be further from my thought.") According to the Popper legend, Sir Karl was a positivist, perhaps even a member of the Vienna Circle; he had been searching for a criterion for meaningfulness, and finally found it in falsifiability.

FIGURE 13.7  Karl Popper turns
a page.

Nothing of that was true! He, Popper, used falsifiability to demarcate scientific from nonscientific statements. He did not use it as a criterion for meaning. Such a criterion did not interest him. As for positivism, it was dead. Stone dead. Killed. Popper readily confessed that he himself had been the perpetrator of the crime, or at least a guilty party. But he had not done it on purpose. Hence he had not committed murder—only manslaughter, at the worst.

The claim that positivism was slain by Sir Karl is in truth vastly exaggerated. But it is always pleasing to be able to end a thrilling tale with a confession. And we have now reached the last page.

Popper wrote that it ought to be seen as a mark of respect that he had devoted most of his first book to criticizing the Vienna Circle. "The Vienna Circle," he added, by way of obituary, "was an admirable institution. Indeed, it was a unique seminar of philosophers working in close cooperation with first-rate mathematicians and scientists. Its dissolution was a most serious loss." It was a body blow not only to philosophy but also, it might be added, to the town of Vienna.

FIGURE 13.8 The magical Strudlhof staircase
close to the place where the Vienna Circle met.

Soon after World War II was over, Viktor Kraft declared: "The work of the Vienna Circle has not been completed; it has been broken off."

In the words of the Viennese art historian and cabaret performer Egon Friedell, whose life ended when he jumped out a window on the day of the *Anschluss:* "The Viennese have always been remarkably talented in getting rid of their teachers."

# *Afterword*

I HAD ALWAYS WANTED TO WRITE SOMETHING ABOUT THE VIENNA Circle; with my background, it was almost inevitable. Already as a schoolboy, I fell for Ludwig Wittgenstein's *Tractatus* and knew many of its pages by heart (which does not mean, of course, that I understood them in the least). As a student at the University of Vienna, I attended lectures by Bela Juhos, the only member of the Vienna Circle who still taught at the institute of philosophy. His colleagues did their best to make life hell for the old gentleman.

Later, as a professor of mathematics, I had my office on the same corridor as the lecture room where the Schlick Circle had held its meetings, and I was a regular in the coffeehouses where they had held their lively discussions. (Today the lecture room is a quantum physics laboratory, and most of the coffeehouses have closed their doors.) Almost every day, I walk by the Berggasse or the Strudlhof staircase, with all their intimate links to Freud's and Doderer's Vienna of yore. When I worked on statistical mechanics, my window opened on the Boltzmanngasse. When I later turned to game theory, my address became Oskar Morgenstern Platz. A few of my elder colleagues were able to tell me about their personal acquaintances with members of the Vienna Circle. (Their stories improved with time.) I spent many hours listening to Paul Neurath, Otto's son. And none less than Sir Karl Popper wrote an essay—his last one ever—for the collected works of Hans Hahn, which I had been editing. I coauthored a pictorial biography of Kurt Gödel. In short, the Vienna Circle had been with me for half a century.

The immediate impetus for writing this book came when I co-curated the exhibition *The Vienna Circle,* which was organized by the University of Vienna on the occasion of its 650th anniversary. I owe many thanks to my co-workers Friedrich Stadler, Christoph Limbeck-Lilienau, Hermann Czech, Bea Laufersweiler, and Peter Weibel, as well as to Dieter Schweizer, Falk Pastner, and Heinz Engl, the rector of the university, for their unstinting help in situations that were not always easy. I benefited greatly from the expertise of Elisabeth Nemeth, Wolfgang Reiter, Josef Hofbauer, Mitchell Ash, Matthias Baaz, Jakob Kellner, Vincent Jansen, and Helmut Veith. I owe much to Klaus Taschwer, as well as to my former student Bernhard Beham. Christian Palmers, Christian Ehalt, and Michael Stampfer were essential, each in his own way. Christos Papadimitriou, Simon Bang, Dirk van Dalen, Helmut Widder, and in particular Bea Laufersweiler helped with the illustrations. Ulrike Schmickler-Hirzebruch from Springer Spektrum was an ideal editor for the German version.

At the Austrian National Library, I was supported by Alfred Schmidt, Andreas Fingernagel, and Julia Kamenicek; at the Archive of the University of Vienna by Thomas Maisel and Kurt Mühlberger; at the Library of the University of Vienna, by Andrea Neidhardt, Peter Graf, Alexander Zartl, and Günter Müller; at the Vienna Library by Julia Danielczyk and Silvia Mattl-Wurm; at the Austrian Academy of Science by Stefan Sienell; in the Museum for Social and Economic Affairs by Hans Hartweger; in Princeton by Don Skemer and Marcia Tucker; at Duke University by Elisabeth Dunn; and in Cambridge by Michael Nedo. I am pleased to thank my employers—the University of Vienna and the Institute for Applied Systems Analysis in Laxenburg—for their generous support. And very special thanks go, of course, to my wife, Anna Maria Sigmund, for her help, expertise, and encouragement. She happens to be a historian.

The German version of the book, originally intended simply to be a catalog of the exhibition, evolved into a companion volume. To my own great surprise, the story fell into place almost as if it were writing itself. The book was well received, and in fact it won the honor of Austrian Science Book of the Year 2016. But my ambition went further. I had always hoped most of all for an Anglo-Saxon audience. After all, the best books on Vienna come from that world. So I embarked on the adventure of rewriting my book in English, with lots of additional material and the foolhardy hope that my Viennese accent would not show too much in print. Camilla Nielsen and Steve MacManus helped a lot.

Thanks to John Brockman (the founder of EDGE), his son Max, and Russell Weinberger, my book proposal was accepted by Basic Books. Here, with the helpful and unerring advice of T. J. Kelleher and Helene Barthelemy, the manuscript got ready for delivery. I am deeply grateful for their work.

And then, just a few weeks before the deadline, something remarkable happened. Out of the blue, I got a message from Douglas Hofstadter, in Bloomington, Indiana. The author of *Gödel, Escher, Bach* is clearly a man of impeccable taste, for he had enjoyed the German original, and he even spontaneously proposed to write a preface for the English version, and on top of that, to help polish the English in my manuscript. I was delighted, and so in the next few weeks, he went ahead, chapter by chapter, paragraph by paragraph, sentence by sentence, word by word, letter by letter (honest!), and as he sent me the revised chapters, I felt like someone watching King Midas and his magic touch. But Douglas Hofstadter did more than just add luster: quite a few pages are entirely due to him (I will of course point them out in the notes to the chapters). It was an incredibly generous action. I am still at a loss to find the proper words of gratitude—and now I feel that I understand a wee bit better what Ludwig Wittgenstein meant by his phrase "the inexpressible."

# Illustration Credits

Figure 12.6: Richard Arens

Figures 1.1, 1.2, 12.1: Archive of the University of Vienna

Figure 2.1: Austrian Academy of Science

Figures 3.2 (Photographer F. Schmutzer), 3.5, 10.1, 13.8: Bildarchiv Österreichische Nationalbibliothek

Figures 5.2, 6.1, 6.2 (Photos T. Fleischmann), 9.4 (Photo T. Fleischmann), 9.5, 12,2, 12.9: Carnap Papers Archive University of Pittsburgh

Figures 4.6, 11.1: Collection van de Velde-Schlick

Figures 9.1, 9.2, 13.7: Estate Sir Karl Popper, University of Klagenfurt

Figure 3.3: Fachbibliothek Wirtschaft und Mathematik Universität Wien

Figures 1.7, 1.8, 1.10, 4.1, 4.5, 7.6, 9.3 (right), 11.3, 13.1: Institut Wiener Kreis University of Vienna

Figures 7.4, 12.7. 12.8 (left): Isotype Collection Department of Typography University of Reading

Figure 12.8 (right): Peter and Richard Kershaw. (The photograph was taken from the deck of HMS VENOMOUS on the 15 May 1940 by Lt Peter Kershaw RNVR and consent is given by his son Richard Kershaw.)

Figures 1.9, 8.4, 8.5, 8.7, 12.3, 12.5, 13.4, 13.5, 13.6: Kurt Gödel papers Princeton University Library. (All rights reserved by Institute for Advanced Study. The Kurt Gödel Papers are on deposit in the Manuscript Division, Department of Rare Books and Special Collections, Princeton University Library)

Figures 1.4, 1.5, 8.1: Photos by Bea Laufersweiler

Figure 2.2: E. Mach (2011)

Figure 13.2: Nachlass Paul Feyerabend, Philosophische Archiv Konstanz

Figure 10.2 Oskar Morgenstern papers, Duke University

Figure 4.3: Paul Neurath Papers University of Vienna

# *Notes*

The following mostly lists historically oriented secondary literature. The philosophical texts of the main protagonists have been published (or are being published) in many new editions, collected works, or working editions, for instance those by Mach (Xenomoi, Vienna), Schlick (Springer, Vienna), Gödel (Princeton University Press), Menger (Springer, Vienna), Hahn (Springer, Vienna), Neurath (Hölder-Pichler-Tempsky, Vienna), Popper (Mohr-Siebeck, Tübingen), Carnap (Open Court, Chicago), Wittgenstein (Suhrkamp, Frankfurt), Russell (Routledge, London) and Einstein (Princeton University Press). Moreover, much of the archival material is online. Of particular use are Archive of the Vienna Circle Foundation in Haarlem (on Schlick and Neurath); Archive and Manuscripts Collection of the University of Pittsburgh (on Carnap and Rand); Wittgenstein Archives in Cambridge, UK, and Bergen, Norway; Department of Typography and Graphic Art of the University of Reading (on Neurath); Russell Archive of McMaster University, Hamilton, Canada; Einstein Archives of Hebrew University, Jerusalem; Deutsches Museum in Munich (on Mach); Duke University Archives (on Morgenstern and Menger); Department of Rare Books and Special Collections of the Princeton University Library (on Gödel); Österreichische Nationalbibliothek (on Neurath and Wittgenstein); Zentralbibliothek für Physik, Vienna (on Boltzmann); Archive of the University of Vienna; and Archive of the Austrian Academy of Science.

## CHAPTER 1

Books on the Vienna Circle: Geier (1992), Haller (1993), Sarkar (1996a, b), Limbeck and Stadler (2015), and Stadler (2015), as well as Thurm and Nemeth (2003), who provide a town guide. The notion of an Austrian philosophy is discussed by Fischer (1991, 1995), Johnston (1983), Juhos (1965), Keyserling

(1965), Uebel (2003). Incidentally, the title "Die wissenschaftliche Weltauffassung" is often translated as "The Scientific World-Conception," with excellent reasons—but "The Scientific Worldview" sounds better.

v *I am sitting* . . . Wittgenstein (1974), On Certainty, see Wittgenstein, Collected Works

v *If we were to open the window* . . . Bergmann (1988)

2 *The scientific worldview* . . . This and all other quotes in Chapter 1 are from "*Die wissenschaftliche Weltauffassung: Der Wiener Kreis*" (1929), reprinted in Stadler and Uebel (2012)

## Chapter 2

On Friedrich Adler: Adler (1918b), Ardelt (1984), Bauer (2004), Maier and Maderthaner (2006). On Mach: Blackmore (1972, 1978, 1992), Blackmore and Hentschel (1985), Blackmore et al. (2001), Bradley (1971), Cohen (1975), Frank (1937/38), Haller and Stadler (1988), Juhos (1972), Kraft (1966), Lampa (1923), Menger (1960). On Boltzmann: Blackmore (1995, 1999), Broda (1955), Cercignani (1998), Fasol-Boltzmann (1990), Höflechner (1994), Lindley (2001). On Gomperz: Gomperz (1936a, b), Kann (1974).

12 *When you gave a talk* . . . Heinrich Gomperz to Mach, February 17, 1908, in Haller and Stadler (1988)

12 *Most respected colleague* . . . Theodor Gomperz to Mach, September 27, 1894, in Haller and Stadler (1988)

13 *My life's task* . . . Mach Notebook 57, October 5, 1902, in Haller and Stadler (1988)

13 *The 15-year old boy* . . . Mach CV, 1913, in Haller and Stadler (1988)

14 *The Austrians figured out* . . . Neurath, O. (1936), Die Entwicklung des Wiener Kreises und die Zukunft des Logischen Empirismus, in Neurath (1981)

14 *The only disagreeable moments* . . . Mach CV, 1913, in Haller and Stadler (1988)

16 *Few great men* . . . Popper (2002), 151–152

17 *The present volume* . . . Mach (2012)

17 *The gist and kernel* . . . Mach (2012), preface

17 *Let us direct our attention* . . . Mach (2012), 92

18 *All of science* . . . Mach (2012), 503

19 *Most researchers ascribe* . . . Mach (2012), 530

19 *Someone who knew the world only through the theater* . . . Mach (2012), 531

19 *Instructing individuals in science* . . . Mach (2012), 504

19 *ingeniously contrived barriers* . . . Mach (2014), 353

19 *I will not meet any opposition* . . . Mach (2014), 336

20 *Without any doubt* . . . Mach (2014), 340

20 *I know of nothing* . . . Mach (1881), Über den relative Bildungswert . . . , see Mach (2014)

20 *Colors, sounds, temperatures* . . . Mach (2008)

21 *A pencil held up* . . . this and the following quotes are from Mach Notebook 23, January 26, 1881, in Haller and Stadler (1988)

21 *One bright summer's day* . . . Mach (2008), 30n

22 "*I experience green* . . . Mach (2008)

23 *When I say* . . . Berta Zuckerkandl (1980), Literatur und Philosophie: Hermann Bahr, Ernst Mach und Emil Zuckerkandl im Gespräch, in *Die Wiener Moderne* (ed. Wunberg G.), Reclam, Stuttgart

23 *In this phrase* . . . Hermann Bahr (1904), *Dialog vom Tragischen*

23 *All our Machists* . . . Lenin, V. (1909), *Materialism and Empiriocriticism,* see The Marxist Internet Archive

24 *Ernst Mach's Victory over Mechanical Materials* . . . see Adler 1918a

25 *It thinks* . . . Lichtenberg G., *Aphorismen (Sudelbücher),* see Database Project Gutenberg

25 *the bizarre opinion* . . . Boltzmann Aphorisms, see Fasol-Boltzmann (1990)

25 *I disdain experiments* . . . Boltzmann Aphorisms, see Fasol-Boltzmann (1990)

25 *His professor Josef Stefan* . . . see Boltzmann (1895), Josef Stefan, in Boltzmann (1905)

26 *No matter how little I believe* . . . Boltzmann to Aigentler, September 27, 1875, in Flamm (1995)

28 *To reconcile the entropy hypothesis* . . . Broda (1955), 85

30 *One usually starts out* . . . Boltzmann, inaugural lecture, October 1902, in Boltzmann (1905)

31 *Both Boltzmann and Mach* . . . Popper (2002)

32 *To praise Mach* . . . Boltzmann, inaugural lecture, October 1902, in Boltzmann (1905)

32 *Up till now* . . . Boltzmann, inaugural lecture, October 1902, in Boltzmann (1905)

33 *Whereas I felt some qualms* . . . Boltzmann, inaugural lecture, October 1902, in Boltzmann (1905)

33 *My lectures on philosophy* . . . Boltzmann to Brentano, December 24, 1904

33 *What the brain is to man* . . . Boltzmann Aphorisms, see Fasol-Boltzmann (1990)

33 *No equation* . . . Boltzmann (1899), Entwicklung der Methoden der theoretischen Physik in neuerer Zeit, in Boltzmann (1905)

34 *We hesitate to ascribe* . . . Broda (1955), 111

34  *A worm stepped on by a man* . . . Graf, P., Zartl, A. (1906), *A Ludwig Boltz-mann 1844–1906 Ausstellung Zentralbibliothek für Physik Vienna* (reported by Karl Przibram)

35  *Metaphysics appears to exert* . . . Boltzmann, inaugural lecture, October 1902, in Boltzmann (1905)

35  *Which definition of philosophy* . . . for this and the following quotes, see Boltzmann (1904), Über eine These Schopenhauers, in Boltzmann (1905)

37  *The irrepressible urge* . . . see Boltzmann (1904), Über eine These Schopen-hauers, in Boltzmann (1905)

37  *I'm sleeping very poorly* . . . Boltzmann to his wife, undated, see Flamm (1995)

38  *In well-informed circles* . . . *Neue Freie Presse*, September 8, 1906

38  *This extremely talented scientist* . . . Brentano to Mach, January 31, 1908

38  *to introduce a new philosophy* . . . Mach (2011)

38  *a scientist can rest content* . . . Mach (2011)

38  *Science has progressed almost more* . . . Mach (2012), 562

39  *The aim of science* . . . Mach (2012), 517

39  *The Ego varies* . . . Mach Notebook 29 (1884/85)

39  *Sensations are the common elements* . . . Mach Notebook 55, November 9, 1900

39  *I faced a saint* . . . Lampa (1923)

39  *I do not think* . . . William to Alice James, in Richardson, R. (2007), *William James*, Harcourt, San Diego

39  *Should this letter be my last* . . . Archive, Austrian Academy of Science

39  *grandiose one-sidedness* . . . Einstein, A. (1916), Nachruf auf Ernst Mach, *Physikalische Zeitschrift* 17(7), 1–2

CHAPTER 3

On the *Urkreis:* Haller (1979, 1993), Stadler (2015), Uebel (1991, 1993). On Richard von Mises: Siegmund-Schultze (2001, 2017). On Viennese Modern-ism: Beller (2008), Janik and Toulmin (1973), Kandel (2012), LeRider (1993), Schorske (1981), Timms and Robertson (1990). On Einstein: Bührke (2004), Clark (1984), Corry et al. (1998), Einstein (1954), Fölsing (1998), Frank (1947), Galison (2003), Hentschel (1990), Hoffmann and Dukas (1972), Holton (1981, 1996), Isaacson (2007), Neffe (2007), Pais (1982), Regis (1987), Renn (2005), Rigden (2005), Schilpp (1944a), Yourgrau (2005). Most of the text on Einstein has kindly been written by D. H. Hofstadter. For Einstein's intriguing Mach connection: Wolters (1987). On Hilbert: Reid (1996). On Russell: Ayer (1972), Clark (1988), Doxiadis and Papadimitriou (2009), Grayling (1992), Monk (1997, 2001a), Russell (1967–1969), Schilpp (1944b). On Musil: Berghahn (1972), Corino (2003). On the Redl affair: Moritz and Leidinger (2012).

41  *Germans! When will you finally give up* . . . Bolzano, B. (1837), *Wissenschafts-lehre,* vol. 4, 590, Sulzbach, Vienna

41  *Demonstration That Schopenhauer* . . . Boltzmann (1904), Über eine These von Schopenhauer, reprinted in Boltzmann (1905)

43  *Strange to say* . . . Frank (1949)

43  *I belonged to a group* . . . Frank (1949)

44  *This treatise* . . . Archive of the Technical University of Vienna: Report on the PhD Thesis of Richard von Mises

44  *Empire of Improbabilities* . . . Brentano to Mach, January 31, 1908, see Brentano (1988)

50  *The fact that sometimes* . . . Letter proposing to elect Einstein in the Prussian Academy of Science, June 12, 1913

50  *Experiment has outrun theory* . . . Millikan, R. A. (1917), *The Electron,* 230, University of Chicago Press, Chicago

57  *I did not bother much* . . . Blumenthal, Lebensgeschichte, in Hilbert (1970)

62  *We developed pretty much in parallel* . . . Friedrich to Viktor Adler, June 19, 1908

63  *the disease of logic* . . . see Adler (1918b)

68  *with the support of a university philosopher* . . . Frank (1949)

68  *The Minute Endangered* . . . Frank (1947)

70  *the science with the evil eye* . . . Musil, R. (1978), *Mann ohne Eigenschaften,* Vol. II, p. 1979, Rowohlt, Hamburg

71  *elected to fall in love* . . . Corino (2003)

71  *"Appreciation of the Teachings of Mach"* . . . see Musil (1982)

73  *something deeply flawed* . . . Hofmannsthal, H. (1907), *Der Dichter und diese Zeit,* Fischer, Berlin

CHAPTER 4

On Hahn: Frank (1934), Popper (1995), Sigmund (1995, 2015). On Neurath: Cohen and Neurath (1973), Hegselmann (1979), Neurath, O. (1981), Neurath P. (1993), Nemeth and Stadler (1996), Sandner (2014). On Schlick: Engler and Iven (2007, 2008), Hentschel (1986), Iven (2008), Menger (1994). On Adler: Adler (1918b), Bauer (2004), Galison (2008), Maier and Maderthaner (2006).

79  *In the last year* . . . Hahn to Ehrenfest, December 26, 1909

79  *opportunity to ponder* . . . Hahn to Ehrenfest, December 26, 1909

81  *slick and inscrutable* . . . Doderer, H. (1956), *Die Dämonen,* Biederstein, München

84  *the diverse views* . . . Neurath, O. (1912), Das Problem des Lustmaximums, in Neurath (1981)

84  *a philosophizing Russian* . . . Schnitzler Diary 23

84  *a science just like ballistics* . . . Neurath, O. (1910), Die Kriegswirtschaft, in Neurath (1981)

86  *the plan of all plans* . . . Neurath, O. (1919), Technik und Wirtschaftsordnung, see Sandner (2014)

87  *With the best of conscience* . . . Otto Bauer, Testimonial for Otto Neurath, June 30, 1919

88  *We are like mariners* . . . Neurath, O. (1920) Anti-Spengler, see Neurath (1981).

89  *a demagogue called in from Austria* . . . Müller, K. A. (1954), *Mars und Venus,* Kilpper, Stuttgart

89  *Has a notebook* . . . Musil Diary 9, 1920 (undated)

90  *If there were no words* . . . Stöhr (1910)

94  *whose selflessness had made him land in the soup* . . . Einstein to Zangger, March 10, 1917

95  *which may be said to be* . . . Friedrich Adler to Viktor and Katia Adler, February 14, 1917

95  *I confided in you as a friend* . . . Friedrich Adler to Viktor Adler, February 25, 1917

97  *a fairly sterile rabbinistic dunderhead* . . . Einstein to Besso, April 29, 1917

97  *I am curious to know* . . . Einstein to Friedrich Adler, October 20, 1918

99  *After thus having sentenced theoretical philosophy to death* . . . Schlick, Lebenslauf I (1900), in Schlick, Collected Works IV/1

99  *a hint of fate* . . . Schlick, inaugural lecture, 1922, Archive of the University of Vienna

99  *I turned to physics* . . . Schlick Lebenslauf III (probably 1915), in Schlick, Collected Works IV/1

100  *My dear Doctor Schlick* . . . Blanche to Moritz Schlick, May 19, 1905

101  *You are well-versed in physics* . . . Laue to Schlick, December 27, 1911

102  *This work is among the best* . . . Einstein to Schlick, December 14, 1915

102  *I have been fighting* . . . Einstein to Hilbert, December 20, 1915

103  *Spacetime tells matter* . . . Wheeler (1990)

103  *The Great War? Ah yes* . . . Schlick (1920), Einsteins Relativitätstheorie, in Schlick, Collected Works I/5

103  *Your presentation is of unsurpassable clarity* . . . Einstein to Schlick, February 6, 1917

104  *Fortunately, one of the most immediate consequences* . . . Laue to Schlick, August 19, 1913

104  *In the meantime* . . . Laue to Schlick, October 7, 1917

104  *We have become a community* . . . Born to Schlick, July 11, 1919

105  *to escape from the sleepiness of Rostock* . . . Schlick to Einstein, February 22, 1920

105  *Our experience has been that one cannot talk about prices* . . . Laue to Schlick, September 3, 1922

106  *It was hard for me, in the end, to move to Vienna* . . . Schlick to Einstein, August 13, 1922

106  *We particularly enjoy* . . . Schlick to Störring, February 5, 1935

107  *Professor Schlick's lectures* . . . Nagel (1936)

107  *He was very sincere* . . . Menger (1994)

108  *He had first studied physics* . . . Menger (1994)

108  *In Vienna a great deal of philosophy* . . . Schlick to Einstein, June 12, 1926

109  *Over the years, the size of the Circle* . . . Menger (1994)

110  *The room was filled* . . . Menger (1994)

110  *People would mill around* . . . Menger (1994)

CHAPTER 5

On Reidemeister: Bachmann et al. (1972), Epple (1995, 1999). On Carnap: Awodey and Klein (2004), Carnap (1963), Carus (2007), Giere and Richardson (1997), Mormann (2000), Richardson (1998), Schilpp (1963). From the enormous literature on Wittgenstein: Bartley (1988), Baum (2014), Engelmann (1967), Grayling (1988), Janik and Toulmin (1973), Klagge (2001), McGuiness (2005), Monk (2001b), Nedo (2012), Wijdeveld (2000), Wuchterl and Hübner (1998).

111  *The interests of the Vienna Circle* . . . Menger (1960)

112  *nothing less than a complete reform of philosophy* . . . Schlick to Einstein, June 5, 1927

112  *No transcendental twaddle* . . . Wittgenstein to Engelmann, January 16, 1918

113  *at no other meeting in living memory* . . . Otto Schreier to Menger, November 21, 1922

116  *It is historically understandable* . . . Carnap (1928a), preface

117  *I had the opportunity to speak* . . . Schlick to Carnap, October 21, 1924

117  *Zilsel, a member of the Vienna Circle* . . . Taschwer (2015)

118  *"The concepts of the scientific domain* . . . Carnap (1928a)

118  *My interests and my basic philosophical views* . . . Carnap (1963)

120  *After all, the Magic Flute* . . . Boltzmann (1884), Über Luftschiffahrt, reprinted in Boltzmann (1905)

121  *a famous series of letters by Russell* . . . Russell to Lady Ottoline Morell, see Nedo (2012) or Monk (2001a)

122  *As soon as I had read the first sentence* . . . Russell, BBC 4 broadcast from 1955, http://www.bbc.co.uk/programmes/b0184rgn

122  *Because Wittgenstein always looks frightfully puzzled* . . . Russell to Lady Ottoline Morell, see Monk (2001a)

122  *We expect the next big step* . . . Hermine Wittgenstein (1944), Familienerinnerungen, in Nedo (2012)

122  *Wittgenstein was perhaps the most perfect example* . . . Russell (1967–1969)

123  *[His critique] was an event of first-rate importance* . . . Russell to Lady Ottoline Morell, March 4, 1916

124  *Hell, buy yourself a gun* . . . Hermine Wittgenstein (1944), Familienerinnerungen, in Nedo (2012)

124  *The war saved my life* . . . McGuiness (2005)

124  *What can be said at all* . . . Wittgenstein, preface to *Tractatus*

125  *I think I have solved the problems once and for all* . . . Wittgenstein to Russell, March 13, 1919

126  *to have found everything Ernst Mach had been looking for* . . . Friedrich Adler to Viktor Adler, February 14, 1917

126  *Most thankful that you are still alive* . . . Russell to Wittgenstein, March 2, 1919

126  *I am sure you are right* . . . Russell to Wittgenstein, June 21, 1919

127  *In order to croak with a good conscience* . . . Somavila I, Begegnungen mit Wittgenstein, in Hänsel (2012)

127  *"Yes," retorted Wittgenstein* . . . Heinz von Förster, personal communication

128  *Clear as crystal* . . . Wittgenstein to Russell, March 13, 1919

128  *The inexpressible is contained* . . . Wittgenstein to Engelmann, April 9, 1917

129  *We would often start class* . . . Leinfellner and Windholz (2005)

129  *and then he got frightfully angry* . . . Leinfellner and Windholz (2005)

129  *I am still at Trattenbach* . . . Wittgenstein to Russell, November 28, 1921

130  *The people here are so narrow-minded* . . . Monk (2001a)

130  *from the three best teachers of my generation* . . . Wijdeveld (2000)

132  *To me, the Tractatus* . . . Menger (1994)

132  *Logic is not a theory about* . . . Hahn (1930), Diskussion zur Grundlegung der Mathematik, in Hahn (1980)

132  *For me Wittgenstein was the philosopher* . . . Carnap (1963)

133  *told him that his interruptions were disturbing* . . . Hempel, Memoirs: The Vienna Circle and Empiricism, in Hempel (2012)

133  *It would give me particular pleasure* . . . Schlick to Wittgenstein, December 25, 1924

134  *He asks me to convey his warmest regards* . . . Stonborough to Schlick, February 19, 1927

134  *Each of us considered the other one to be mad* . . . Monk (2001b), 242

134  *the Viennese Ludwig Wittgenstein* . . . Schlick to Einstein, July 14, 1927

135  *Before the first meeting* . . . Carnap (1963)

135 *When at last I met Wittgenstein* . . . Carnap (1963)

135 *It was fascinating to behold* . . . see Nedo (2012)

135 *Recently, Brouwer held two lectures* . . . Schlick to Carnap, March 27, 1928

136 *Wittgenstein had rapidly come to the conclusion* . . . Neider (1977)

136 *the rigorous and responsible attitude of scientific research* . . . Carnap (1928a), preface

136 *By assigning to each individual* . . . Carnap (1928a), preface

137 *I cannot found a school* . . . Wittgenstein MS 134

137 *When Wittgenstein talked about philosophical problems* . . . Carnap (1963)

CHAPTER 6

Mulder (1968) describes the origins of the manifesto; see also Stadler and Uebel (2012). Ayer (1936), Feigl (1969), Frank (1941), Mann (1986), Menger (1994), and Neider (1977) produce firsthand reports. On Heidegger: Biemel (1973). On his Carnap connection: Friedman (1999, 2000). On Neurath: Nemeth (1981), Nemeth and Stadler (1996), Reidemeister-Neurath (1980). On Schlick: Iven (2008), McGuiness (1985).

139 *All that the Austrian ministry was prepared to offer* . . . Schlick to Majer, May 29, 1929

139 *Were Professor Schlick to leave Vienna* . . . Verein Ernst Mach to Schlick, April 2, 1929

139 *This would be a severe blow* . . . Carnap to Schlick, April 6, 1929

140 *How great was our joy* . . . Menger (1982)

141 *Neurath advises us* . . . Carnap Diary, June 17, 1929

141 *Several of us ardently wished* . . . Frank, Introduction: Historical Background, in Frank (1949)

141 *The sour duty and the sweet right* . . . Carnap to Neurath, July 26, 1929

142 *Finished typing the brochure* . . . Carnap Diary, July 25, 1929

142 *The German Physical Society* . . . Frank (1949)

143 *fell upon the audience* . . . Frank (1949)

143 *And now, out of the blue* . . . Carnap to Schlick, September 30, 1929

144 *one of those compromises* . . . Menger (1980)

145 *This is a deeply troubling affair* . . . Wittgenstein to Waismann, July 1929 (undated)

146 *From time to time* . . . Schlick (1930), Die Wende in der Philosophie, *Erkenntnis* 1, 4–11

146 *This peculiar fate* . . . Schlick (1930), Die Wende in der Philosophie, *Erkenntnis* 1, 4–11

146 *I allow myself* . . . Schlick (1930), Die Wende in der Philosophie, *Erkenntnis* 1, 4–11

147  *Philosophy is not a set of statements* . . . Schlick (1930), Die Wende in der Philosophie, *Erkenntnis* 1, 4–11

147  *It is easy to see* . . . Schlick (1930), Die Wende in der Philosophie, *Erkenntnis* 1, 4–11

147  *The difference between the task of the scientist* . . . Schlick (1986)

148  *It helps us to confess our faith* . . . Hahn (1930), Die Bedeutung der Wissenschaftlichen Weltauffassung, *Erkenntnis* 1, 96–105; see also Hahn (1980)

148  *This term serves to contrast* . . . Hahn (1930), Die Bedeutung der Wissenschaftlichen Weltauffassung, *Erkenntnis* 1, 96–105, see also Hahn (1980)

149  *in the systems of German idealism* . . . Hahn (1930), Occams Rasiermesser, in Hahn (1980)

150  *Away with them all!* . . . Hahn (1930), Occams Rasiermesser, in Hahn (1980)

150  *I have received more than a few compliments* . . . Hahn to Ehrenfest, March 30, 1912

150  *How is the empiricist position* . . . Hahn (1930), Diskussion zu den Grundlagen, see Hahn (1980)

151  *Logic does not say anything about the world* . . . Hahn (1933), Logik Mathematik und Naturerkennen, see McGuiness (1987)

151  *Whoever does not accept logical inference* . . . Hahn (1933), Logik Mathematik und Naturerkennen, see McGuiness (1987)

151  *And indeed, it hardly seems believable* . . . Hahn (1933), Logik Mathematik und Naturerkennen, see McGuiness (1987)

151  *An omniscient subject needs no logic* . . . Hahn (1930), Bedeutung der Wissenschaftlichen Weltauffassung, see Hahn (1980)

152  *It could be said that Hahn was in a certain sense* . . . Frank (1934)

152  *He proceeded by taking almost imperceptible steps* . . . Menger (1994)

152  *chilly clarity* . . . Auguste Dick, personal communication

152  *The personal impression I gained* . . . Popper (1995)

153  *immediately pending* . . . *Personalakte* Hahn, Archive of the University of Vienna

153  *My daughter has become an actress* . . . Hahn to Ehrenfest, December 8, 1932

153  *He based this opinion on two arguments* . . . Menger, Introduction, in Hahn (1980)

154  *I came to hold the view* . . . Carnap (1963)

155  *If someone asserts "There is a God"* . . . Carnap (1963)

155  *Everything that is beyond the factual* . . . Carnap (1928), Scheinprobleme in der Philosophie, see Stadler and Uebel (2012)

155  *In science, there are no depths* . . . Stadler and Uebel (2012)

156  *Yet there still remains* . . . Carnap (1963)

156 *Making itself intelligible is suicide for philosophy* . . . Heidegger, Beiträge zur Philosophie (Zum Ereignis), see Heidegger, Collected Works

157 *Where do we look for the nothing?* . . . Heidegger (1929), Sein und Zeit, see Heidegger, Collected Works

157 *so instructive* . . . Hilbert (1931)

157 *The supposed sobriety and superiority of science* . . . Heidegger (1929), Sein und Zeit, see Heidegger, Collected Works

157 *the primitive traits* . . . Hahn (1929), Occam's Rasiermesser, see Hahn (1980)

158 *one of the most stimulating* . . . Carnap (1963)

158 *Schlick could not be sure of the privacy of his correspondence* . . . Schlick to Carnap, August 5, 1927

159 *for technical reasons* . . . Ina to Rudolf Carnap, January 18, 1930, and Carnap to Ina, January 22, 1930

159 *She has two faces* . . . Carnap Diary, March 6, 1930

160 *So I arrived at Neurath's house* . . . Neider (1977)

160 *Don't read any Kant* . . . Mann (1986)

161 *Guess what—I kissed Mietze today!* . . . Reidemeister-Neurath (1980)

162 *I can't have a man with such a loud voice* . . . Neider (1977)

162 *Herr Neurath has declared himself ready* . . . Neider (1977)

162 *It was only at the end of the nineteenth* . . . Neurath, O. (1929), Wissenschaftliche Weltauffassung, Arbeiterzeitung, October 13, reprinted in Neurath (1981), 347

163 *through the often distorting lens* . . . Menger (1994)

163 *The pathway of the socialist proletariat* . . . Neurath, O. (1929), Bertrand Russell, der Sozialist, *Der Kampf* 22, 234–238, reprinted in Neurath (1981)

163 *For Russell, Marxism means* . . . Neurath, O. (1929), Bertrand Russell, der Sozialist, *Der Kampf* 22, 234–238, reprinted in Neurath (1981)

CHAPTER 7

On the political situation at the University of Vienna: Taschwer (2015, 2016). On adult education: Kadrnoska (1981), Kutalek and Fellinger (1969). On pictorial statistics: Arntz (1976, 1982), Burke et al. (2013), Neurath (1993). On the Bettauer case: Hall (1978). On architecture: Blau (1999), Dahms (2004), Frank J (1931), Galison (1990), Rukschcio and Schachel (1982), Welzig (1998). On literature: Berghahn (1972), Corino (2003), Eckert and Müller (1989), Lützeler (1985), Müller (1992).

165 *All of us in the Circle* . . . Carnap (1963)

166 *Of course, not every single adherent* . . . Wissenschaftliche Weltauffassung, see Stadler and Uebel (2012)

166  *The Vienna Circle is not satisfied* . . . Wissenschaftliche Weltauffassung, see Stadler and Uebel (2012)

167  *We are witnessing* . . . Wissenschaftliche Weltauffassung, see Stadler and Uebel (2012)

169  *societies for the deflagration of corpses* . . . Weissensteiner (1990)

173  *abandon all your hopes* . . . Weber (1919)

173  *welded the anti-Semitic groups* . . . Taschwer (2015)

174  *Zilsel was a militant leftist* . . . Menger (1994)

178  *The ultimate consequence of empiricism* . . . Neurath (1981)

178  *The proletariat, as a class, is eager* . . . Neurath, O. (1926), Statistik und Proletariat Kulturwille 4, see Neurath (1981)

178  *Statistics is pure joy* . . . Neurath, O. (1926), Statistik und Proletariat Kulturwille 4, in Neurath (1991)

178  *Contemporary people receive a great deal* . . . Neurath, O. (1930), Statistische Hieroglyphen, see Neurath (1981). For this and the following, see Neurath and Nemeth (1993)

179  *The more abstract an art becomes* . . . Musil (1923), Schwarze Magie, in *Nachlass zu Lebenszeiten* (1936); in Musil, Collected Works (1978)

180  *A larger quantity is represented* . . . Neurath, O. (1930), Gesellschaft und Wirtschaft, see Neurath (1991)

180  *An adult education institute for social enlightenment* . . . Neurath, O. (1929), Bildstatistik und Arbeiterbewegung, see Neurath (1991)

181  *Whatever can be shown by a picture* . . . see Neurath, O. (1991), 243

181  *Words divide* . . . Neurath, O. (1931), Bildstatistik nach der Wiener Methode, see Neurath (1991), 190

182  *Whoever leaves the most out is the best teacher* . . . Isotype and Graphics, see Neurath (1991), 343

182  *A pictorial survey of the world economy* . . . Neurath, O. (1931), Pictorial Statistics in Economic Planning, in Neurath (1991)

183  *Loos has become an insufferable philistine* . . . Wittgenstein to Engelmann, September 2, 1919

183  *Great architects for small houses* . . . Blau (1999)

184  *All small-scale gardeners* . . . Neurath, O., Planmässige Siedlungs-, Wohnungs-, und Kleingartenorganisation, in Neurath (1991)

184  *palaces for people's apartments* . . . Frank, J. (1926), Der Volkswohnpalast, *Aufbau* 7, 1

185  *I couldn't stand living in an apartment* . . . Musil, R. (1930), Der Mann ohne Eigenschaften, in Musil, Collected Works (1978)

185  *What is happening here* . . . Achleitner (1996)

185 *how people will live in a maximally happy way* . . . Neurath, O. (1932), Die Internationale Werkbundsiedlung Wien 1932, in Neurath (1981)

186 *There are fourteen unemployment offices* . . . Brunngraber (1932)

187 *exactly according to the quantities of raw materials* . . . Brunngraber (1932)

188 *the very picture of a condottiere* . . . Brunngraber (1949)

189 *Philosophers are terrorists* . . . Musil, R. (1930), Der Mann ohne Eigenschaften, in Musil, Collected Works

189 *The one thing that can be said with certainty* . . . Musil, R. (1930), Der Mann ohne Eigenschaften, in Musil, Collected Works

191 *Whenever there is talk of algebra* . . . Musil, R. (1930), Der Mann ohne Eigenschaften, in Musil, Collected Works

194 *It seems to me that there are points in common* . . . Musil, R. (1981), *Briefe* (ed. Filsé, A.), Rowohlt, Hamburg

194 *as a copy of his own enterprise* . . . Canetti (1980)

196 *Two young mathematicians* . . . Eckert and Müller (1989)

197 *I read your* Pseudo-Problems . . . Schlick to Carnap, August 30, 1928

197 *Some time ago, this paper published an anecdote* . . . *Prager Presse*, April 9, 1926

198 *We may already today have reached the point* . . . Musil, R. (1932), Der Mann ohne Eigenschaften, Vol. II, Rowohlt, Hamburg

CHAPTER 8

The PhD thesis of Bernhard Beham (2013) uncovered new material; for other sources, see Geymonat (1991), Menger (1994), Sigmund (1998, 2002). On Carl Menger: Streissler (1994). On Brouwer: van Dalen (2002, 2005). Part of the text on the excluded middle is by Douglas R. Hofstadter. On Gödel: Baaz et al. (2011), Badesa et al. (2009), Dawson (1997), Dawson and Sigmund (2006), DePauli-Schimanovich and Weibel (1997), Fefermann (1986, 1996), Franzen (2005), Goldstein (2005), Hintikka (1999), Hofstadter (1979), Kreisel (1980), Köhler et al. (2002), Mancosu (2010), Nagel and Newman (2005), Regis (1987), Sigmund et al. (2006), Stewart (2013), van Atten and Kennedy (2003), van Heijenoort (2002). Again, the reader will not fail to recognize the Hofstadter touch. On Hilbert's program: Mancosu (1998, 2010), Sieg (2013), Zach (2006). On Taussky-Todd: Davis (1996), Taussky-Todd (1985). On Abraham Wald and mathematical economy: Menger (1952, 1972).

200 *Man values goods* . . . Wieser (1923)

201 *There is no metaphysics* . . . Menger (1994)

205 *Horrible trouble and toil* . . . Karl Menger Diary, see Beham (2013)

205 *We are publishing an article* . . . *Neue Freie Presse*, November 17, 1921

206 *It is not true, as Kant maintained* ... Hahn (1933), The Crisis in Intuition, in Hahn (1980)

208 *At the institute, there was also Karl Menger* ... Popper (1995)

209 *I will of course tell you everything in detail* ... Schreier to Menger, September 24, 1923

214 *I have always tried to avoid* ... Menger (1994)

216 *He was a slim, unusually quiet young man* ... Menger (1994)

225 *But Münchhausen was a liar* ... Enzensberger, Hommage à Gödel, in Enzensberger (2002), *Elixiere der Wissenschaft*, Suhrkamp, Frankfurt

227 *Gödel's theorem forces us* ... Wittgenstein MS 163, July 11, 1941, see Nedo (2012), 378

229 *Today you and I have once again* ... Menger (1994)

230 *small and thin, obviously poor* ... Menger (1994)

232 *My dear Göderl* ... Natkin to Gödel, May 27, 1931

233 *A scientific breakthrough of the first order* ... *Habilitationsakte* Gödel, Archive of the University of Vienna

CHAPTER 9

On Popper: Edmonds and Eidinow (2001), Geier (1994), Hacohen (2000), Kraft (1974), Magee (1973), Popper (1995, 2002). On the dispute between Schlick and Neurath: Cartwright et al. (1996), Haller (1979, 1982), Hegselmann (1979), Uebel (1993). On Wittgenstein, Waismann, and the Vienna Circle: Baker (1979, 2003), Manninen (2014), McGuiness (1979), Nagel (1936). On Carnap: Awodey and Carus (2007), Awodey and Klein (2004), Carnap (1963), Krauth (1970), Mormann (2000).

235 *None did so much* ... Popper (2002)

236 *It was in the middle of the winter* ... Popper (1995)

237 *Only the department of mathematics* ... Popper (2002)

238 *a kind of hasty last-minute affair* ... Popper (2002)

238 *Popper's work is clearly of a secondary and literary nature* ... Popper PhD thesis, University of Vienna, 1928

238 *I could hardly believe my ears* ... Popper (2002)

240 *It was only after my Ph.D. examination* ... Popper (2002)

240 *Induction is merely guesswork* ... Schlick (1934), Fundamente der Erkenntnis, in Schlick (2008), Collected Works I/6

242 *Never let yourself be goaded* ... Popper (2002)

242 *Gomperz invited me from time to time* ... Popper (2002)

242 *the meeting became decisive for my whole life* ... Popper (2002)

243 *But once I started the book* ... Popper (2002)

243 *From the beginning* ... Popper (2002)

244 *I was never invited* . . . Popper (2002)

244 *I never was a member* . . . Popper, in Schilpp (1974)

244 *Do you think he is any good* . . . Gödel to Menger, August 4, 1932

245 *the whole thing leaves an unpleasant aftertaste* . . . Feigl to Schlick, September 14, 1933

245 *Gomperz says that on the points* . . . Carnap to Schlick, January 19, 1933

245 *Popper is in a terrible hurry* . . . Schlick to Carnap, June 18, 1933

246 *It is an exceptionally intelligent work* . . . Schlick to Carnap, November 1, 1934

246 *I started reading it with great hopes* . . . For this and the following, Schlick to Frank, July 16, 1930

247 *Almost every argument* . . . Schlick to Frank, July 16, 1930

247 *Unfortunately, when I apprised Neurath of my decision* . . . Report attached to Schlick's letter to Frank, July 16, 1930

248 *Neurath would merely claim* . . . Report attached to Schlick's letter to Frank, July 16, 1930

248 *was planning to submit a completely new manuscript* . . . Schlick to Carnap, August 7, 1930

249 *I do not believe that tactfulness and good taste* . . . Schlick to Carnap, August 7, 1930

250 *I greatly like his careful reasoning* . . . Manninen (2014)

250 *Waismann has written up Wittgenstein's basic ideas* . . . Schlick to Carnap, January 4, 1928

250 *Unfortunately, Waismann's text is not finished yet* . . . Schlick to Carnap, January 29, 1928

250 *He [Waismann] waited with the greatest patience* . . . Wittgenstein to Schlick, May 6, 1932

251 *Except to small, exclusive groups* . . . Nagel (1936)

252 *The solutions to philosophical questions* . . . Waismann, conversation with Wittgenstein, December 9, 1931

253 *I cannot bear* . . . Schlick to Carnap, August 7, 1930

253 *My life in Prague* . . . Carnap (1963)

253 *On February 8 [1933], we were married by civil law* . . . Carnap to Schlick, March 5, 1933

254 *After I had been thinking about these problems* . . . Carnap (1963)

255 *Our business is not to set up prohibitions* . . . Carnap (1937)

255 *He is one of those few* . . . Nagel (1936)

256 *cognitively unimportant* . . . Carnap (1928b)

256 *Neurath even once tried to translate* . . . Carnap (1963)

257 *with outstanding conscientiousness* . . . Wittgenstein to Schlick, March 6, 1932

257  *One must be a great deal more profound* . . . Wittgenstein to Schlick, August 8, 1932

257  *only a psychoanalyst* . . . unpublished part of Carnap's autobiography; see Stadler (2015)

258  *Enclosed with this letter* . . . Schlick to Carnap, August 24, 1932

258  *This, of course, would be highly undesirable for me* . . . Wittgenstein to Schlick, May 5, 1932

259  *To command and to obey an order* . . . Wittgenstein, *Philosophical Investigations*, §23

259  *logical analysis brings hidden things to light* . . . Wittgenstein, Big Typescript TS, 213

259  *A proposition should be thought of as being like a tool* . . . Wittgenstein, *Philosophical Investigations* §421

260  *to show the fly an escape route* . . . Wittgenstein, *Philosophical Investigations*, §309

260  *A philosopher treats a question* . . . Wittgenstein, *Philosophical Investigations*, §255

260  *make the subconscious conscious* . . . Wittgenstein Diary, October 29, 1930

260  *Now I am charged with constructing a series of examples* . . . Waismann to Menger, summer 1936 (undated)

260  *the difficulty of a joint work* . . . Waismann to Schlick, August 9, 1934

261  *Are the words of the Anointed One* . . . Neurath to Neider, June 8, 1934

261  *Waismann's book is indeed close to completion* . . . Schlick to Carnap, March 12, 1934

261  *I have to correct my last note* . . . Schlick to Carnap, August 8, 1934

261  *The most recent development* . . . Schlick to Carnap, July 24, 1934

261  *It is hell to work with me* . . . Waismann to Schlick, August 9, 1934

CHAPTER 10

On Dollfuss: Portisch (1989), Weissensteiner (1990). On Menger and Morgenstern: Leonard (1998), Rellstab (1992). For a commented reprint of the *Ergebnisse eines Mathematischen Kolloquiums:* Dierker and Sigmund (1998). On Richard von Mises: Basch (1953), Föllmer and Küchler (1998), Frank (1954), Siegmund-Schultze (2017), Stadler (1990). On the mathematicians of the Viennese Colloquium: Pinl and Dick (1974). On ethics in the Vienna Circle: Siegetsleitner (2014). On Nelböck: Lotz-Rimbach (2009), Malina (1988).

264  *Nothing unnerves the Sozis more* . . . Dollfuss, May 3, 1933, see Hanisch (2011)

266  *In 1933, the year of Hitler's coming to power* . . . Menger (1994)

267  *No, the institute is dead* . . . Reid (1996)

267 *A Small Textbook of Positivism* . . . see Mises, R. v. (1956)

267 *Adolf Hitler, our great leader and chancellor* . . . Heidegger, early August 1933, in Collected Works, vol. 16, 151

267 *Led by an inner urge* . . . Schlick to Dollfuss, May 21, 1933

268 *devoid of cognitive meaning* . . . Carnap (1963)

269 *Act only according to that maxim* . . . Kant (1788), *Kritik der praktischen Vernunft*

269 *While the political situation* . . . Menger (1994)

270 *Morals will, so to speak* . . . Menger (1974)

270 *Groups similar to those considered in my notes* . . . Menger (1974)

272 *Could he by any chance be a Jew* . . . Morgenstern Diary, September 12, 1923

272 *an aristocrat of the old school* . . . Morgenstern (1927)

272 *Morgenstern (Aryan)* . . . Rosenstein-Rodan to Morgenstern, May 26, 1928

273 *This event was all to my credit* . . . Mises, L. v. (1978)

274 *vicious circle* . . . Morgenstern (1928)

274 *truly exact thinking* . . . Morgenstern Diary, April 19, 1936

275 *Yesterday noon, lunch with Karl Menger* . . . Morgenstern Diary, September 11, 1935

275 *He read them the mathematical riot act* . . . Schumpeter (1954)

277 *Popper in Menger's Mathematical Colloquium* . . . this was on February 6, 1934

277 *Yet one more mathematics lesson* . . . Morgenstern Diary, November 8, 1935

278 *I like to be of help to my friends* . . . Friedrich Schiller, Distichon Gewissensskrupel

278 *Morality, lying at the core of our sense of free will* . . . Kant I (1788), *Kritik der praktischen Vernunft*

278 *a shopkeeper's determination to recognize virtue* . . . Schlick (1934), Fragen der Ethik, see Schlick (1984)

279 *According to the empiricist criterion of meaning* . . . Carnap (1963)

279 *less as an invitation to read it* . . . Schlick to Wittgenstein, November 27, 1930

280 *Whoever has understood, as we do* . . . Schlick (1934), Fragen der Ethik VIII.7, see Schlick (1984)

280 *I am always amazed by the superficiality* . . . Schlick (1934), Fragen der Ethik VIII.9, see Schlick (1984)

281 *Thank God that there are still some psychologists* . . . Schlick to Lorenz, November 17, 1935

281 *Here we find . . . a most remarkable hint* . . . Schlick (1934), Fragen der Ethik VIII.9, see Schlick (1984)

281 *Be ready for happiness* . . . Schlick (1934), Fragen der Ethik VIII.10, see Schlick (1984).

281 *Moral behavior springs from pleasure and pain* . . . Schlick (1934), Fragen der Ethik VIII.10, see Schlick (1984).

281 *[Questions of Ethics] is a book breathing its author's spirit* . . . Menger (1994)

282 *This seemed paradoxical even to Max Planck* . . . Planck to Schlick, August 21, 1927

282 *Only in play* . . . Schlick (1927), Vom Sinn des Lebens, Symposion 1, see Schlick (2008) Collected Works I/6

282 *Man only plays* . . . Friedrich Schiller, *Briefe über die ästhetische Erziehung des Menschen*

284 *A nervous girl with a somewhat strange character* . . . Pötzl to the Rector of the University of Vienna, October 22, 1931; see also Stadler (2015)

284 *A psychopath with bizarre and megalomaniacal ideas* . . . see Stadler (2015)

CHAPTER 11

On the civil war of 1934: Maimann (1988), Weinzierl and Skalnik (1983), Weissensteiner (1990). On Neurath: Sandner (2014). On the dissolution of the Verein Ernst Mach: Stadler (2015). On the Circle during the Schuschnigg years: Ayer (1959, 1977), Bergmann (1988), Hegselmann (1979), Naess (1993), Nagel (1936). Obituaries for Hahn: Frank (1934), Menger (1934). On the persecution and assassination of Schlick: Lotz-Rimbach (2009), Malina (1988), Menger (1994), Stadler (2015). On Turing: Hodges (1983). Most of the text on Turing is due to Douglas R. Hofstadter. On Quine: Keil (2011), Quine (1985). On Tarski: Fefermann and Fefermann (2004). On the emigration of mathematicians and members of the Vienna Circle: Bergmann (1988), Dahms (1987, 1988), Hacohen (2000), Popper (1995), Siegmund-Schultze (2009), Sigmund (2001), Thiel (1984). On the protocol debate: Haller (1979, 1982), Uebel (1991, 1993).

288 *If you agree with Karl Marx's slogan* . . . Trotsky, L. (1931), *History of the Russian Revolution,* Haymarket Books, Chicago

288 *We are known for our great cunning* . . . Otto Neurath to Martha Tausk, Easter Monday 1932 or 1933

289 *Through Marie Reidemeister* . . . Reidemeister-Neurath (1980)

291 *it has been brought to our attention* . . . Sicherheitskommissar to Schlick, March 20, 1934

291 *true to the spirit* . . . Schlick to Sicherheitskommissar, March 23, 1934

291 *Not for a minute* . . . Schlick to Sicherheitskommissar, March 23, 1934

291 *pure chance* . . . Schlick to Sicherheitskommissar, March 23, 1934

292 *It is somewhat tragicomical . . .* Schlick to Sicherheitskommissar, March 23, 1934

292 *So the Ernst Mach Society has really been disbanded . . .* Schlick to Carnap, March 25, 1934

292 *Up with Dollfuss . . .* Neurath to Carnap, July 7, 1934

294 *It occurred to me . . .* Nagel (1936)

294 Language, Truth and Logic . . . see Ayer (2001)

294 *the city of my dreams . . .* Quine (1985) and Creath (2007)

295 *With Hans Hahn's premature death . . .* Nagel (1936)

295 *As for myself . . .* Hahn to Ehrenfest, December 8, 1932

296 *a letter to the ministry . . . Personalakte* Schlick, Archive of the University of Vienna

297 *such a catholic Interest . . .* Nagel (1936)

297 *There will be no meetings . . .* Schlick to Rynin, November 4, 1933

297 *We naturally feel very intensely and painfully . . .* Schlick to Carnap, January 20, 1935

298 *Schlick gently objected that this definition seemed circular . . .* Schlick to Carnap, May 29, 1932

298 *Otto's protocol . . .* Neurath, O. (1932), Protokollsätze, *Erkenntnis* 3, see Neurath (1981)

299 *Which is exactly what Carnap did . . .* Carnap, R. (1932), Über Protokollsätze, *Erkenntnis* 3, 215–228

299 *reduced the danger that younger persons would unwittingly stray . . .* Carnap, R. (1932), Über Protokollsätze, *Erkenntnis* 3, 215–228

299 *The next task . . .* Carnap, R. (1932), Über Protokollsätze, *Erkenntnis* 3, 215–228

299 *beside the point and of little interest . . .* Schlick to Carnap, April 6, 1935

300 *in exactly the fashion that they popped into my mind . . .* Schlick to Carnap, April 16, 1935

300 *they do not lie at the foundations of science . . .* Schlick (1934), Über das Fundament der Erkenntnis, *Erkenntnis* 4, see Schlick (2008), Collected Works, I/6

301 *Some may appreciate such lyricism . . .* Neurath, O. (1934), Radikaler Physikalismus und "Wirkliche Welt," *Erkenntnis* 4, see Neurath (1981)

301 *A representative of the Vienna Circle . . .* Neurath, O. (1934), Radikaler Physikalismus und "Wirkliche Welt," *Erkenntnis* 4, see Neurath (1981)

301 *As you can imagine . . .* Schlick to Carnap, April 6, 1935

301 *I was a little surprised . . .* Schlick (1935), Facts and Propositions, see Schlick (2008), Collected Works I/6

302  *If anyone were to tell me* . . . Schlick (1935), Tatsachen und Aussagen, see Schlick (2008), Collected Works I/6

303  *And yet I must confess* . . . Carnap to Schlick, November 8, 1935

303  *your mathematician's mind-set* . . . Schlick to Carnap, November 14, 1935

303  *This discussion was just one of the many* . . . Menger (1994)

303  *The trouble arises entirely* . . . Schlick to Carnap, June 5, 1934

304  *You cannot claim that I faithfully cling* . . . Schlick to Carnap, January 20, 1935

304  *It has recently been very hot* . . . Schlick to Carnap, July 24, 1934

304  *Wittgenstein wants to prevent the publication* . . . Neider to Neurath, July 27, 1934

304  *This masterstroke planned by the master himself* . . . Neurath to Neider, August 1934

304  *Only dark reports on the Schlick Circle* . . . Neurath to Carnap, January 28, 1935

305  *To my dismay* . . . Neider to Neurath, February 8, 1935

305  *holy sanctuary for Wittgenstein's philosophy* . . . Bergmann to Neurath, see Bergmann (1988)

306  *would be very grateful if you could provide us* . . . Carnap to Rand, May 31, 1935

306  *Rose Rand just informed me* . . . Carnap to Neurath, June 22, 1935

307  *for your kindness in permitting me* . . . Schlick to Pötzl, January 8, 1935

307  *He is a mathematician of the highest caliber* . . . Schlick to Pötzl, January 8, 1935

309  *The greatest* sane *logician* . . . Fefferman and Fefferman (2004)

312  *The whole is more* . . . Schlick (1935/36), Über den Begriff der Ganzheit, see Schlick (2008), Collected Works I/6

314  *That's not a friend* . . . Menger (1994)

318  *It is not often* . . . *Neue Freie Presse,* June 24, 1936

319  *The philosophy chairs at the University of Vienna* . . . *Schönere Zukunft,* July 12, 1936

319  *"But a cat always lands on its feet* . . . Frank to Einstein, August 4, 1936

319  *We claimed merely that he was a* friend *of the Jews* . . . *Schönere Zukunft,* August 1936

319  *For a full fourteen years* . . . see Malina (1988)

320  *far removed from all the centers of production* . . . Menger to Alt, December 31, 1937

320  *I am deeply saddened* . . . Menger to Alt, December 31, 1937

321  *the ground is burning under your feet* . . . Carnap to Popper, October 3, 1936

322  *halfway to the moon* . . . Popper to Kaufmann, October 1936, see Hacohen
(2000), 324

323  *One could live here* . . . Wittgenstein to Pattisson, see Nedo (2012), 330

CHAPTER 12

Zuckmayer (1984) and Rudin (1997) were eyewitnesses of the *Anschluss*. On the
cultural exodus: Siegmund-Schultze (2009), Stadler (1987), Stadler and Weibel
(1995). On Gödel: Dawson (1997), Kennedy (2014), Mancosu (2010), Russell
(1967–1969), Sigmund (2011), Sigmund et al. (2006), Stewart (2013, 2017),
Thiel (1984), van Atten and Kennedy (2003), Yourgrau (2005). The text on Turing is mostly due to Douglas Hofstadter. On Neurath: Kinross (2002), Sandner
(2014). The *Zeemanshoop* tale is at http://www.holywellhousepublishing.co.uk
/Zeemanshoop.html. On the University of Vienna during the Third Reich: Heiss
et al. (1989), Reiter (2001). On Adler: Bauer (2004).

325  *That night hell broke loose* . . . Zuckmayer (1984)

325  *Cabled today* . . . *Personalakte* Menger, Archive of the University of Vienna

326  *Everything has changed now* . . . Archive of the University of Vienna

327  *In a sharp and united manner* . . . Sigmund et al. (2006), 56

327  *In one respect, we were better off* . . . Rudin (1997)

330  *strongly motivated by national sentiments* . . . This and the following quotes
are in Stadler (2015)

331  *It was probably the last meeting ever* . . . Dawson (1997), 125

331  *Her type: Viennese washerwoman* . . . Morgenstern Diary, July 4, 1940

332  *When such "states" were particularly intense* . . . Aydelotte to Selective Service
Board, April 14, 1943, IAS Archive, Princeton

339  *endless errands* . . . Gödel to Menger, August 30, 1939

339  *Obscure offices that no one had ever heard of before* . . . Perutz (1996)

339  *generally, mathematics in the Vienna of the System Era* . . . *Personalakte* Gödel,
Archive of the University of Vienna

340  *Gödel is in a class by himself* . . . von Neumann to Flexner, October 7, 1939

340  *Gödel is absolutely irreplaceable* . . . von Neumann to Flexner, September 27,
1939

341  *The coffee is wretched* . . . Morgenstern Diary, March 10, 1940

342  *As already reported several times* . . . *Personalakte* Gödel, Archive of the University of Vienna

343  *Why did Einstein enjoy talking with me?* . . . Gödel to Seelig, September 7,
1955

343  *The one man who has certainly been by far Einstein's best friend* . . . Straus,
Reminiscences, see Pais (1982)

343  *I go to my office* . . . Morgenstern, letter to Austrian government, 1965

344  *and do something to that person* . . . Gödel, A Remark About the Relationship Between Relativity Theory and Idealistic Philosophy, in Schilpp (1944a), see Gödel (1990), Collected Works II

345  *the German penchant for metaphysics* . . . Russell (1967–1969)

348  *If we cannot find a boat* . . . Reidemeister-Neurath (1980)

350  *You can swim if you like* . . . http://www.holywellhousepublishing.co.uk/ Zeemanshoop.html

350  *We were received with bananas, kindness, and tea* . . . Neurath to P. J. de Kanter, July 12, 1945

351  *neither persecuted nor aggrieved* . . . Neurath to Jette Pront, July 24, 1945; see also Sandner (2014), 265

351  *I have always been interested* . . . Neurath to Kaufmann, April 20, 1942

CHAPTER 13

The Menger episode is documented in Sigmund (1998). On the postwar appointments (and nonappointments) at the University of Vienna: Stadler (2015). On Viktor Kraft: Kainz (1976). On Feyerabend: Feyerabend (1975, 1994). The essays by Bachmann are reedited in Bachmann (2005). On the Popper-Wittgenstein debate (and a lot about the Vienna Circle): Edmonds and Eidinow (2001). On Carnap: Carus (2007), Cirera (1994), Creath (1990, 2007, 2012), Irzik and Grünberg (1995), Wagner (2009). For the later years of Kurt Gödel: Awodey and Carus (2010), Dawson (1997), Feferman (1986), Kennedy (2014), Sigmund et al. (2006), van Atten and Kennedy (2003). On the Circle in the United States: Bergmann (1954), Feigl (1969), Feyerabend (1966), Holton (1993), Zilsel (1992, 1988). On philosophy of science: Kuhn (1962), Nagel (1961), Reisch (2005), Richardson (2008), Salmon and Wolters (1994). Karl Popper (1995) had the last word.

356  *A typical episode centered on Karl Menger* . . . *Personalakte* Menger, Archive of the University of Vienna

360  *a former student, a paranoid psychopath* . . . Kraft (1953)

361  *Already at that time* . . . Wiener Wochenausgabe 4 (1951)

362  *In Vienna itself, the Vienna Circle is dead* . . . Bachmann (2005)

364  *Erich Heintel was given the job* . . . Archive of the University of Vienna

364  *The man has grown fat* . . . McGuiness, B., Waismann the Wandering Scholar, in McGuiness (2011)

365  *burnt him in his pocket* . . . Wittgenstein to Sraffa, March 14, 1938

365  *no disgrace in living from manual work* . . . Iven (2004)

365  *wet and cold psychic climate* . . . Wittgenstein, April 13, 1947

369 *a struggle against the bewitching of our reason* . . . Wittgenstein, *Philosophical Investigations,* §109

369 *language taking a holiday* . . . Wittgenstcin, *Philosophical Investigations,* §38

369 *I still believe that the surest way to perdition* . . . Popper, in Schilpp (1974)

372 *Peace in thinking* . . . Wittgenstein, *Vermischte Bemerkungen* 87 (1944)

372 *Where to position the lever today?* . . . Bachmann (2005)

376 *With his usual thoroughness, Gödel prepared* . . . Morgenstern Papers, History of the Naturalization of Kurt Gödel, March 13, 1971

376 *How can any of us be a professor* . . . Ulam (1976); see also Dawson (1997), 201

377 *I mean the view that mathematics describes* . . . Gödel, K. (1951), Some Basic Theorems on the Foundation of Mathematics and Their Implications, in Gödel, Collected Works, Vol. III

378 *in view of widespread prejudices* . . . Gödel to Schilpp, February 3, 1959

378 *What Newton had done for physics* . . . Wang (1987)

379 *Of course a scientific foundation* . . . Gödel to his mother, October 6, 1961

381 *The objection that it is impossible* . . . Gödel to his mother, September 12, 1961

381 *For we understand neither* . . . Gödel to his mother, February 27, 1950

381 *But who would have had an interest* . . . Menger, Memories of Kurt Gödel, in Menger (1994)

381 *Gödel clings to me* . . . Morgenstern papers, folder on Kurt Gödel, Morgenstern Archive, Duke University

383 *Popper was never a member* . . . Kraft, in Schilpp (1974)

384 *The Vienna Circle was an admirable institution* . . . Popper, in Schilpp (1974)

385 *The work of the Vienna Circle* . . . Kraft (1953)

385 *The Viennese have always been remarkably talented* . . . Friedell, E. (1922), preface to Nestroy, J., "Das ist klassisch," n.p., Vienna

# Bibliography

Achleitner, F. (1996): *Wiener Architektur*. Böhlau, Vienna.

Adler, F. (1918a): *Ernst Machs Überwindung des mechanischen Materialismus*. Brand, Vienna.

Adler, F. (1918b): *Vor dem Ausnahmegericht*. Reprinted in 2016 (eds. Maier M., Spitaler, G.). Promedia, Vienna.

Ardelt, R. G. (1984): *Friedrich Adler: Probleme einer Persönlichkeitsentwicklung um die Jahrhundertwende*. Österr. Bundesverlag, Vienna.

Arntz, G. (1976): *Kritische Grafik und Bildstatistik*. Werkkatalog, The Hague.

Arntz, G. (1982): Otto Neurath, Ich und die Bildstatistik. In Stadler, F. (ed.), *Arbeiterbildung in der Zwischenkriegszeit*, 31–34. Jugend und Volk, Vienna-Munich.

Awodey, S., Carus, A. W. (2007): Carnap's Dream: Gödel, Wittgenstein and "Logical Syntax." *Synthese* 159, 23–45.

Awodey, S., Carus, A. W. (2010): Gödel and Carnap. In Feferman, S., et al., *Kurt Gödel: Essays for His Centennial*. Cambridge University Press, New York.

Awodey, S., Klein, C. (eds.) (2004): *Carnap Brought Home: The View from Jena*. Open Court, Chicago.

Ayer, A. J. (1956): The Vienna Circle. In Ryle, G. (ed.), *Revolution in Philosophy*, 70–87. Fontana, London.

Ayer, A. J. (1959): *Logical Positivism*. Free Press, New York.

Ayer, A. J. (1972): *Russell*. Fontana, London.

Ayer, A. J. (1977): *Part of My Life*. Collins, London.

Ayer A. J. (2001): *Language, Truth and Logic*. Penguin, Harmondsworth, UK.

Baaz M., Papadimitriou, C. H., Putnam, H. W., Scott, D. S., Harper, J. L. (2011): *Kurt Gödel and the Foundations of Mathematics: Horizons of Truth*. Cambridge University Press, New York.

Bachmann, F. et al. (1972): Nachruf auf Kurt Reidemeister: Math. *Annalen* 199, 1–11.

Bachmann, I. (2005): *Kritische Schriften.* Piper, Munich.

Badesa, C., Mancosu, P., Tach, R. (2009): The Development of Mathematical Logic from Russell to Tarski, 1900–1935. In Haaparanta, L. (ed.), *The History of Modern Logic.* Oxford University Press, Oxford, UK.

Baker, G. (1979): Verehrung und Verkehrung: Waismann und Wittgenstein. In Luckhardt, C. G. (ed.), *Wittgenstein, Sources and Perspectives,* 243–285. Cornell University Press, Ithaca, NY.

Baker, G. (ed.) (2003): *The Voices of Wittgenstein.* Routledge, London.

Bartley, W. W. III (1988): *Wittgenstein.* Cresset, London.

Basch, A. (1953): Richard von Mises zum 70. Geburtstag. *Osterreichisches Ingenieur-Archiv* 7, 73–76.

Bauer, M. (2004): *Friedrich Adler—Rebell der Einheit.* Trotzdem, Vienna.

Baum, J. (2014): *Wittgenstein im Ersten Weltkrieg: Die "Geheimen Tagebücher" und die Erfahrungen an der Front (1914–1919).* Kitab, Klagenfurt.

Beham, B. (2013): Karl Menger 1903–1925. PhD thesis, University of Vienna.

Beller, S. (2008): *Vienna and the Jews 1867–1938: A Cultural History.* 2nd ed. Cambridge University Press, New York.

Berghahn, W. (1972): *Robert Musil.* Rowohlt, Reinbek.

Bergmann, G. (1954): *The Metaphysics of Logical Positivism.* Longmans Green, New York.

Bergmann, G. (1988): Erinnerungen an den Wiener Kreis. Letter to Otto Neurath. In Stadler, F. (ed.), *Kontinuität und Bruch 1938–1945–1955,* 171–180. Jugend und Volk, Vienna.

Biemel, W. (1973): *Martin Heidegger.* Rowohlt, Reinbek.

Blackmore, J. T. (1972): *Ernst Mach: His Work, Life, Influence.* University of California Press, Berkeley.

Blackmore, J. T. (1978): Three Autobiographical Manuscripts by Ernst Mach. *Annals of Science* 35, 401–418.

Blackmore, J. T. (ed.) (1992): *Ernst Mach—A Deeper Look: Documents and New Perspectives.* Kluwer, Dordrecht.

Blackmore, J. T. (ed.) (1995): *Ludwig Boltzmann—His Later Life and Philosophy 1900–1906.* 2 vols. Kluwer, Dordrecht.

Blackmore, J. T. (ed.) (1999): Ludwig Boltzmann—Troubled Genius as Philosopher. *Synthese* 119, 1–232.

Blackmore, J. T., Hentschel, K. (eds.) (1985): *Ernst Mach als Aussenseiter.* Braumüller, Vienna.

Blackmore, J. T., Itagaki, R., Tanaka, S. (eds.) (2001): *Ernst Mach's Vienna 1895–1930.* Kluwer, Dordrecht.

Blau, E. (1999): *The Architecture of Red Vienna.* MIT Press, Cambridge, MA.

Boltzmann, L. (1905): *Populäre Schriften.* Barth, Leipzig.

Bradley, J. (1971): *Mach's Philosophy of Science.* Collins, London.

Brentano, F. (1988): Über Ernst Machs "Erkenntnis und Irrtum" (eds. Chisholm, R., Marek, J.). Rodopi, Vienna.

Broda, E. (1955): *Ludwig Boltzmann: Mensch, Physiker, Philosoph.* Springer, Vienna.

Brunngraber, R. (1932): *Karl und das 20. Jahrhundert.* Zsolnay, Vienna.

Brunngraber, R. (1949): *Der Weg durchs Labyrinth.* Zsolnay, Graz.

Bührke, T. (2004): *Albert Einstein.* dtv, Munich.

Burke, C., Kindel, E., Walker, S. (2013): *Isotype: Design and Contexts 1925–1971.* Hyphen Press, London.

Canetti, E. (1980): *Die Fackel im Ohr.* Hanser, Munich.

Carnap, R. (1928a): *Der logische Aufbau der Welt.* Meiner, Frankfurt.

Carnap, R. (1928b): *Scheinprobleme der Philosophie.* Weltkreis, Berlin.

Carnap, R. (1937): *The Logical Syntax of Language.* Kegan Paul, London.

Carnap, R. (1963): Intellectual Autobiography. In Schilpp, P. A. (ed.), *The Library of Living Philosophers—Carnap.* Northwestern University Press, Chicago.

Cartwright, N., Cat, J., Fleck, L., Uebel, T. E. (eds.) (1996): *Otto Neurath: Philosophy Between Science and Politics.* Cambridge University Press, New York.

Carus, A. (2007): *Carnap and Twentieth-Century Thought.* Cambridge University Press, New York.

Cercignani, C. (1998): *Ludwig Boltzmann—The Man Who Trusted Atoms.* Oxford University Press, Oxford, UK.

Cirera, R. (1994): *Carnap and the Vienna Circle: Empiricism and Logical Syntax.* Rodopi, Atlanta.

Clark, R. W. (1988): *The Life of Bertrand Russell.* Weidenfeld and Nicolson, London.

Clark, R. W. (1984): *Einstein—The Life and Times.* HarperCollins, New York.

Cohen, R. S. (ed.) (1975): *Ernst Mach: Physicist and Philosopher.* Kluwer, Dordrecht.

Cohen, R. S., Neurath, M. (1973): *Otto Neurath: Empiricism and Sociology.* Kluwer, Dordrecht.

Corino, K. (2003): *Robert Musil: Eine Biographie.* Rowohlt, Reinbek.

Corry, L., Renn, J., Stachel, J. (1998): Belated Decision in the Hilbert-Einstein Priority Dispute. *Science* 278, 1270–1273.

Creath, R. (ed.) (1990): *Dear Carnap, Dear Van: The Quine-Carnap Correspondence and Related Work.* University of California Press, Berkeley.

Creath, R. (2007): Vienna, the City of Quine's Dreams. In Richardson, A., Uebel, T. (eds.), *The Cambridge Companion to Logical Empiricism,* 332–345. Cambridge University Press, New York.

Creath, R. (ed.) (2012): *Rudolf Carnap and the Legacy of Logical Empiricism.* Springer, Dordrecht.

Dahms, H. J. (1987): Die Emigration des Wiener Kreises. In Stadler, F. (ed.), *Vertriebene Vernunft,* 66–122. Jugend und Volk, Vienna.

Dahms, H. J. (1988): Die Bedeutung der Emigration des Wiener Kreises für die Entwicklung der Wissenschaftstheorie. In Stadler, F. (ed.), *Kontinuität und Bruch 1938–1945–1955,* 155–168. Jugend und Volk, Vienna.

Dahms, H. J. (2004): Neue Sachlichkeit in the Architecture and Philosophy in the 1920s. In Awodey, S., Klein, C. (eds.), *Carnap Brought Home: The View from Jena.* Open Court, Chicago.

Davis, C. (1996): Remembering Olga Taussky-Todd. *AWM Newsletter* 26, 7–9.

Dawson, J. W. (1997): *Logical Dilemmas: The Life and Work of Kurt Gödel.* AK Peters, Wellesley, MA.

Dawson, J. W., Sigmund, K. (2006): Gödel's Vienna. *Mathematical Intelligencer* 28, 44–55.

DePauli-Schimanovich, W., Weibel, P. (1997): *Kurt Gödel: Ein mathematischer Mythos.* Hölder-Pichler-Tempsky, Vienna.

Dierker, E., Sigmund, K. (eds.) (1998): *Karl Menger—Ergebnisse eines Mathematischen Kolloquiums.* Springer, Vienna.

Doxiadis, A., Papadimitriou, C. H. (2009): *Logicomix—An Epic Search for Truth.* Bloomsbury, London.

Eckert, B., Müller, H. H. (1989): *Leo Perutz 1882–1957. Eine Ausstellung der deutschen Bibliothek, Frankfurt.* Zsolnay, Vienna.

Edmonds, D. J., Eidinow, J. A. (2001): *Wittgenstein's Poker: The Story of a Ten-Minute Argument Between Two Great Philosophers.* HarperCollins, New York.

Einstein, A. (1954): *Ideas and Opinions.* Crown, New York.

Engelmann, P. (1967): *Letters from Ludwig Wittgenstein.* Blackwell, Oxford, UK.

Engler, F. O., Iven, M. (2007): *Moritz Schlick in Rostock.* Weiland, Rostock.

Engler, F. O., Iven, M. (eds.) (2008): *Moritz Schlick: Leben, Werk und Wirkung.* Parerga, Berlin.

Epple, M. (1995): Kurt Reidemeister (1893–1971): Kombinatorische Topologie und exaktes Denken. In Rauschning, D., von Nerée, D. (eds.), *Die Albertus-Universität zu Königsberg und ihre Professoren,* 567–575. Duncker and Humblot, Berlin.

Epple, M. (1999): *Die Entstehung der Knotentheorie.* Vieweg, Braunschweig.

Fasol-Boltzmann, I. M. (ed.) (1990): *Ludwig Boltzmann: Principien der Naturfilosofi, Lectures on Natural Philosophy 1903–1906.* Springer, Heidelberg.

Feferman, A. S., Feferman, S. (2004): *Alfred Tarski: Life and Logic.* Cambridge University Press, New York.

Feferman, S. (1986): Gödel's Life and Work. In Feferman, S., et al. (eds.), *Gödel's Collected Works, Vol. 1,* 1–36. Princeton University Press, Princeton.

Feferman, S. (1996): *In the Light of Logic.* Oxford University Press, Oxford, UK.

Feferman, S., Parsons, C., Simpson, S. G. (eds.) (2010): *Kurt Gödel: Essays for His Centennial.* Cambridge University Press, New York.

Feigl, H. (1969): Der Wiener Kreis in Amerika. In Fleming, D., Bailyn, B. (eds.), *The Intellectual Migration: Europe and America, 1930–1960,* 630–673. Belknap Press, Cambridge, MA.

Feyerabend, P. (1966): Herbert Feigl. A Biographical Sketch. In Feyerabend, P. K., Maxwell, G. (eds.), *Mind, Matter, Method: Essays in the Philosophy of Science in Honor of Herbert Feigl,* 3–13. University of Chicago Press, Chicago.

Feyerabend, P. (1975): *Against Method.* Verso, New York.

Feyerabend, P. (1994): *Killing Time.* University of Chicago Press, Chicago.

Fischer, R. (1991): *Philosophie aus Wien: Aufsätze zur analytischen und österreichischen Philosophie, zu den Weltanschauungen des Wiener Fin-de-Siècle und Biographisches aus Berkeley, Shanghai und Wien.* UVW, Vienna.

Fischer, R. (ed.) (1995): *Das goldene Zeitalter der Österreichischen Philosophie.* UVW, Vienna.

Flamm, D. (ed.) (1995): *Hochgeehrter Herr Professor! Innig geliebter Louis! Ludwig Boltzmann, Henriette von Aigentler.* Briefwechsel. Böhlau, Vienna.

Föllmer, H., Küchler, K. (1998): Richard von Mises. In Begehr, H., et al., *Mathematics in Berlin, Catalogue of an Exhibition,* 55–60. Birkhäuser, Basel.

Fölsing, A. (1998): *Einstein: A Biography.* Penguin, Harmondsworth, UK.

Frank, J. (1931): *Architektur als Symbol* (reprinted 2005). Löcker, Vienna.

Frank, P. (1934): Hans Hahn. *Erkenntnis* 4, 315–316.

Frank, P. (1937/38): Ernst Mach—The Centenary of His Birth. *Erkenntnis* 7, 247–256.

Frank, P. (1941): *Between Physics and Philosophy.* Harvard University Press, Cambridge, MA.

Frank, P. (1947): *Einstein: His Life and Times.* Knopf, New York.

Frank, P. (1949): *Modern Science and Its Philosophy.* Harvard University Press, Cambridge, MA.

Frank, P. (1954): The Work of Richard von Mises: 1883–1953. *Science* 119, 823–824.

Franzen, T. (2005): *Gödel's Theorem: An Incomplete Guide to Its Use and Abuse.* AK Peters, Wellesley, MA.

Friedman, M. (1999): *Reconsidering Logical Positivism.* Cambridge University Press, New York.

Friedman, M. (2000): *A Parting of the Ways: Carnap, Cassirer and Heidegger.* Open Court, Chicago.

Gadol, E. T. (ed.) (1982): *Rationality and Science: A Memorial Volume for Moritz Schlick on the Occasion of His 100th Birthday.* Springer, Vienna.

Galison, P. (1990): Aufbau/Bauhaus: Logical Positivism and Architectural Modernism. *Critical Inquiry* 16, 709–752.

Galison, P. (2003): *Einstein's Clocks, Poincaré's Maps.* Norton, New York.

Galison, P. (2008): The Assassin of Relativity. In Holton, G., et al. (eds.), *Einstein for the 21st Century: His Legacy in Science, Art, and Modern Culture,* 185–204. Princeton University Press, Princeton.

Geier, M. (1992): *Der Wiener Kreis.* Rowohlt, Reinbek.

Geier, M. (1994): *Karl Popper.* Rowohlt, Reinbek.

Geymonat, L. (1991): Persönliche Erinnerungen an den Wiener Kreis. In Kruntorad, P. (ed.), *Jour Fixe der Vernunft,* 42–48. Hölder-Pichler-Tempsky, Wien.

Giere, R. N., Richardson, A. W. (1997): *Origins of Logical Empiricism.* University of Minnesota Press, Minneapolis.

Gödel, K.: Collected Works. Princeton University Press, Princeton.

Goldstein, R. (2005): *Incompleteness: The Proof and Paradox of Kurt Gödel.* Norton, New York.

Gomperz, H. (1936a): Philosophy in Austria During the Last Sixty Years. *Personalist* 17, 307–311.

Gomperz, H. (ed.) (1936b): *Theodor Gomperz (1832–1912). Briefe und Aufzeichnungen. Eingeleitet, erläutert und zu einer Darstellung seines Lebens verknüpft von Heinrich Gomperz, Vol. I.* Kuppitsch, Vienna.

Grayling, A. C. (1988): *Wittgenstein: A Very Short Introduction.* Oxford University Press, Oxford, UK.

Grayling, A. C. (1992): *Russell: A Very Short Introduction.* Oxford University Press, Oxford, UK.

Hacohen, M. H. (2000): *Karl Popper 1902–1945: The Formative Years.* Cambridge University Press, New York.

Hahn, H. (1980): *Empiricism, Logic and Mathematics* (ed. McGuiness, B.). Reidel, Dordrecht.

Hall, M. (1978): *Der Fall Bettauer.* Löcker, Vienna.

Haller, R. (1979): *Studien zur österreichischen Philosophie: Variationen über ein Thema.* Kluwer, Dordrecht.

Haller, R. (1982): Das Neurath-Prinzip: Grundlagen und Folgerungen. In Stadler, F. (ed.), *Arbeiterbildung in der Zwischenkriegszeit,* 79–87. Jugend und Volk, Vienna.

Haller, R. (1993): *Neopositivismus: Eine historische Einführung in die Philosophie des Wiener Kreises.* Wissenschaftliche Buchgesellschaft, Darmstadt.

Haller, R., Stadler, F. (eds.) (1988): *Ernst Mach—Werk und Wirkung.* Hölder-Pichler-Tempsky, Vienna.

Hanisch, E. (2011): *Der grosse Illusionist: Otto Bauer.* Böhlau, Vienna.

Hänsel, L. (2012): *Tagebücher.* Haymon, Innsbruck.

Hegselmann, R. (1979): Otto Neurath—Empiristischer Aufklärer und Sozialreformer. In Hegselmann, R. (ed.), *Otto Neurath: Wissenschaftliche Weltauffassung, Sozialismus und logischer Empirismus,* 7–78. Suhrkamp, Frankfurt.

Heidegger, M. (1975–): Collected Works. Klostermann, Frankfurt.

Heiss, G., Mattl, S., Meissl, S., Stuhlpfarrer, K. (1989): *Willfährige Wissenschaft: Die Universität Wien 1938–1945.* UVW, Vienna.

Hempel, C. G. (2012). *Selected Philosophical Essays* (ed. Jeffrey, R.). Cambridge University Press, New York.

Hentschel, K. (1986): Die Korrespondenz Einstein–Schlick: Zum Verhältnis der Physik zur Philosophie. *Annals of Science* 43, 475–488.

Hentschel, K. (1990): *Interpretationen und Fehlinterpretationen der speziellen und allgemeinen Relativitätstheorie durch Zeitgenossen Albert Einsteins.* Birkhäuser, Basel.

Hilbert, D. (1931): Die Grundlagen der elementaren Zahlentheorie. *Mathematische Annalen* 104, 485.

Hilbert, D. (1970): *Gesammelte Abhandlungen.* Springer, Berlin.

Hintikka, J. (1999): *On Gödel.* Wadsworth, Belmont, CA.

Hodges, A. (1983): *Alan Turing: The Enigma.* Simon and Schuster, New York.

Hoffmann, B., Dukas, H. (1972): *Albert Einstein: Creator and Rebel.* Viking Press, New York.

Höflechner, W. (ed.) (1994): *Ludwig Boltzmann—Leben und Briefe.* Akademische Druck- und Verlagsanstalt, Graz.

Hofstadter, D. R. (1979): *Gödel, Escher, Bach: An Eternal Golden Braid.* Basic Books, New York.

Holton, G. (1993): From the Vienna Circle to Harvard Square: The Americanization of a European World Conception. In Stadler, F. (ed.), *Scientific Philosophy: Origins and Developments,* 47–74. Kluwer, Dordrecht.

Holton, G. (1981): *Thematische Analyse der Wissenschaft: Die Physik Einsteins und seine Zeit.* Suhrkamp, Frankfurt.

Holton, G. (1996): *Einstein, History, and Other Passions: The Rebellion Against Science at the End of the Twentieth Century.* Addison-Wesley, Reading, MA.

Holton, G., Galison, P., Schweber, S. (2008): *Einstein for the 21st Century: His Legacy in Science, Art, and Modern Culture.* Princeton University Press, Princeton.

Irzik, G., Grünberg, T. (1995): Carnap and Kuhn: Arch Enemies or Close Allies? *British Journal for the Philosophy of Science* 46, 285–307.

Isaacson, W. (2007): *Einstein: His Life and Universe.* Simon and Schuster, New York.

Iven, M. (2004): *Wittgenstein und Rand: Versuch einer Annäherung.* Wittgenstein-Studien 9. Lang, Berlin.

Iven, M. (2008): *Moritz Schlick: Die frühen Jahre (1882–1907).* Parerga, Berlin.

Janik, A., Toulmin, S. (1973): *Wittgenstein's Vienna.* Weidenfeld and Nicolson, London.

Johnston, W. M. (1983): *The Austrian Mind: An Intellectual and Cultural History 1848–1938.* University of California Press, Berkeley.

Juhos, B. (1965): Gibt es in Österreich eine wissenschaftliche Philosophie? In Kadrnoska, F. (ed.), Österreich—Geistige Provinz?, 232–244. Braumüller, Vienna.

Juhos, B. (1972): Mach Ernst, Physiker und Philosoph. In Österreichisches Biographisches Lexikon 1815–1950 Wien, Vol. 5, 388–389. Österreichische Akademie der Wissenschaften, Wien.

Kadrnoska, F. (ed.) (1981): *Aufbruch und Untergang: Österreichische Kultur zwischen 1918 und 1938.* Europaverlag, Vienna.

Kainz, F. (1976): Viktor Kraft. *Almanach der Österreichische Akademie der Wissenschaften,* 519–557.

Kandel, E. R. (2012): *The Age of Insight.* Random House, New York.

Kann, R. A. (1974): *Theodor Gomperz: Ein Gelehrtenleben im Bürgertum der Franz-Josephs-Zeit.* Löcker, Vienna.

Keil, G. (2011): *Quine.* Reclam, Stuttgart.

Kennedy, J. (2014): Gödel's 1946 Princeton Bicentennial Lecture: An Appreciation. In Kennedy, J. (ed.), *Interpreting Gödel: Critical Essays,* 109–130. Cambridge University Press, New York.

Keyserling, A. (1965): *Der Wiener Denkstil: Mach, Carnap, Wittgenstein.* Akademische Druck- und Verlagsanstalt, Graz.

Kinross, R. (2002): Marie Neurath 1898–1986. In Kinross, R. (ed.), *Unjustified Texts,* 12–25. Hyphen Press, London.

Klagge, J. (2001): *Wittgenstein: Biography and Philosophy.* Cambridge University Press, New York.

Köhler, E., et al. (ed.) (2002): *Kurt Gödel—Wahrheit und Beweisbarkeit.* Hölder-Pichler-Tempsky, Vienna.

Kraft, V. (1953): *The Vienna Circle: The Origin of Neo-Positivism, a Chapter in the History of Recent Philosophy.* Greenwood Press, New York.

Kraft, V. (1966): Ernst Mach als Philosoph. *Almanach der Österreichischen Akademie der Wissenschaften* 116, 373–387.

Kraft, V. (1974): Popper and the Vienna Circle. In Schilpp, P. A. (ed.), *The Library of Living Philosophers—Popper,* 185–204. Northwestern University Press, Chicago.

Krauth, L. (1970): *Die Philosophie Carnaps.* Springer, Vienna.

Kreisel, G. (1980): Kurt Gödel. *Biographical Memoirs of Fellows of the Royal Society* 26, 148–224.

Kuhn, T. S. (1962): The Structure of Scientific Revolutions. In *International Encyclopedia of Unified Science*, 2nd ed. (1970). University of Chicago Press, Chicago.

Kutalek, N., Fellinger, H. (1969): *Zur Wiener Volksbildung.* Jugend und Volk, Vienna.

Lampa, A. (1923): Ernst Mach. In *Neue Österreichische Biographie, Vol. I,* 93–102. ÖAW, Vienna.

Leinfellner, E., Windholz, S. (2005): *Ludwig Wittgenstein: Ein Volksschullehrer in Niederösterreich.* Sutton, Vienna.

Leonard, R. J. (1998): Ethics and the Excluded Middle: Karl Menger and Social Science in Interwar Vienna. *Isis* 89, 1–26.

LeRider, J. (1993): *Modernity and Crises of Identity: Culture and Society in Fin-de-Siecle Vienna.* Wiley, New York.

Limbeck, C., Stadler, F. (2015): *Der Wiener Kreis: Texte und Bilder einer Ausstellung.* LIT, Vienna.

Lindley, D. (2001): *Boltzmann's Atom.* Free Press, New York.

Lotz-Rimbach, R. (2009): Mord verjährt nicht: Psychogramm eines politischen Mordes. In Stadler, F., Engler, H. (eds.), *Stationen: Dem Philosophen und Physiker Moritz Schlick zum 125. Geburtstag,* 81–104. Springer, Vienna.

Lützeler, P. M. (1985): *Hermann Broch.* Suhrkamp, Frankfurt.

Mach, E. (2008): Die Analyse der Empfindungen. In *Ernst Mach Studienausgabe.* Xenomoi, Berlin.

Mach, E. (2011): *Erkenntnis und Irrtum.* Xenomoi, Berlin.

Mach, E. (2012): Die Mechanik in ihrer Entwicklung. In *Ernst Mach Studienausgabe.* Xenomoi, Berlin.

Mach, E. (2014): Populär-Wissenschaftliche Vorlesungen. In *Ernst Mach Studienausgabe.* Xenomoi, Berlin.

Magee, B. (1973): *Karl Popper.* Routledge, London.

Maier, M., Maderthaner, W. (eds.) (2006): *Physik und Revolution: Friedrich Adler—Albert Einstein.* Briefe—Dokumente—Stellungnahmen. Löcker, Vienna.

Maimann, H. (ed.) (1988): *Die ersten 100 Jahre: Österreichische Sozialdemokratie 1888–1988.* N.p., Vienna-Munich.

Malina, P. (1988): Tatort: Philosophenstiege. In Benedikt, M., Burger, R. (eds.), *Bewusstsein, Sprache und Kunst,* 231–253. Hölder-Pichler-Tempsky, Vienna.

Mancosu, P. (1998): *From Brouwer to Hilbert: The Debate on the Foundations of Mathematics in the 1920s.* Oxford University Press, Oxford, UK.

Mancosu, P. (2010): *The Adventure of Reason, 1900–1940.* Oxford University Press, Oxford, UK.

Mann, G. (1986): *Erinnerungen und Gedanken: Eine Jugend in Deutschland.* Fischer, Frankfurt.

Manninen, J. (2014): *Wittgenstein's Virtual Presence in the Vienna Circle 1931–1937.* Preprint.

McGuiness, B. (1979): *Wittgenstein and the Vienna Circle: Conversations Recorded by Friedrich Waismann.* Barnes and Noble, New York.

McGuiness, B. (1985): Moritz Schlick. *Synthese* 64.

McGuiness, B. (ed.) (1987): *Unified Science.* Reidel, Dordrecht.

McGuiness, B. (2005): *Wittgenstein: A Life: Young Ludwig: 1889–1921.* Clarendon Press, Oxford, UK.

McGuiness, B. (2011): *Friedrich Waismann: Causality and Logical Positivism.* Springer, New York.

Menger, K. (1934): Hans Hahn. *Ergebnisse eines Mathematischen Kolloquiums* 6, 40–44.

Menger, K. (1952): The Formative Years of Abraham Wald and His Work in Geometry. *Annals of Mathematical Statistics* 23, 14–20.

Menger, K. (1960): Introduction. In Mach, E., *The Science of Mechanics,* iii–xxi. Open Court, La Salle, IL.

Menger, K. (1972): Österreichischer Marginalismus und mathematische Ökonomie. *Zeitschrift für Nationalökonomie* 32, 14–20.

Menger, K. (1974): *Morality, Decision and Social Organization.* Reidel, Dordrecht.

Menger, K. (1980): Introduction. In *Hans Hahn: Empiricism, Logic and Mathematics,* ix–xviii. Vienna Circle Collection 13. Reidel, Dordrecht. Reprinted in Menger, K., *Selecta Mathematica* (eds. Schweitzer, B., et al.), Vol. 2, 565–568. Springer, New York.

Menger, K. (1982): Memories of Moritz Schlick. In Gadol, E. T. (ed.), *Rationality and Science: A Memorial Volume for Moritz Schlick on the Occasion of His 100th Birthday.* Springer, Vienna. Reprinted in Menger, K., *Selecta Mathematica* (eds. Schweitzer, B., et al.), Vol. 2, 569–589. Springer, New York.

Menger, K. (1994): *Reminiscences of the Vienna Circle and the Mathematical Colloquium.* Reidel, Dordrecht.

Mises, L. v. (1978): *Erinnerungen.* Fischer, Stuttgart.

Mises, R. v. (1956): *Positivism: A Study in Human Understanding.* Braziller, New York.

Monk, R. (1997): *Bertrand Russell: 1872–1920. The Spirit of Solitude.* Vintage, London.

Monk, R. (2001a): *Bertrand Russell: 1921–1970. The Ghost of Madness*. Vintage, London.

Monk, R. (2001b): *Ludwig Wittgenstein: The Duty of Genius*. Penguin, Harmondsworth, UK.

Morgenstern, O. (1927): Friedrich von Wieser. *American Economic Review* 17, 669–674.

Morgenstern, O. (1928): *Wirtschaftsprognose*. Springer, Vienna.

Moritz, V., Leidinger, H. (2012): *Oberst Redl: Der Spionagefall, der Skandal, die Fakten*. Residenz, Vienna.

Mormann, T. (2000): *Rudolf Carnap*. Beck, Munich.

Mulder, H. L. (1968): Wissenschaftliche Weltauffassung: Der Wiener Kreis. *Journal of the History of Philosophy* 6, 368–390.

Müller, H. H. (1992). *Leo Perutz*. Beck, Munich.

Musil, R.: Collected Works. Reinbek, Hamburg.

Musil, R. (1982): *On Mach's Theories* (trans. Mulligan, K.). Catholic University of America Press, Washington, DC.

Naess, A. (1993): Logical Empiricism and the Uniqueness of the Schlick Seminar: A Personal Experience with Consequences. In Stadler, F. (ed.), *Scientific Philosophy. Origins and Developments,* 11–26. Kluwer, Dordrecht.

Nagel, E. (1936): Impressions and Appraisals of Analytic Philosophy in Europe. *Journal of Philosophy* 33, 191–246.

Nagel, E. (1961): *The Structure of Science*. Harcourt, New York.

Nagel, E., Newman, J. (2005): *Gödel's Proof*. Routledge, New York.

Nedo, M. (ed.) (2012): *Ludwig Wittgenstein: Ein biographisches Album*. Beck, Munich.

Neffe, J. (2007): *Einstein: A Biography*. Farrar, Straus and Giroux, New York.

Neider, H. (1977): Gespräch mit Heinrich Neider: Persönliche Erinnerungen an den Wiener Kreis. In Marek, K., et al. (eds.), Österreichische Philosophen und ihr Einfluss auf die analytische Philosophie der Gegenwart, 21–42. Conceptus, Innsbruck.

Nemeth, E. (1981): *Otto Neurath und der Wiener Kreis*. Suhrkamp, Frankfurt.

Nemeth, E., Stadler, F. (eds.) (1996): *Otto Neurath (1882–1945): Encyclopedia and Utopia*. Kluwer, Dordrecht.

Neurath, O. (1981): *Gesammelte philosophische und methodologische Schriften* (eds. Haller, R., Rutte, H.). Hölder-Pichler-Tempsky, Vienna.

Neurath, O. (1991): *Gesammelte bildpädagogische Schriften* (eds. Haller, R., Kinross, R.). Hölder-Pichler-Tempsky, Vienna.

Neurath, P. (1993): Otto Neurath (1882–1945) Leben und Werk. In Neurath, P., Nemeth, E. (eds.), *Otto Neurath oder die Einheit von Wissenschaft und Gesellschaft,* 13–96. Böhlau, Vienna.

Neurath, P., Nemeth, E. (eds.) (1993): *Otto Neurath oder die Einheit von Wissenschaft und Gesellschaft.* Böhlau, Vienna.

Pais, A. (1982): *Subtle Is the Lord: The Science and the Life of Albert Einstein.* Oxford University Press, Oxford, UK.

Perutz, L. (1996): *Mainacht in Wien.* Zsolnay, Graz.

Pinl, M., Dick, A. (1974): Kollegen in einer dunklen Zeit. *Jahresbericht DMV* 75, 166–208.

Popper, K. (1995): Hans Hahn—Memories of a Grateful Student. In Schmetterer, L., Sigmund, K. (eds.), *Hans Hahn: Collected Works, Vol. 1,* 1–19, Springer, Vienna.

Popper, K. (2002): *Unended Quest: An Intellectual Autobiography.* Routledge, New York.

Portisch, H. (1989): *Österreich I: Die unterschätzte Republik.* Kremsmayer und Scheriau, Vienna.

Quine, W. v. O. (1985): *The Time of My Life.* MIT Press, Cambridge, MA.

Regis, E. (1987): *Who Got Einstein's Office? Eccentricity and Genius at the Institute for Advanced Study.* Addison-Wesley, Boston.

Reid, C. (1996): *Hilbert.* Copernicus, New York.

Reidemeister-Neurath, M. (1980): *An was ich mich erinnere* (ed. Mulder, H.). Unpublished manuscript.

Reisch, G. (2005): *How the Cold War Transformed Philosophy of Science: To the Icy Slopes of Logic.* Cambridge University Press, New York.

Reiter, W. (2001): Die Vertreibung der jüdischen Intelligenz: Verdoppelung eines Verlustes, 1938–1945. *Internationale Mathematische Nachrichten* 187, 1–20.

Rellstab, U. (1992): Ökonomie und Spiele. PhD thesis, St. Gallen.

Renn, J. (2005): *Albert Einstein: Ingenieur des Universums.* 3 vols. Wiley-VCH, Weinheim.

Richardson, A. (1998): *Carnap's Construction of the World.* Cambridge University Press, New York.

Richardson, A., Uebel, T. (eds.) (2007): *The Cambridge Companion to Logical Empiricism.* Cambridge University Press, New York.

Richardson, R. (2008): Scientific Philosophy as a Topic for History of Science. *Isis* 99, 88–96.

Rigden, J. (2005): *Einstein 1905: The Standard of Greatness.* Harvard University Press, Cambridge, MA.

Rudin, W. (1997): *The Way I Remember It.* American Mathematical Society, Providence, RI.

Rukschcio, B., Schachel, R. (1982): *Adolf Loos Leben und Werk.* Residenz, Salzburg.

Russell, B. (1967–1969): *The Autobiography of Bertrand Russell.* 3 vols. Allen and Unwin, London.

Salmon, W. C., Wolters, G. (eds.) (1994): *Logic, Language and the Structure of Scientific Theories*. University of Pittsburgh Press, Pittsburgh.

Sandner, G. (2014): *Otto Neurath: Eine politische Biographie*. Zsolnay, Vienna.

Sarkar, S. (1996a): *The Emergence of Logical Empiricism: From 1900 to the Vienna Circle*. Routledge, New York.

Sarkar, S. (1996b): *The Legacy of the Vienna Circle: Modern Reappraisals*. Garland, New York.

Schilpp, P. A. (ed.) (1944a): *The Library of Living Philosophers—Einstein*. Northwestern University Press, Chicago.

Schilpp, P. A. (ed.) (1944b): *The Library of Living Philosophers—Russell*. Northwestern University Press, Chicago.

Schilpp, P. A. (ed.) (1963): *The Library of Living Philosophers—Carnap*. Northwestern University Press, Chicago.

Schilpp, P. A. (ed.) (1974): *The Library of Living Philosophers—Popper*. Northwestern University Press, Chicago.

Schlick, M.: Collected Works. Springer, Vienna.

Schlick, M. (1984): *Fragen der Ethik*. Suhrkamp, Frankfurt.

Schlick, M. (1986): *Die Probleme der Philosophie in ihrem Zusammenhang*. (Vorlesungen 1933/34). Suhrkamp, Frankfurt.

Schorske, C. E. (1981): *Fin-de-Siècle Vienna: Politics and Culture*. Vintage, New York.

Schumpeter, J. (1954): *History of Economic Analysis*. Oxford University Press, Oxford, UK.

Schweitzer, B., Sklar, A., Sigmund, K., Schmetterer, L. Gruber, P., Hlawka, E., Reich, L. (eds.) (2002): *Karl Menger Selecta Mathematica, Vols. I and II*. Springer, Vienna.

Sieg, W. (2013): *Hilbert's Program and Beyond*. Oxford University Press, Oxford, UK.

Siegetsleitner, A. (2014): *Moral und Ethik im Wiener Kreis*. Böhlau, Vienna.

Siegmund-Schultze, R. (2001): Richard von Mises. *Internationale Mathematische Nachrichten* 187, 21–32.

Siegmund-Schultze, R. (2009): *Mathematicians Fleeing from Nazi Germany: Individual Fates and Global Impact*. Princeton University Press, Princeton.

Siegmund-Schultze, R. (2017): *Richard von Mises*. Forthcoming.

Sigmund, K. (1995): A Philosopher's Mathematician—Hans Hahn and the Vienna Circle. *Mathematical Intelligencer* 17, 16–29.

Sigmund, K. (1998): Menger's Ergebnisse: A Biographical Introduction. In Dierker, E., Sigmund, K. (eds.), *Karl Menger—Ergebnisse eines Mathematischen Kolloquiums*, 5–31. Springer, Vienna.

Sigmund, K. (2001): *Kalter Abschied von Europa*: Österreichische Math. Gesellschaft, Vienna.

Sigmund, K. (2002): Karl Menger and Vienna's Golden Autumn. In Schweitzer, B., et al. (eds.), *Karl Menger Selecta Mathematica, Vol. I,* 7–21. Springer, Vienna.

Sigmund, K. (2011): Dozent Gödel Will Not Lecture. In Baaz, M., et al. (eds.), *Kurt Gödel and the Foundations of Mathematics: Horizons of Truth,* 75–95. Cambridge University Press, New York.

Sigmund, K. (2015): Mathematik an der Universität Wien. In Fröschl, T., et al. (eds.), *650 Jahre Universität Wien,* 459–473. UVW, Vienna.

Sigmund, K., Dawson, J., Mühlberger, K. (2006): *Kurt Gödel—The Album.* Vieweg, Wiesbaden.

Stadler, F. (1987): *Vertriebene Vernunft.* Jugend und Volk, Vienna.

Stadler, F. (ed.) (1988): *Kontinuität und Bruch 1938–1945–1955.* Jugend und Volk, Vienna.

Stadler, F. (1990): Richard von Mises: Wissenschaft im Exil. In Mises, R. v., *Kleines Lehrbuch des Positivismus,* 7–48. Suhrkamp, Frankfurt.

Stadler, F. (ed.) (1993): *Scientific Philosophy: Origins and Developments.* Kluwer, Dordrecht.

Stadler, F. (2015): *The Vienna Circle: Studies in the Origins, Development and Influence of Logical Empiricism.* Suhrkamp, Frankfurt.

Stadler, F., Engler, H. (ed.) (2009): *Stationen: Dem Philosophen und Physiker Moritz Schlick zum 125. Geburtstag.* Springer, Vienna.

Stadler, F., Uebel, T. (eds.) (2012): *The Scientific World-Conception: Wissenschaftliche Weltauffassung.* Springer, Vienna.

Stadler, F., Weibel, P. (eds.) (1995): *The Cultural Exodus from Austria.* Springer, Vienna.

Stewart, I. (2013): *Visions of Infinity: The Great Mathematical Problems.* Basic Books, New York.

Stewart, I. (2017): *Infinity: A Very Short Introduction.* Oxford University Press, Oxford, UK.

Stöhr, A. (1910): *Lehrbuch der Logik in psychologisierender Darstellung.* Deuticke, Vienna.

Streissler, E. (1994): Menger, Carl. In *Neue Deutsche Biographie (NDB), Vol. 17.* Duncker und Humblot, Berlin.

Taschwer, K. (2015): *Hochburg des Antisemitismus.* Czernin, Vienna.

Taschwer, K. (2016): *Der Fall Paul Kammerer.* Brandstätter, Vienna.

Taussky-Todd, O. (1985): An autobiographical essay. In Albers, D. J., Alexanderson, G. L. (eds.), *Mathematical People: Profiles and Interviews,* 309–336. Birkhäuser, Boston.

Thiel, C. (1984): Folgen der Emigration deutscher und österreichischer Wissenschaftstheoretiker und Logiker zwischen 1933 und 1945. *Berichte zur Wissenschaftsgeschichte* 7, 227–256.

Thurm, V., Nemeth, E. (2003): *Wien und der Wiener Kreis*. Facultas, Vienna.

Timms, E., Robertson, R. (eds.) (1990): *Vienna 1900: From Altenberg to Wittgenstein*. Edinburgh University Press, Edinburgh.

Uebel, T. (ed.) (1991): *Rediscovering the Forgotten Vienna Circle: Austrian Studies on Otto Neurath and the Vienna Circle*. Kluwer, Dordrecht.

Uebel, T. (1993): Vernunftkritik und Wissenschaft: Otto Neurath und der erste Wiener Kreis. *Vienna Circle Yearbook* 1, 47–74.

Uebel, T. (2003): On the Austrian Roots of Logical Empiricism. In Parrini, P., et al. (eds.), *Logical Empiricism: Historical and Contemporary Perspectives*, 67–93. University of Pittsburgh Press, Pittsburgh.

Ulam, S. (1976): *Adventures of a Mathematician*. Scribner, New York.

Van Atten, M., Kennedy, J. (2003): Gödel's Philosophical Developments. *Bulletin of Symbolic Logic* 9, 470–492.

Van Dalen, D. (2002, 2005): *Mystic, Geometer, and Intuitionist: The Life of L. E. J. Brouwer*. 2 vols. Clarendon Press, Oxford, UK.

Van Heijenoort, J. (ed.) (2002): *From Frege to Gödel. A Source Book in Mathematical Logic 1897–1931*. Harvard University Press, Cambridge, MA.

Wagner, P. (ed.) (2009): *Carnap's Logical Syntax of Language*. Palgrave Macmillan, Basingstoke, UK.

Wang, H. (1987): *Reflections on Kurt Gödel*. MIT Press, Cambridge.

Weber, M. (1919): *Wissenschaft als Beruf*. Duncker und Humblot, Berlin.

Weinzierl, E., Skalnik, K. (eds.) (1983): Österreich 1918–1938. 2 vols. Styria, Graz.

Weissensteiner, F. (1990): *Der ungeliebte Staat: Österreich zwischen 1918 und 1938*. ÖBV, Vienna.

Welzig, M. (1998): *Josef Frank 1885–1967: Das architektonische Werk*. Böhlau, Vienna.

Wheeler, J. A. (1990): *A Journey into Gravity and Spacetime*. Vol. 31. Scientific American Library, New York.

Wieser, F. (1923): Biographie von Carl Menger. In Wieser, F., *Gesammelte Abhandlungen* (1929), iii–xviii. Mohr, Tübingen.

Wijdeveld, P. (2000): *Ludwig Wittgenstein, Architekt*. Wiese, Basel.

Wittgenstein, L.: Collected Works, Electronic Edition. ISBN: 978-1-57085-203-9.

Wolters, G. (1987): *Mach I, Mach II, Einstein und die Relativitätstheorie*. De Gruyter, Berlin.

Wuchterl, K., Hübner, A. (1998): *Wittgenstein. Mit Selbstzeugnissen und Bilddokumenten*. Rowohlt, Reinbek.

Yourgrau, P. (2005): *A World Without Time: The Forgotten Legacy of Gödel and Einstein*. Basic Books, New York.

Zach, R. (2006): Hilbert's Program Then and Now. In Jacquette, D. (ed.), *Philosophy of Logic: Handbook of the Philosophy of Science, Vol. 5*, 411–447. Elsevier, Amsterdam.

Zilsel, E. (1992): *Wissenschaft und Weltanschauung: Aufsätze 1929–1933*. Europaverlag, Vienna.

Zilsel, P. (1988): Über Edgar Zilsel. In Stadler, F. (ed.), *Kontinuität und Bruch 1938–1945–1955, Vol. 2*, 929–932. Jugend und Volk, Vienna.

Zuckmayer, C. (1984): *A Part of Myself: Portrait of an Epoch*. Carroll and Graf, New York.

# Index

Bea Laufersweiler

**KARL SIGMUND** s a professor emeritus of mathematics at the University of Vienna and researcher at IIASA, Laxenburg. One of the pioneers of evolutionary game theory, he lives in Vienna.